Advanced Topics in Energy

Edited by **George Thomson**

NY RESEARCH PRESS
New York

Published by NY Research Press,
23 West, 55th Street, Suite 816,
New York, NY 10019, USA
www.nyresearchpress.com

Advanced Topics in Energy
Edited by George Thomson

International Standard Book Number: 978-1-63238-024-1 (Hardback)

Printed in China.

Contents

Preface

Energy is the most essential part and basic unit of our universe. It's a very common saying - Energy can neither be destroyed nor be created; it's very true because it's a never ending process that makes energy change its forms from one to another. Energy is used to transform efforts into work, energy illuminates our life with lights, and energy provides power to our transportation system such as cars, railways, aircrafts, jets, etc. It is also used to heat down our homes, cook food, play music systems, and for powering machinery in manufacturing units. The most important source of energy is sun, which provides light, heat, biochemical processes for plants growth. If we consider the food cycle sun brings energy to plants, which is consumed by herbivores, and finally consumed by carnivores.

Energy is the fuel of every process in our universe; when in stored form it is called potential energy, and moving energy is called kinetic energy. There are many more forms of energy such as; Mechanical energy, sound energy, chemical energy, electrical energy, magnetic energy, nuclear energy, elastic energy, gravitational pull, thermal energy, and many more. This book is a collection of chapters that discusses some of the crucial aspects of energy and energy conservation as well.

I would like to express my sincere appreciation to the authors of the chapters in this book for their excellent contributions and for their efforts involved in the publication process. I do believe that the contents in this book will be helpful to many researchers in this field around the world. I would also like to thank my publisher for giving me this unmatched opportunity of editing the work of renowned experts.

Editor

1

Experimental Evaluation of Supercapacitor-Fuel Cell Hybrid Power Source for HY-IEL Scooter

Piotr Bujlo,[1,2] Grzegorz Pasciak,[1] Jacek Chmielowiec,[1] and Andrzej Sikora[1]

[1] *Electrotechnical Institute Wroclaw Division of Technology and Materials Science, M. Sklodowskiej-Curie 55/61, 50-369 Wroclaw, Poland*
[2] *Hydrogen South Africa (HySA) Systems and Validation Centre, University of the Western Cape, Robert Sobukwe Road, Bellville, Cape Town 7535, South Africa*

Correspondence should be addressed to Piotr Bujlo; pbujlo@gmail.com

Academic Editor: Mattheos Santamouris

This paper presents the results of development of a hybrid fuel cell supercapacitor power system for vehicular applications that was developed and investigated at the Energy Sources Research Section of the Wroclaw Division of Electrotechnical Institute (IEL/OW). The hybrid power source consists of a polymer exchange membrane fuel cell (PEMFC) stack and an energy-type supercapacitor that supports the system in time of peak power demands. The developed system was installed in the HY-IEL electric scooter. The vehicle was equipped with auxiliary components (e.g., air compressor, hydrogen tank, and electromagnetic valves) needed for proper operation of the fuel cell stack, as well as electronic control circuits and a data storage unit that enabled on-line recording of system and vehicle operation parameters. Attention is focused on the system energy flow monitoring. The experimental part includes field test results of a vehicle powered with the fuel cell-supercapacitor system. Values of currents and voltages recorded for the system, as well as the vehicle's velocity and hydrogen consumption rate, are presented versus time of the experiment. Operation of the hybrid power system is discussed and analysed based on the results of measurements obtained.

1. Introduction

Hydrogen becomes a more reliable future energy carrier and its usage as fuel for fuel cells becomes a more and more probable future method for electric energy production. This statement can be supported by a large number of scientific papers that present the latest developments on fuel cell technology [1–3], innovative fuel cell stationary [4, 5], and mobile and automotive applications [6–8], as well as considerations on hydrogen economy prospects [9–11].

Polymer exchange membrane fuel cell (PEMFC) is the most promising fuel cell type for future commercialization and application. Although PEMFC has suitable parameters for the previously mentioned applications, among others low operating temperature and high current density, it cannot serve as the only energy source in the power system. Hybridization of power sources is needed to provide immediate power supply for auxiliary devices during the fuel cell start-up and to supply power during peak demands. The latter is important especially in the case of automotive applications where load changes are sudden and the power train requires an immediate power supply for a comfortable drive. There are literature reports concerning battery/fuel cell hybrid systems [12, 13]. Because the battery presents some drawbacks, such as low life cycle, long recharging time and low power density, it is more reasonable to use supercapacitors that overcome the previously mentioned problems with batteries.

In the literature there are many papers describing the projects of fuel cell powered lightweight vehicles, such as and Pios—hydrogen powered tricycle [17], HYSYRIDER—fuel cell based electric scooter [18], and MHV and HFCHV—hydrogen fuel cell vehicle [15, 16], a few concepts of fuel cell powered electric bicycles [14] and hydrogen fuel cell hybrid vehicles [20, 21] and scooters [19, 22]. Typically, batteries are applied as an energy buffer because a standalone fuel cell system integrated into an electric drive is not always sufficient to provide the load demands of a vehicle [23]. In order to provide the peak power during such transients as start-up,

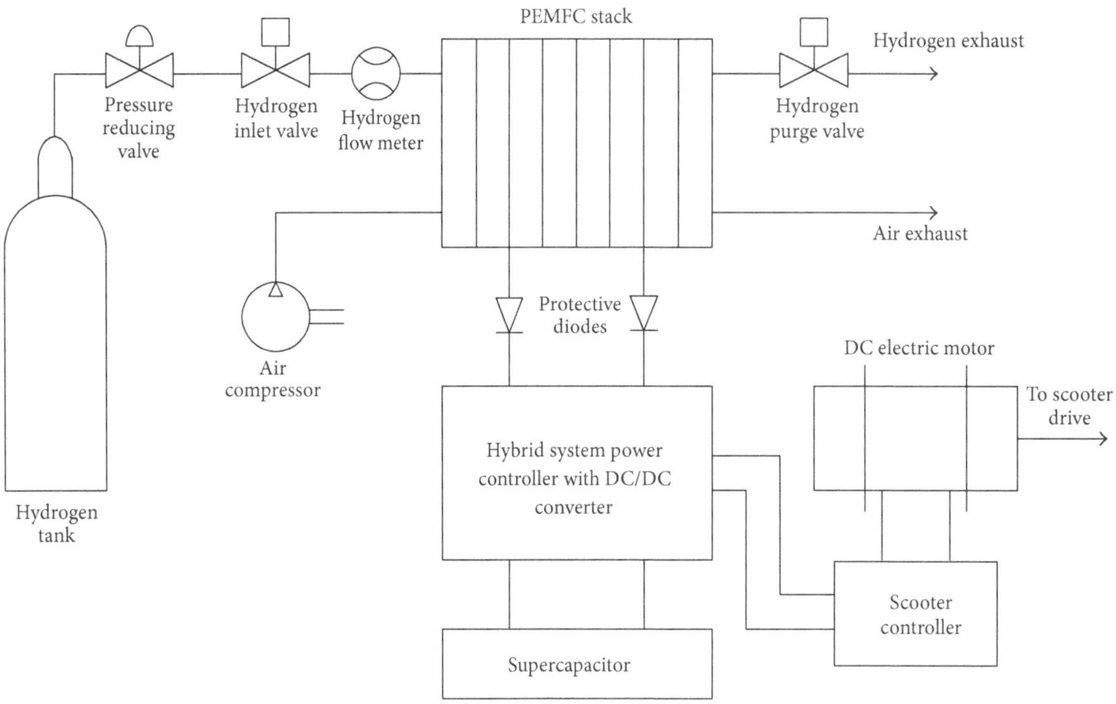

FIGURE 1: Component layout diagram of the Hy-IEL scooter showing hybrid system components and electrical drive train.

acceleration or sudden changes in load and also to take advantage of the regenerative power of an electric vehicle during braking, it is much more suitable to use a supercapacitor in addition to the fuel cell stack [20, 21]. The primary purpose of using a supercapacitor instead of batteries as the energy buffer is that a supercapacitor provides better power densities than conventional batteries and better energy densities than conventional capacitors. The results of research and measurements have proved that a supercapacitor can be interchangeably used instead of batteries for vehicular applications.

This paper deals with an experimental evaluation of the supercapacitor-fuel cell hybrid power source implemented on an Hy-IEL electric vehicle. Operating parameters of the hybrid power source and the vehicle are measured and stored during field tests. Operation of the hybrid power system is discussed and analysed based on the results of measurements obtained.

2. Hybrid Power Source Integration in a Vehicle

System technical specification and the components were selected taking into account available fuel cell stack and its parameters as well as vehicle electric motor requirements. The construction of the developed hybrid system consists of the ANUVU fuel cell stack with 500 W of nominal power, operating with the efficiency of up to 40% at 70 A and two MAXWELL supercapacitors with a nominal capacity of 59 F (50 F measured) and nominal voltage of 15 V each, operating with the efficiency of 98%. The two mentioned energy storage sources were connected in parallel. The supercapacitor was chosen in order to smooth out peak power demands of the vehicle during acceleration and to supply power for balance of plant (BoP) components during fuel cell start-up. Based on the well-known parameters of energy sources and vehicle motor requirements a hybrid system power controller was constructed. The electronic circuit was integrated with energy sources. The hybrid system power controller input voltage of 10–20 V DC and maximum input current of 50 A was adjusted to fuel cell stack electrical parameters and output parameters of 40 V DC and 15 A were imposed by the scooter electric motor. Figure 1 presents integration layout of the fuel cell hybrid system components. It was assumed that the power source should be able to deliver 300 W of power, which is enough to supply the electric scooter power trains which are usually equipped with motors up to 250 W power. In order to ensure a continuous operation of the fuel cell stack it was necessary to install some auxiliary components. Oxygen is delivered to the stack cathode with the aid of the air compressor, and hydrogen is stored in the form of compressed gas. The compressor was selected based on stack air flow rate requirements and it delivers oxidant at the flow rate of 30 dm^3/min and 0.5 bar pressure. Other stack working parameters (temperature, pressure, and humidity) are maintained at optimum values that were determined at the time of the stack testing and described elsewhere [24]. High pressure (125 bar) stainless steel Swagelok gas container with the capacity of 2.5 dm^3 allows to store 35 g of hydrogen, which ensures fuel supply for 1 hour of operation. The complete hybrid power source, integrated in the construction of Hy-IEL scooter, is shown in Figure 2.

With the use of a specially designed data storage unit, eleven parameters can be measured in order to provide information about the energy flow and behaviour of the powering

FIGURE 2: Photograph of Hy-IEL electric scooter powered with the hybrid fuel cell-supercapacitor energy source. Components marked in the photograph: 1: fuel cell stack, 2: hydrogen tank, 3: air compressor, 4: supercapacitor, 5: batteries, 6: control panel, 7: data storage unit.

modules, as well as the driving conditions. The voltages and currents of the battery, supercapacitor module and fuel cell are measured, as well as the powering condition of the compressor. Additionally, the hydrogen flow and the velocity of the scooter and its inclination along the driving direction are recorded. The data are stored on a flash memory device (SD card) and can be downloaded to a PC with the use of standard software. More detailed description of the data storage unit topography can be found elsewhere [25].

3. Experimental Results

Values of currents and voltages of the hybrid system, as well as the velocity of the vehicle, hydrogen consumption rate, and inclination were measured, and the data stored, during the scooter's field test operation. Data were recorded at a sampling rate of 8 measurements per second. An example of measurement results is presented in Figure 3.

The top characteristic presents voltage values of motor supply, supercapacitor and fuel cell. The voltage that supplies the electric motor of the vehicle is constant although there are many transient disturbances observed which are probably caused by the original electronic control system installed in the scooter. The value of motor supply voltage equals 40 V and is imposed by the electric motor demands. The voltage of fuel cell and supercapacitor changes depending on the load, within the range from 18 to 10 V. At this condition the power generated by those sources changes from 0 to 250 W. During the start-up and acceleration of the vehicle the energy flows at first from the supercapacitor, which is easily seen in the current characteristic plotted versus time of the trial. The middle characteristic presents current values and its flow direction in the system. In comparison with the supercapacitor, the fuel cell needs more time to generate the required power and both energy sources complement each other. When the supercapacitor is discharged, the fuel cell stack is ready to deliver electric energy for the drivetrain and the surplus of energy is used to charge supercapacitor, which can be observed as negative current values measured for the supercapacitor. The maximum electric motor current

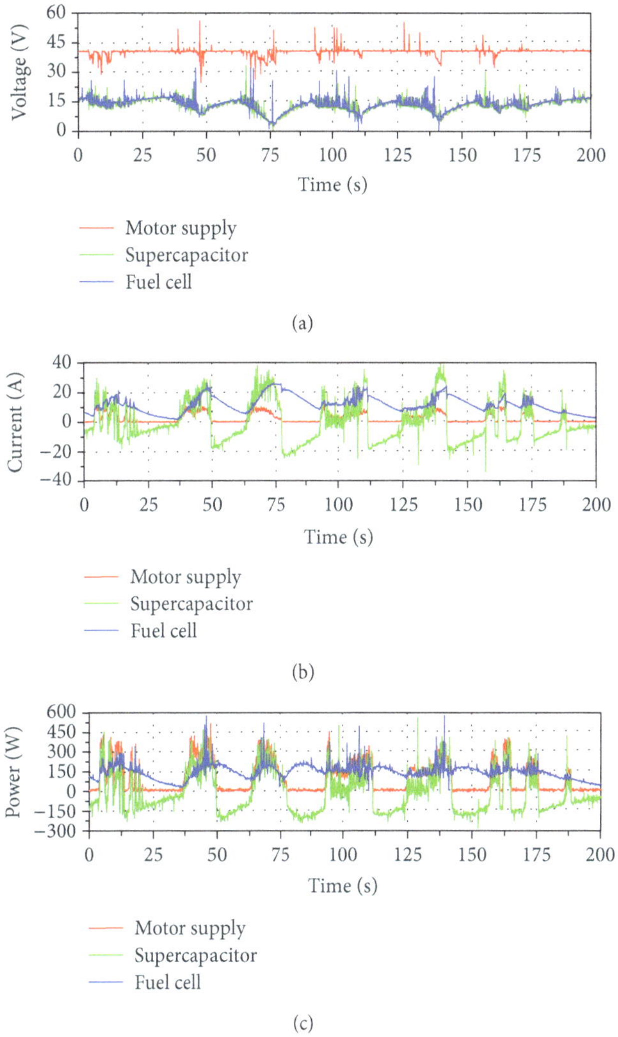

FIGURE 3: Electrical parameters (voltage, current, and power) of the hybrid system measured during field test of the vehicle.

is about 12 A but such high values are measured only during the start-up and acceleration of the vehicle. The figure in which power is plotted versus time shows that the power demand from the motor (350 W in peak) is higher than the power output from the PEMFC stack (200 W during the test), and the extra power required comes from the supercapacitor. The graph presented in Figure 4 includes results of the measured vehicle velocity, hydrogen consumption rate, and inclination that provide information as to whether the vehicle is moving uphill or downhill. The maximum velocity that the vehicle reached over a short distance of the trial exceeded 10 km/h. During the start-up and acceleration, when energy is needed, the hydrogen consumption rate increased up to 3 SLPM. Although the test was performed on a horizontal path, without elevations, the installed inclinometer was very sensitive to the vehicle's position and measured changes during braking, which is seen, for example, in the 50th and the 75th seconds of the trial. The previously described parameters provide information about some very important

TABLE 1: Comparison of fuel cell vehicles.

Name of vehicle	Vehicle type	Motor power (W)	PEMFC stack power (W)	Second energy storage	H_2 Consumption	H_2 storage method	Vehicle speed (km/h)
HFCHV [14]	Cart	4000	1200	Lead acid battery	10 g/km	Compressed cylinder 350 bar, 600 g H_2	25
MHV [15]	Cart	4000	3200	—	—	High pressure cylinder 10 L at 200 bar, 1400 Nl H_2	—
PAB [16]	Bicycle	230	300	—	—	Metal hydride tank 6.8 g of H_2 at <10 bar	25
PIOS [17]	Tricycle	300	300	Lithium polymer battery	3.7 SLPM	Composite pressurized tank 26 L at 350 bar	40–65
HYSYRIDER [18]	Scooter	250	300	Lead acid battery	—	Low pressure metal hydride tank, 200 Nl H_2	20
HFCHS [19]	Scooter	800	500	Sealed lead acid battery	2-3 SLPM	Metal hydride tank, 54 g of H_2 at 15 bar	10–12
Hy-IEL	Scooter	250	500	Supercapacitor	3.0 SLPM	Stainless steel high pressure tank 2.5 L at 125 bar, 34 g H_2	10–15

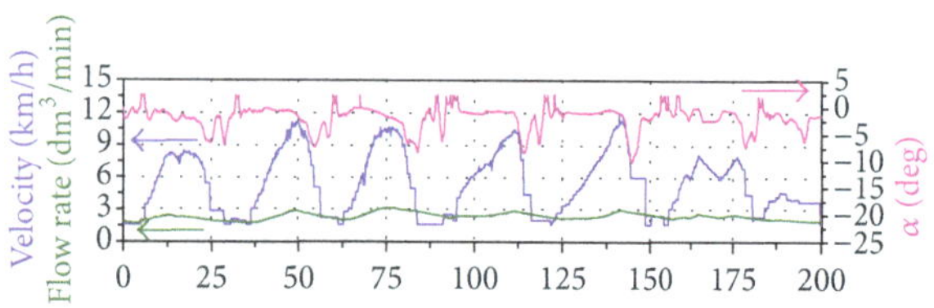

FIGURE 4: Results of the measured vehicle velocity, hydrogen consumption rate, and inclination versus time.

parameters of the hybrid power supply unit and will be useful for energy balance calculation during future investigations.

In Table 1 there are presented parameters of electric vehicles (carts, bicycles, and scooters) in the construction of which the fuel cell was used to deliver required power to the electric drivetrain. As can be noticed the bigger vehicles like carts are using fuel cell stacks with the power above 1000 W while the bicycles and scooters have lower power demands and can be powered with smaller units up to 500 W, as in the case presented in the paper on hybrid power source applied in Hy-IEL electric scooter. Different battery types were used for hybridization and most commonly lead acid batteries were applied. The disadvantage of application of the lead acid batteries is their weight and long charging time while supercapacitor used in the Hy-IEL scooter offers much higher energy density. Moreover the charging time of the supercapacitor is very short and it can be charged with high currents which enables application of breaking energy recuperation. Different hydrogen storage methods were used for fuel supply but similar amounts of gas were stored and similar hydrogen consumption for scooter type vehicles was measured. It was reported by authors of HFCHS vehicle that application of metal hydride tank caused problem and sufficient hydrogen flow rate that could not be achieved because of sluggish hydrogen release mechanism [22]. The usage of the pressurised tank seems to be the best option for hydrogen storage in automotive applications. The maximum speed of the scooter class fuel cell vehicles reported in the literature varies from 10 to 65 km/h but most of them drive at the speed of about 20 km/h. This makes the described Hy-IEL scooter comparable with similar vehicles in this class. The results of the performed field tests of the Hy-IEL scooter were recorded during test on the short distance which can have influence on low vehicle speed that was reported in the paper. Maximum speed of the scooter declared by the producer equals 25 km/h, but installation of the additional weight decreased the maximum speed to 20 km/h that was recorded but not reported in the figures.

4. Conclusions

The hybrid fuel cell-supercapacitor power supply system was developed and installed in the electric scooter named Hy-IEL. The construction of the vehicle was tested during field tests. Experimental results indicate that the developed hybrid power system with a fuel cell-supercapacitor power source is suitable for powering electric vehicles.

The performance of the Hy-IEL electric scooter operating with integrated supercapacitor-fuel cell hybrid power source is similar to previously reported in the literature similar developments. The advantage of the developed power source is application of the supercapacitor which improves the dynamics of energy flow and reduces the weight of the second energy source of the hybrid system. Application of the supercapacitor gives also opportunity for easier braking energy recuperation which will be studied in the future project.

The power system designed and implemented on the vehicle will be modified and used for future research on hybrid systems. The research work will be focused on developing a more sophisticated power control unit, as well as on comparing the experimental results with the calculated energy balance model.

Acknowledgments

The authors wish to thank the Polish Ministry of Science and Higher Education for financial support. The presented work was completed under the national Project no., COST/261/2006 carried out within the framework of COST Action 542.

References

[1] D. Ye and Z. Zhan, "A review on the sealing structures of membrane electrode assembly of proton exchange membrane fuel cells," *Journal of Power Sources*, vol. 231, pp. 285–292, 2013.

[2] A. Morin, F. Xu, G. Gebel, and O. Diat, "Influence of PEMFC gas flow configuration on performance and water distribution studied by SANS: evidence of the effect of gravity," *International Journal of Hydrogen Energy*, vol. 36, no. 4, pp. 3096–3109, 2011.

[3] F. Alcaide, G. Álvarez, J. A. Blázquez, P. L. Cabot, and O. Miguel, "Development of a novel portable-size PEMFC short stack with electrodeposited Pt hydrogen diffusion anodes," *International Journal of Hydrogen Energy*, vol. 35, no. 11, pp. 5521–5527, 2010.

[4] H. S. Han, C. Cho, S. Y. Kim, and J. M. Hyun, "Performance evaluation of a polymer electrolyte membrane fuel cell system for powering portable freezer," *Applied Energy*, vol. 105, pp. 125–137, 2013.

[5] H. S. Chu, F. Tsau, Y. Y. Yan, K. L. Hsueh, and F. L. Chen, "The development of a small PEMFC combined heat and power system," *Journal of Power Sources*, vol. 176, no. 2, pp. 499–514, 2008.

[6] S. F. Tie and C. W. Tan, "A review of energy sources and energy management system in electric vehicles," *Renewable and Sustainable Energy Reviews*, vol. 20, pp. 82–102, 2013.

[7] F. Wang and Y. Chiang, "Design and control of a PEMFC powered electric wheelchair," *International Journal of Hydrogen Energy*, vol. 37, no. 15, pp. 11299–11307, 2012.

[8] A. Veziroglu and R. MacArio, "Fuel cell vehicles: state of the art with economic and environmental concerns," *International Journal of Hydrogen Energy*, vol. 36, no. 1, pp. 25–43, 2011.

[9] T. Larriba, R. Garde, and M. Santarelli, "Fuel cell early markets: techno-economic feasibility study of PEMFC-based drivetrains in materials handling vehicles," *International Journal of Hydrogen Energy*, vol. 38, no. 5, pp. 2009–2019, 2013.

[10] W. G. Colella, M. Z. Jacobson, and D. M. Golden, "Switching to a U.S. hydrogen fuel cell vehicle fleet: the resultant change in emissions, energy use, and greenhouse gases," *Journal of Power Sources*, vol. 150, no. 1-2, pp. 150–181, 2005.

[11] C. Wang, S. Zhou, X. Hong, T. Qiu, and S. Wang, "A comprehensive comparison of fuel options for fuel cell vehicles in China," *Fuel Processing Technology*, vol. 86, no. 7, pp. 831–845, 2005.

[12] L. Barelli, G. Bidini, and A. Ottaviano, "Optimization of a PEMFC/battery pack power system for a bus application," *Applied Energy*, vol. 97, pp. 777–784, 2012.

[13] O. Erdinc and M. Uzunoglu, "Recent trends in PEM fuel cell-powered hybrid systems: Investigation of application areas, design architectures and energy management approaches," *Renewable and Sustainable Energy Reviews*, vol. 14, no. 9, pp. 2874–2884, 2010.

[14] L. Cardinali, S. Santomassimo, and M. Stefanoni, "Design and realization of a 300 W fuel cell generator on an electric bicycle," *Journal of Power Sources*, vol. 106, no. 1-2, pp. 384–387, 2002.

[15] K. Kendall, B. G. Pollet, A. Dhir, I. Staffell, B. Millington, and J. Jostins, "Hydrogen fuel cell hybrid vehicles (HFCHV) for Birmingham campus," *Journal of Power Sources*, vol. 196, no. 1, pp. 325–330, 2011.

[16] J. J. Hwang, D. Y. Wang, and N. C. Shih, "Development of a lightweight fuel cell vehicle," *Journal of Power Sources*, vol. 141, no. 1, pp. 108–115, 2005.

[17] J. Weigl and H. Saidi, "Pios hydrogen fuel cell tricycle," *International Journal of Hydrogen Energy*, vol. 30, no. 9, pp. 1035–1036, 2005.

[18] E. Novo, L. Baldini, L. Borello, P. Cherchi, G. Gianolio, and N. Valtuegna, "Development of fuel cell hybrid scooter equipped with an integrated metal hydride tank," in *HYSYDAYS 1st World Congress of Young Scientists on Hydrogen Energy Systems*, pp. 405–407, 2006.

[19] N. Shih, C. Lin, C. Chang, and D. Wang, "Experimental tests of an air-cooling hydrogen fuel cell hybrid electric scooter," *International Journal of Hydrogen Energy*, 2013.

[20] P. Thounthong, S. Raël, and B. Davat, "Control strategy of fuel cell/supercapacitors hybrid power sources for electric vehicle," *Journal of Power Sources*, vol. 158, no. 1, pp. 806–814, 2006.

[21] P. Rodatz, G. Paganelli, A. Sciarretta, and L. Guzzella, "Optimal power management of an experimental fuel cell/supercapacitor-powered hybrid vehicle," *Control Engineering Practice*, vol. 13, no. 1, pp. 41–53, 2005.

[22] J. L. Shang and B. G. Pollet, "Hydrogen fuel cell hybrid scooter (HFCHS) with plug-in features on Birmingham campus," *International Journal of Hydrogen Energy*, vol. 35, no. 23, pp. 12709–12715, 2010.

[23] A. Jossen, J. Garche, H. Doering, M. Goetz, W. Knaupp, and L. Joerissen, "Hybrid systems with lead-acid battery and proton-exchange membrane fuel cell," *Journal of Power Sources*, vol. 144, no. 2, pp. 395–401, 2005.

[24] P. Bujlo, *Polimeric, superionic membranes for PEM fuel cells [Ph.D. thesis]*, Wroclaw University of Technology, Wroclaw, Poland, 2006, http://www.dbc.wroc.pl/Content/1539/Bujlo+Piotr+Rozprawa+Doktorska.pdf.

[25] P. Bujło, A. Sikora, G. Paściak, and J. Chmielowiec, "Energy flow monitoring unit for Hy-IEL hybrid (PEM fuel cellsupercapacitor) electric scooter," *Przeglad Elektrotechniczny*, vol. 86, no. 3, pp. 271–273, 2010.

2

Development of an Analytically Described Pitch Regulator for a Wind Turbine to Be Used for Grid Disturbance Studies

Pinar Tokat, Torbjörn Thiringer, and Peiyuan Chen

Division of Electric Power Engineering, Department of Energy and Environment, Chalmers University of Technology, 412 58 Gothenburg, Sweden

Correspondence should be addressed to Pinar Tokat; tokat@chalmers.se

Academic Editor: Onder Ozgener

In this paper, a pitch controller for a variable-speed wind turbine to be used in the high wind speed region is derived. The pitch regulator parameters are determined using analytical expressions and are compared with numerical calculations. In order to derive the pitch regulated wind turbine model, blade element momentum theory is utilized and reformulated analytically. Appropriate simplifications are made and, finally, the analytically derived pitch regulated wind turbine model is tested under grid disturbances such as voltage dips and spinning reserve provision. From this work it was found that by linearizing the blade profiles, one can analytically derive a fully functioned pitch regulator. In spite of all nonlinearities, a single pitch controller setting which is valid for the whole operation region is shown to be sufficient. This system is tested under grid disturbances and it is proven that the system is capable of operating well during a 0% remaining voltage dip and also during the voltage recovery back to the rated voltage level. Accordingly, grid codes commonly referred to can be handled with this simply derived pitch regulator. Moreover, it is shown that the derived system works well for a spinning reserve application using a 90% spinning reserve ability and still maintains a robust turbine control.

1. Introduction

Due to the increasing concern about the emissions caused by the usage of fossil fuels and the security issues regarding nuclear power, wind energy has nowadays become a popular source of electric power production. Wind energy is one of the fastest growing energy sources, which means that a great number of wind power plants have been installed and many more are to be installed further. Accordingly, the impact of wind energy units on the grid is of utmost importance to be studied.

Many wind turbine studies require modeling and simulation of wind turbines. Today, most wind turbines are of the variable-speed type and pitching is an essential feature of the variable-speed operation. In [1], pitch controller is explained but not derived, and in [2] pitch regulator is derived using a fairly extensive procedure. In [3, 4], the pitch regulator is designed using complex control methods. However, a simple design of a pitch regulator based on an analytical wind turbine model is not found in the literature. Such a "simple and analytically described" pitch regulator design would be useful while performing and in particular while comparing different studies.

Motivated by the reasons above, the purpose of this paper is to provide a fully traceable and analytically expressed derivation of a pitch regulator for a power system generic wind turbine, so that it can be used for complex studies. One of the goals of this work is to present the required steps such as modifying the blade element momentum theory and linearizing the blade profile of the wind turbine. Furthermore, a target is to demonstrate that the turbine through the pitching activity can handle a voltage dip without the need for a braking chopper. Finally, another aim is to test the system under load balancing disturbances, that is, to demonstrate a spinning reserve application.

2. Noniterative Blade Element Momentum Theory

A wind turbine is a device that converts the energy in the wind to electrical energy. The main idea in this conversion is to produce a driving torque by the turbine blades, in order to create a torque on the electric generator. For this study, the production of the driving torque is expressed using the blade element momentum theory (BEM) [5].

The rated power of the wind turbine used in this work is 2 MW and the rated rotor speed is 19 rpm. The blades of the wind turbine are nonuniformly shaped and have a length of approximately 35 meters for this case. Detailed information about the blade geometry can be found in [6].

Due to the fact that the torque formation is not uniformly distributed along the blade, the turbine blade is divided into smaller pieces during the analysis, where the wind is considered to be homogenous on each piece. These pieces are called the blade segments and on each blade segment the effect of the wind is different.

The blade segment is subjected to two forces created by the wind. These are called the edge and the flap forces; their directions are axial and tangential to the rotor plane, respectively. The edge force, F_{edge}, is the source of the rotation, hence the source of the output torque. The torque produced by each blade segment is called the torque contribution (ΔT) of the respective segment. The relation between F_{edge} and ΔT is

$$F_{\text{edge}} = \frac{1}{2}\rho W^2 c \Delta r \left(c_L \sin\phi - c_D \cos\phi\right), \quad (1)$$

$$\Delta T = F_{\text{edge}} r = \frac{1}{2}\rho W^2 c \Delta r \left(c_L \sin\phi - c_D \cos\phi\right) r, \quad (2)$$

where ρ is the air density, and W is the relative velocity composed by the incoming wind speed and the wind speed due to the rotation of the turbine. c is the chord, and Δr is the radial length of the blade segment. c_L and c_D represent the lift and drag coefficients, respectively, and ϕ is the relative flow angle.

The summation of the torque contribution of each segment determines the torque output of one turbine blade. The turbine used in this study has three turbine blades. Therefore, the total output torque of the wind turbine is

$$T = n_b \sum_{i=1}^{m} \Delta T_i. \quad (3)$$

Here, T indicates the output torque of the wind turbine, n_b is the number of blades, and m is the number of blade segments.

A key point of the BEM for this work is the derivation of the relative flow angle, ϕ. It is given as

$$\phi = \tan^{-1}\frac{v_\omega\left(1 + a'\right)}{r\omega_r\left(1 - a\right)}, \quad (4)$$

where ω_r is the angular rotor speed and r is the radius of the segment [7]. a and a' are called axial and tangential induction factors, respectively [8]. The problematic point about the ϕ calculation is that a and a' are calculated iteratively. This, in principle, makes it very difficult to derive an analytical model of the turbine blades and complicates the simulations. However, these factors are of less importance in the high wind speed region. Therefore, in the high speed region, ϕ can be simplified to

$$\phi = \tan^{-1}\frac{v_\omega}{r\omega_r}, \quad (5)$$

where a and a' are neglected. For more detailed information about the effects of a and a', please refer to [6]. Using this expression for ϕ, the BEM becomes a noniterative method. This is essential for the upcoming derivation of the pitch regulator in analytical terms.

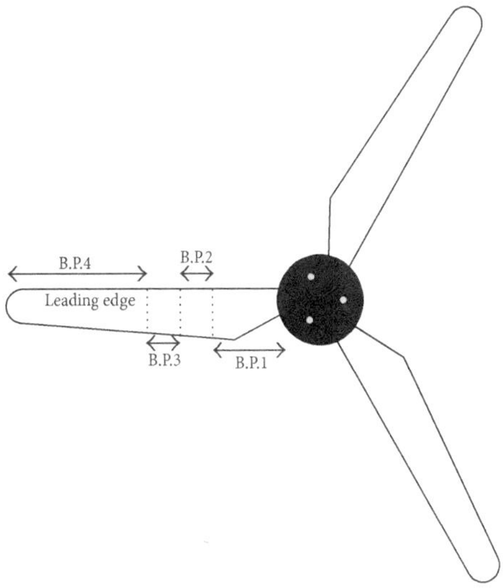

FIGURE 1: The blade profile distribution along the wind turbine blade.

3. Linearizing the Blade Profile

The wind acting on the blade has different directions according to the position of the blade. The angle between the incoming wind and the chord line of the turbine blade is called the angle of attack. The angle of attack is an important parameter of the wind turbine, since the lift and drag coefficients are depending on it. Initially, for each angle of attack, there is a set of c_L and c_D. The table which shows the lift and drag coefficient values for different angle of attacks gives the blade profile, as shown in Figure 3. Unfortunately, the relation between these coefficients and the angle of attack is non-linear. It would be convenient to have a linear relation between them in order to build the analytical model.

Due to the nonuniform shape of the turbine blades, different blade profiles need to be used for different parts of the blade. Figure 1 shows the blade profile distribution along the turbine blade.

This wind turbine has four blade profiles. They are distributed as shown in Figure 1 and as it can be observed, blade profile 4 is the dominating blade profile. Furthermore, it has been mentioned that each blade segment has a different contribution to the output torque.

Figure 2 shows the comparison between the lift forces in the utilized operational range [6] obtained from the different blade profiles, which is -10° to 10° since high wind speed

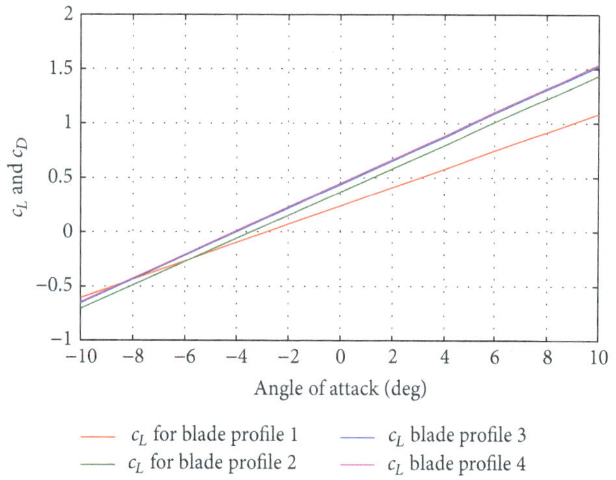

FIGURE 2: The blade profiles for the operation range of the wind turbine system. For the whole wind speed range, refer to [4].

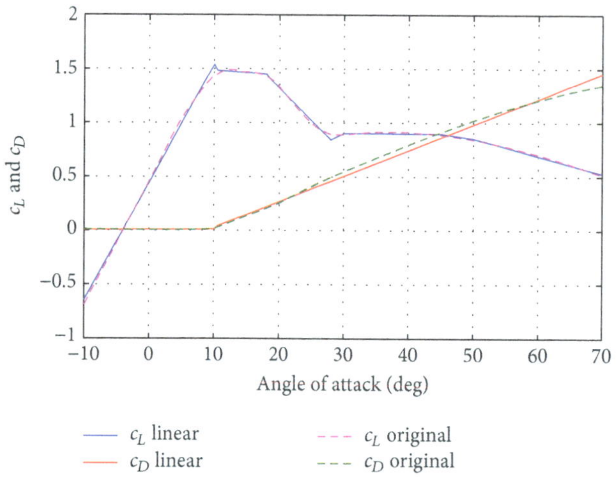

FIGURE 3: Original and linearized blade profiles. As it is observed the linearization is made piecewise and the error between the original and the linearized data is minimized.

TABLE 1: Linearized lift and drag coefficient expressions for different angle of attack regions.

Angle of attack (α)	Lift coefficient (c_L)	Drag coefficient (c_D)
−10°–10°	$0.1092 * \alpha + 0.4466$	$2.23e^{-4} * \alpha + 0.0077$
10°–18°	$-0.0043 * \alpha + 1.5293$	$0.0238 * \alpha - 0.2076$
18°–28°	$-0.0597 * \alpha + 2.5141$	$0.0238 * \alpha - 0.2076$
28°–45°	$-1.08e^{-4} * \alpha + 0.9029$	$0.0238 * \alpha - 0.2076$
45°–70°	$-0.0164 * \alpha + 1.6722$	$0.0238 * \alpha - 0.2076$

region, where the pitch regulator is active, is the focus of this paper. The drag coefficients for this range are fairly close to zero. Due to the above reasons and for simplicity, blade profile 4 is used to represent the whole turbine blade.

In the whole wind speed region, for different wind speeds, it is observed that angle of attack varies in the range of −10° to 70°. Although the relation is non-linear in between these limits, c_L and c_D can be easily linearized as

$$c_{Li} = a_i\alpha + b_i,$$
$$c_{Di} = d_i\alpha + e_i, \tag{6}$$

where i represents the number of the blade profile.

Due to the piecewise linearization, the constants a, b, d, and e would change for each angle of attack region. Table 1 shows the expressions of c_L and c_D for different regions. Figure 3 shows the original and the linearized data of the lift and drag coefficients for the operating range of the angle of attack.

4. Derivation of the Pitch Regulator

The wind turbine considered in this paper is a variable-speed wind turbine with pitching ability. Operation of the variable speed wind turbine differs according to the wind speed. Due to the operation principles, the wind speed range is divided into three regions. These are named the low, the middle, and the high wind speed regions. Figure 4 shows the power output and the rotor speed behavior of the wind turbine in the whole wind speed range. It is observed, that for the low wind speed region, the power output and the rotor speed increase with the wind speed. For the middle wind speed region, the rotor speed has reached the rated value, while the power output still increases. And lastly, for the high wind speed region, the output power reaches its rated value as well. It is desirable that both the rotor speed and the electric power are kept close to their corresponding rated values.

According to the different operation principles, different control strategies ought to be developed in order to achieve the optimal operation. The level of the captured power from the wind turbine can be easily understood from the behavior of the power coefficient, C_p. Figure 5 presents the behavior of C_p according to the tip speed ratio, λ, and the pitch angle, β. The "optimal operation path" of the turbine is depicted as well, which is marked with the stars. It is shown in this figure that C_p decreases with the increasing pitch angle.

It is observed from Figure 5 that the pitch angle is not changed for the low and the middle speed regions; however, it constantly increases at the high speed region. In this way, an increase in the rotor speed and the power output is prevented. Furthermore, for the low speed range, C_p can be maximized due to the variable-speed operation.

A wind turbine controller which could maximize C_p for the low wind speed operation, has a robust control in the middle wind speed operation and keeps the output power and the rotor speed constant for high wind speeds. A suitable design is shown in Figure 6. It is worth emphasizing that this paper focuses on the pitch controller; therefore, the only part that is explained further is the controller for the high wind speed region. For the information of the whole system, please refer to [6].

Since the study focuses on the wind turbine operation at high wind speeds, the pitching is active. In order to control the pitch angle of the wind turbine for different wind speeds, a pitch regulator is necessary. An analytically derived regulator

TABLE 2: Different K values calculated through the different blade profiles. All the K values are similar to each other.

Wind speed	K (blade profile 1)	K (blade profile 2)	K (blade profile 3)	K (blade profile 4)
16	−142150	−182420	−185040	−186990
17	−152560	−195590	−198440	−200430
18	−163170	−209040	−212120	−214170
19	−174010	−222760	−226080	−228180
20	−185080	−236780	−240340	−242490
21	−196380	−251090	−254910	−257100
22	−207930	−265710	−269780	−272040
23	−219720	−280650	−284980	−287280
24	−231770	−295910	−300510	−302880
25	−244080	−311500	−316370	−318780

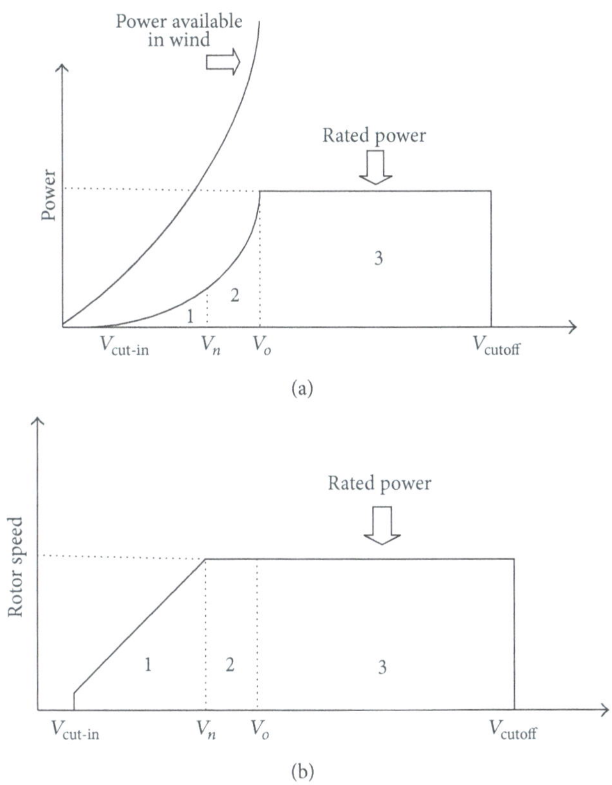

FIGURE 4: Wind turbine operation in the whole wind speed range.

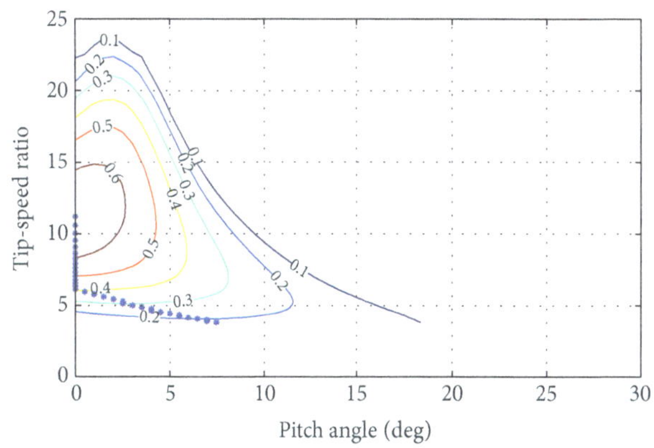

FIGURE 5: The contour plot shows the C_p as a function of λ and β and the star indicates the optimal operation for the wind turbine.

would be convenient to have and for this, there preferably should be a linear relation between the pitch angle and the output torque. It is assumed that the torque is directly proportional to the pitch angle and the relation between them is

$$T = f(\beta) = K\beta, \tag{7}$$

where β is the pitch angle and K is a constant that represents the relation between pitch angle and output torque. Table 2 shows the different K values obtained for different blade profiles.

The pitch regulator is designed according to this assumption using the internal mode control method [9]. Figure 7 shows the controller scheme of the pitch regulated wind turbine model.

If the pitch regulator is named F, then the system can be represented as

$$\frac{\omega}{\omega^*} = \frac{F(K/sJ)}{1 + F(K/sJ)}. \tag{8}$$

Since the torque-pitch angle relation is assumed as in (7), the pitch regulator is designed as a P-controller. Therefore, a pitch regulator can be expressed as

$$F = \frac{\alpha J}{K} = k_p, \tag{9}$$

where k_p is the proportional parameter of the pitch regulator and α is the controller bandwidth. The next step of the controller development is to determine K in terms of the turbine data and blade profile as well as the wind and rotor speed.

5. Model Development

The general expression of the torque contribution from a single blade segment is as in (2). Here, all the lift and drag coefficients are constant values for an individual situation of

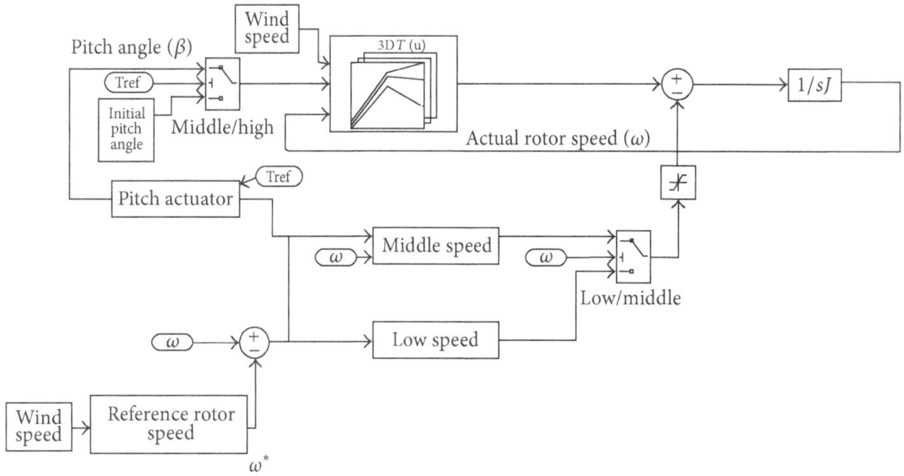

FIGURE 6: The controller scheme that is valid for the whole operational range.

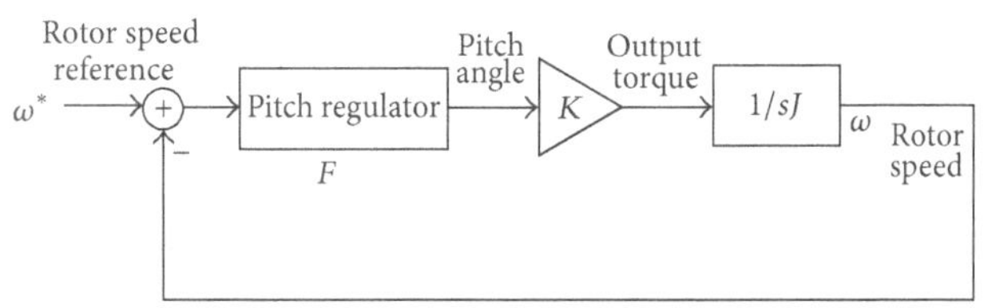

FIGURE 7: The general scheme of the pitch regulated wind turbine system.

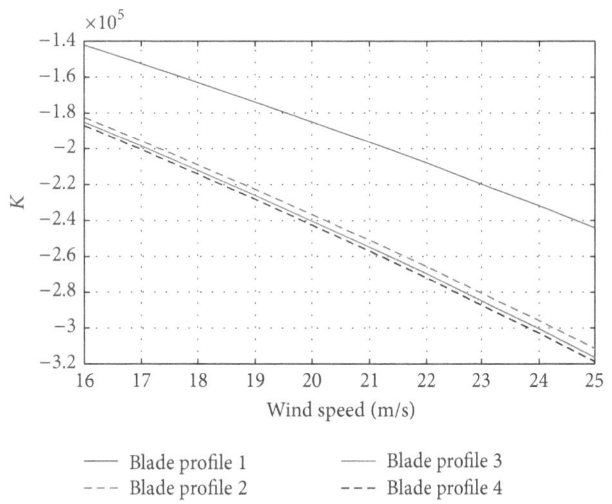

FIGURE 8: K parameter obtained by using different blade profiles.

the wind and the wind turbine. This means that, in order to have the output torque as a function of pitch angle, the lift and drag coefficient should be a function of it as well. Equation (6) show that c_L and c_D are functions of the angle of attack. The angle of attack is expressed as

$$\alpha = \phi - \beta \tag{10}$$

which means that α and therefore c_L and c_D are functions of the pitch angle.

The pitch regulated system is designed to operate in the high wind speed region. According to [6], the angle of attack in this region is between -10° and 10°. Table 1 shows that, for this α interval, the slope and the intercept values for c_L and c_D equations can be set to constants. The torque contribution is now expressed as

$$\Delta T = \frac{1}{2}\rho W^2 c\Delta r\left(\left(a\alpha + b\right)\sin\phi - \left(d\alpha + e\right)\cos\phi\right)r. \tag{11}$$

In the light of the explanations above, it is understood that the torque contribution of a blade segment is actually a function of the pitch angle. The relation can be expressed as the partial derivation of ΔT according to β

$$K = \frac{\partial T}{\partial \beta} = \frac{1}{2}\rho W^2 c\Delta r\left(a\sin\phi - d\cos\phi\right)r. \tag{12}$$

This expression of K is needed in order to determine the controller parameter k_p, as presented in (9). As shown in (12), K is depending on the wind speed and Figure 8 shows the value of K for different wind speed with different blade profiles. As a consequence of this, there is a different k_p parameter for each wind speed. However, due to practicality, one single value will be chosen to be used for the whole operating range, which is the value obtained at a 20 m/s wind speed. Furthermore, blade profile 4 is considered to be the only profile for the whole turbine blade. While blade profiles 2 and 3 give quite similar results to the ones that are obtained by using blade profile 4, blade profile 1 contributes to an approximately 20% of error. Fortunately, the part that has blade profile 1 contributes very little to the output torque [6]; therefore, considering blade profile 4 as the sole profile for the turbine does not affect the overall torque output of the system, significantly.

Another important point regarding the pitch regulator design is that the output torque expression is more complicated than that assumed in (7). Although the torque-pitch relation still can be expressed as a first-order function, there is an offset. This offset that varies with wind speed and the blade segment would accumulate into a large steady state error, if

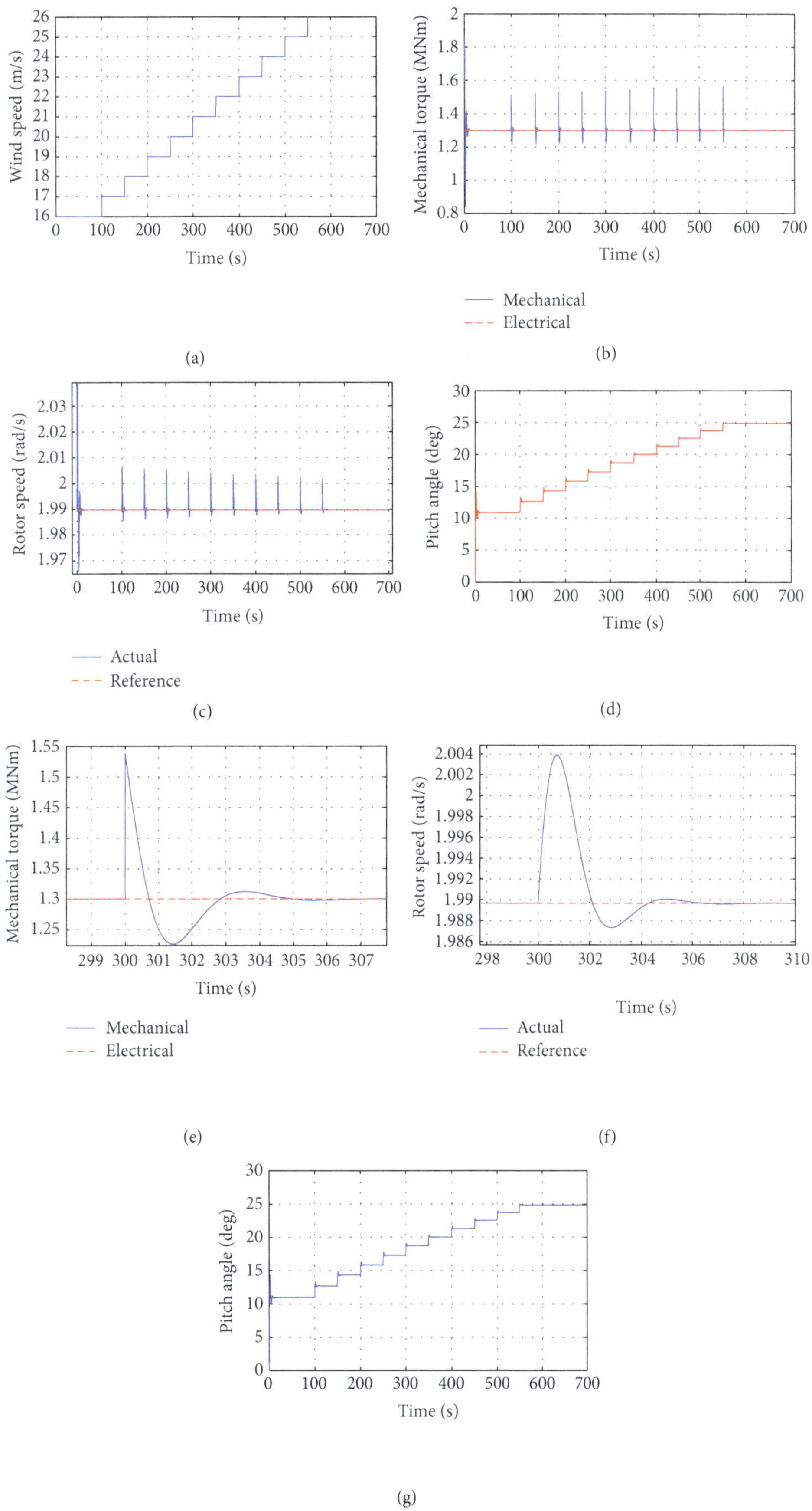

FIGURE 9: Step response of the pitch regulated WT system under ideal grid conditions: (a) wind speed, (b) output torque, (c) rotor speed, (d) pitch angle for the whole high speed region and (e) output torque, (f) rotor speed, and (g) pitch angle for one wind step.

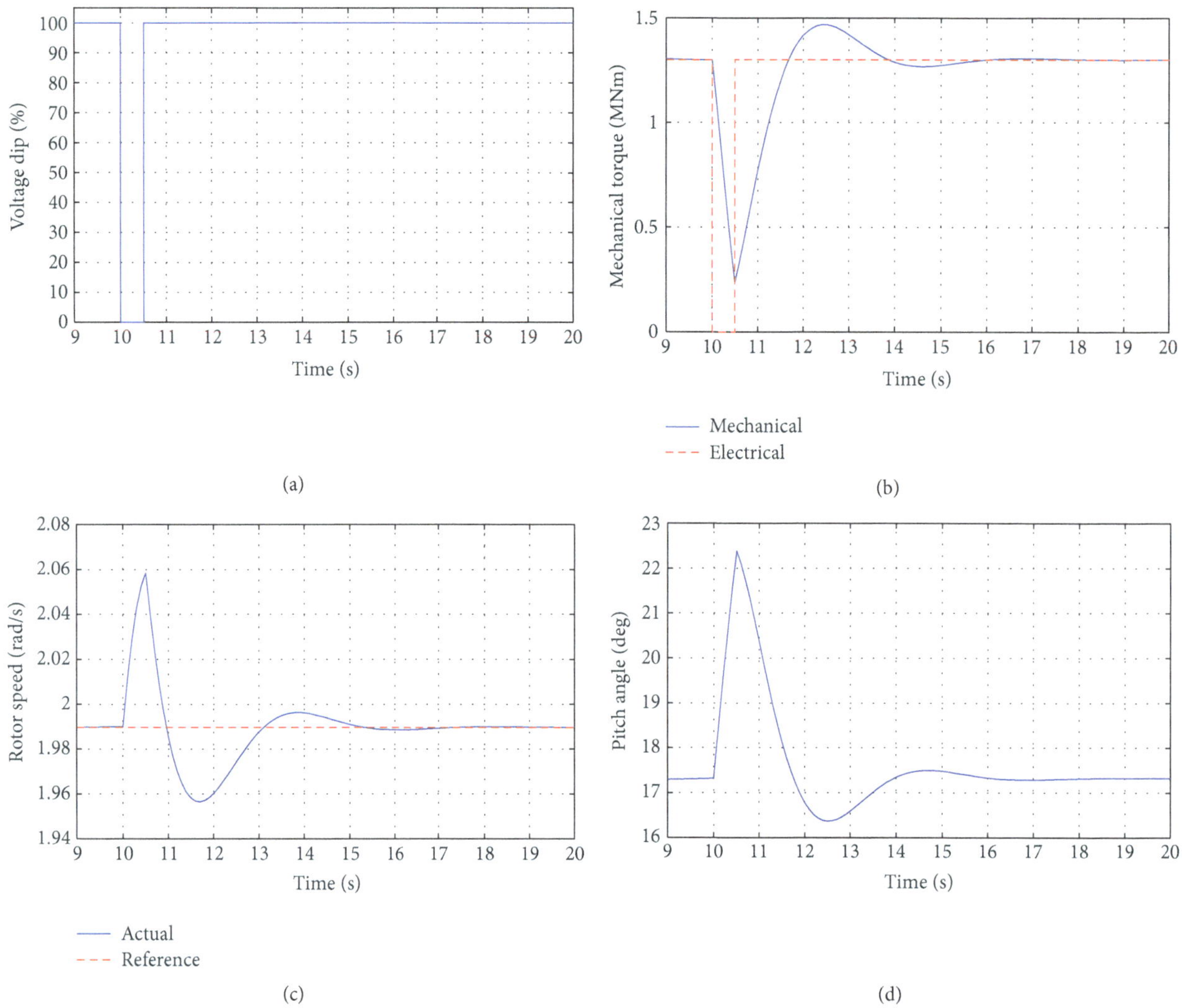

FIGURE 10: Step response of the pitch regulated WT system under 100% voltage dip: (a) power level, (b) output torque, (c) rotor speed, and (d) pitch angle.

not taken into consideration. Therefore, in order to prevent a large steady state error, the pitch controller needs to have an integral part in addition to the proportional part.

The integral parameter of the pitch regulator, in the absence of a natural damping, is calculated through the active damping method [10] as

$$k_i = \frac{\alpha^2 J}{K}. \quad (13)$$

The analytically expressed pitch regulated speed controller of the wind turbine system is hence derived.

6. Investigating the Performance of the Pitch Regulator

6.1. Verifying Normal Operation Behavior. The pitch regulated WT system is now studied when operating in the high speed region, which means that the output torque and the rotor speed of the turbine are operating close to their rated values. The nominal rotor speed of the wind turbine is 19 rpm. The nominal output power that this turbine can provide is 2 MW. Firstly, the response of the pitch regulated system is simulated for the wind increases. While the WT operates in steady state, the wind speed would be changing stepwise, and each step is initiated 50 seconds after the previous value. The results obtained from these conditions are shown in Figure 9.

When the WT system is subjected to a step wise increase in the wind speed, the shaft torque and rotor speed have slight overshoots. This is due to the integral part of the pitch regulator, which has been calculated through the active damping method. Although the instantaneous wind speed increase is slightly unrealistic, the overshoot of the rotor speed is 0.7%, which is an acceptable value, due to wind turbine regulations. It is worth noting that the wind steps simulated in this paper are instantaneous, which is a very drastic event compared to the realistic winds. The increase in the shaft torque is balanced by the pitching action. The pitch angle is increased in order to keep the shaft torque constant at the rated value for the wind turbine.

6.2. Voltage Dip. According to IEEE 1159, a reduction in the voltage is considered as a voltage dip when the RMS voltage

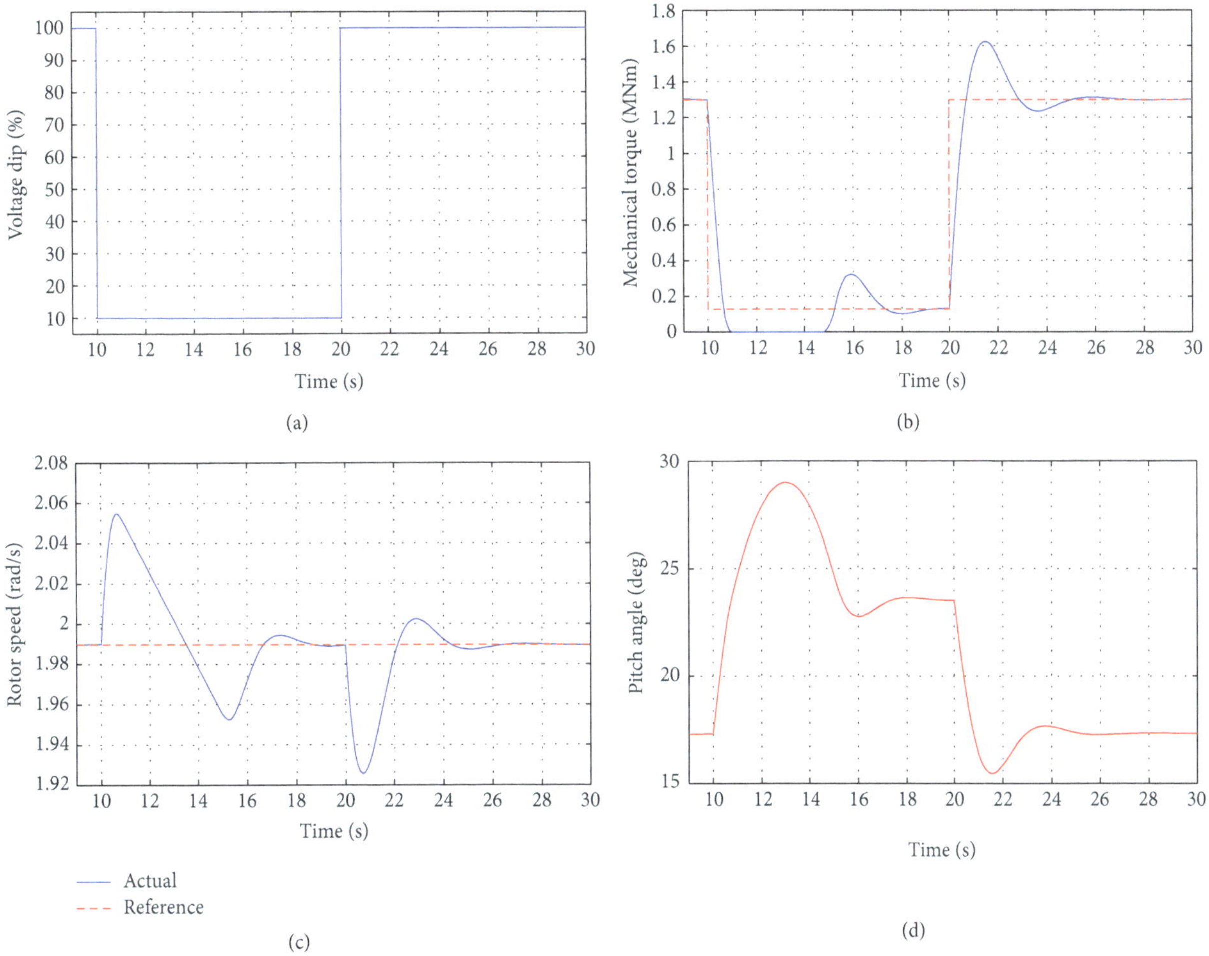

FIGURE 11: Step response of the pitch regulated WT system when the spinning reserve ability is activated and the power is reduced with 90%: (a) power level, (b) output torque, and (c) rotor speed, (d) pitch angle.

decreases between 0.1 and 0.9 p.u of the nominal voltage for a duration of half a cycle up to one minute [11].

Voltage dips are undesired incidents and might cause disturbances for the wind turbine system. There are several solutions for protecting the WT from the DC-link overvoltages that occur during a voltage dip and still keeping them online as stated by the grid codes, one of which is to use a braking chopper. A braking chopper is a device that discharges the DC-link of the wind turbine during a voltage dip and in this way prevents the DC-link voltage from being too high. When a braking chopper is used, the turbine operates undisturbedly, since all the excess energy is dissipated in the braking chopper. However, in this paper, pitch regulation is used in order to cope with the problem that all energies cannot be sent to the grid during the dip.

The system is tested for the most severe case, when the system experiences a voltage dip with 0% remaining voltage. The duration of the voltage dip is 500 ms. The simulation results are presented in Figure 10. When the dip occurs, the reference torque directly decreases to 0 from its prefault level. Generally, at a full power converter system, the wind turbine is not affected by the grid disturbances directly; however, in order to be able to control the WT system with the pitch regulator, the reference torque must be adjusted likewise. The rotor speed of the wind turbine accelerates and results in a 3.5% overshoot.

In order to be able to reduce the shaft torque by 100%, the pitch angle increases to 5°. This is expected due to the significant reduction in the voltage level. The effect of the pitch actuator is ignored in this paper, for simplicity.

All the parameters have overshoots but since the voltage dip is a short duration disturbance, the time to reach the steady state is not affected. As noted from Figure 10, the pitch regulated WT system is able to operate under severe voltage dips, up to 100% reduction.

6.3. Spinning Reserve Ability. A loss of load or generation in a power system leads to a frequency disturbance. These disturbances can be mitigated to a slight degree by changing the power level of wind turbines connected to the power system. Therefore, the pitch regulated WT system developed in this paper should be able to operate when the power needed by the system changes.

Of course, mainly the case with a required power increase is of the greatest interest; however, in order to broaden the investigation, the power level is first reduced and then increased back to its nominal level. The instant increase of the power level will show the spinning reserve ability of the pitch regulated WT system.

The spinning reserve action in this paper is simulated as the severe case scenario and for a longer duration than the voltage dip. The power level is decreased to 10% and the duration of it is 10 seconds. The simulation results can be observed from Figure 11.

When the power level is reduced, the reference torque directly decreases by 90%, instantly. Due to the insufficient electric loading of the generator for the present operating conditions, the rotor speed starts to increase. The rotor speed increase is detected by the controller and it is pushed back to its rated level, having a 3% overshoot when the controlling process starts.

The pitch angle difference adjusts the torque behavior of the WT system, and to be able to reduce the output torque by 90%, the pitch angle increases from its prefault value, 17.3° to 28° at first, but reaches a steady-state value of 23° for the new operation level. After the power is restored, the pitch angle decreases to its prefault value.

When the power level is restored, the rotor speed is decreased due to the increased electrical load; however, it reaches steady state shortly. From Figure 11 it can be observed that the pitch regulated WT system is able to operate when there is a drastic change in the power level, both for an increase and for a reduction.

7. Conclusion

A pitch regulator for a wind turbine system expressed in analytical terms has been designed on the basis of a simplified aerodynamic model of the wind turbine. These simplifications are made in order to calculate the torque produced by the wind turbine in an easily described way. This modified torque calculation allows us to extract a linear, torque-pitch angle relation, which is used while deriving the pitch regulator. Moreover, one single setting of the regulator parameters, which have been obtained using blade profile 4 at 20 m/s, is shown to be sufficient for the whole operating region.

The procedure presented in this paper is also applicable to other turbines. Using the aerodynamic data of that wind turbine, a pitch regulator can easily be derived.

The operation of the pitch regulated wind turbine system is shown to be robust and stable according to the simulation results. Since the offset of the pitch angle-mechanical torque relation is neglected while designing the regulator, the excess part of the equation is balanced by deriving an integrator, according to the *"active damping"* method. The integrator constant calculated through this method results in overshoots of the controlled parameters. However, since these overshoots are in the 1% range, accordingly the performance of the system is almost not affected. If desired, the integrator can be enhanced in order to have a smoother control.

The pitch regulated wind turbine system is able to operate even when a 100% voltage dip occurs. Furthermore, the system is tested for the spinning reserve ability and the system can at least handle a 90% change in the power level. After both disturbances, the pitch regulated WT system can easily return to steady state operation. The system is able to satisfy the grid codes which are explained in [12].

References

[1] E. Muljadi, "Pitch-controlled variable-speed wind turbine generation," *IEEE Transactions on Industry Applications*, vol. 37, no. 1, pp. 240–246, 2001.

[2] S. H. Lee, Y. J. Joo, J. Back, and J. H. Seo, "Sliding mode controller for torque and pitch control of wind power system based on PMSG," in *International Conference on Control, Automation and Systems (ICCAS '10)*, pp. 1079–1084, October 2010.

[3] J. Jonkman, S. Butterfield, W. Musial, and G. Scott, "Definition of a % MW reference wind turbine for offshore system development," Tech. Rep. NREL/TP-500-38060, National Renewable energy Laboratory, February 2009.

[4] F. Bianchi, H. De Battista, and R. J. Mantz, *Wind Turbine Control Systems-Principles, Modelling and Gain Scheduling Design*, Springer, London, UK, 2007.

[5] D. Eggleston and F. Stoddard, *Wind Turbine Engineering Design*, Van Nostrand Reinhold, 1987.

[6] P. Tokat, *Development and Evaluation of a Pitch Regulator for a Variable Speed Wind Turbine*, Chalmers University of Technology, March 2012.

[7] S. Heier, *Wind Energy Conversion Systems*, John Wiley & Sons, 2006.

[8] G. Ingram, *Wind Turbine Blade Analysis Using the Blade Element Momentum Method*, Durham University, 2011.

[9] T. Thiringer and A. Petersson, *Control of a Variable-Speed Pitch-Regulated Wind Turbine*, Chalmers University of Technology, 2005.

[10] L. Harnefors, *Control of Variable-Speed Drives*, Mälardalen University, 2002.

[11] IEEE Standard 1159–2009 (Revision of IEEE Standard 1159–1995), "IEEE Recommended Practica for Monitoring Electric Power Quality".

[12] M. Tsili and S. Papathanassiou, "A review of grid code technical requirements for wind farms," *IET Renewable Power Generation*, vol. 3, no. 3, pp. 308–332, 2009.

3

Life-Cycle Analyses of Energy Consumption and GHG Emissions of Natural Gas-Based Alternative Vehicle Fuels in China

Xunmin Ou[1,2] and Xiliang Zhang[1,2]

[1] *Institute of Energy, Environment and Economy (3E), Tsinghua University, Beijing 100084, China*
[2] *China Automotive Energy Research Center, Tsinghua University, Beijing 100084, China*

Correspondence should be addressed to Xunmin Ou; ouxm@mail.tsinghua.edu.cn

Academic Editor: Qilian Liang

Tsinghua life-cycle analysis model (TLCAM) has been used to examine the primary fossil energy consumption and greenhouse gas (GHG) emissions for natural gas- (NG-) based alternative vehicle fuels in China. The results show that (1) compress NG- and liquid NG-powered vehicles have similar well-to-wheels (WTW) fossil energy uses to conventional gasoline- and diesel-fueled vehicles, but differences emerge with the distance of NG transportation. Additionally, thanks to NG having a lower carbon content than petroleum, CNG- and LNG-powered vehicles emit 10–20% and 5–10% less GHGs than gasoline- and diesel-fueled vehicles, respectively; (2) gas-to-liquid- (GTL-) powered vehicles involve approximately 50% more WTW fossil energy uses than conventional gasoline- and diesel-fueled vehicles, primarily because of the low efficiency of GTL production. Nevertheless, since NG has a lower carbon content than petroleum, GTL-powered vehicles emit approximately 30% more GHGs than conventional-fuel vehicles; (3) The carbon emission intensity of the LNG energy chain is highly sensitive to the efficiency of NG liquefaction and the form of energy used in that process.

1. Background

1.1. Alternative Vehicle Fuels in China

1.1.1. General Background. During the period of the eleventh five-year (2006–2010) plan, the Chinese automobile market experienced strong growth. Vehicle ownership increased 20% annually during that period, amounting to 87 million by the end of 2010. This strong growth promoted a steadily increasing demand for fuels and, accordingly, a substantial escalation in petroleum prices. Meanwhile, pollutant and carbon dioxide (CO_2) emissions associated with petroleum combustion also posed major environmental concerns. Consequently, alternative vehicle fuels are gaining increasing interest.

Currently, several nonconventional fuels have been marketed as alternatives to regular vehicle fuels (i.e., gasoline and diesel); these include vehicular natural gas—compressed natural gas (CNG) and liquefied natural gas (LNG)—biodiesel, methanol-gasoline blends, ethanol-gasoline blends, and coal-derived fuels. Electric vehicles have also been introduced as a potential solution to the disadvantages of vehicles consuming conventional fuels.

The study of CAERC (2012) [1] found that the ownership of natural gas vehicles in China exceeded 500,000 as of 2010; annually 4.6 million tons of vehicular gasoline and diesel are being replaced by natural gas in 2010.

Moreover, the annual consumption of methanol-gasoline blends, biofuels, and coal-derived fuels was 200 million tons, 1.83 million tons, and 33,000 tons, respectively. The annual saving of gasoline and diesel by electric vehicles is less than 40,000 tons. In total, these alternative fuels replace only a fraction (3.3%) of China's vehicular gasoline and diesel consumption. However, the acceptance of alternative vehicle fuels is profoundly affected by industrial productivity and governmental policies. Furthermore, their acceptance may have a substantial impact on local liquid fuel markets.

1.1.2. Current Status. Two types of natural gas products have received attention as alternative vehicle fuels—CNG and LNG.

CNG Vehicle Fuel. According to the NGV Global (2012) [4] statistic, as of June of 2010, the ownership of natural gas vehicles in China exceeded 1,104,000; 98.9% of them are CNGV. Currently, CNG vehicles are in use in 30 provinces (cities or autonomous regions) of China. As of 2010, the officially registered CNG vehicle ownership amounted to 20,000. Taking unregistered vehicles into account (i.e., CNG vehicles converted from regular-fuel autos), the total number of CNG vehicles exceeded 500,000. Meanwhile, CNG station development has made considerable progress. As of 2009, there were 1,055 CNG stations across China, which was 500 more than there were at the end of 2007.

Statistical data of China LNG Vehicle Net (CLNGVN, 2012) [5] in 2010 from several regions where CNG vehicles have been relatively popular (i.e., Sichuan, Chongqing, Harbin, Urumqi, and Xi'an) show that CNG vehicles are primarily used as city buses (replacing diesel buses), taxicabs (replacing gasoline taxis), and governmental automobiles. Approximately 80% of these vehicles were used for commercial purposes.

According to the NGV Global (2012) [4], as of June of 2012, the ownership of natural gas vehicles in China exceeded 1.1 million; 98.9% of them are CNGV.

Current high gasoline prices reinforce the economic advantages of CNG vehicles. As a result, these vehicles have been well marketed in regions with abundant natural gas sources. However, owing to technical limitations (e.g., fuel availability and difficulties in fuel station construction), CNG vehicles are primarily suitable for city buses and short-distance transport. As of 2010, there were 600,000 CNG vehicles in China, which consumed 6 billion cubic meters of CNG annually as a replacement for 2.76 million tons of gasoline and 260,000 tons of diesel.

LNG Vehicle Fuel. LNG vehicles have not been widely utilized in China, primarily owing to the lack of LNG sources, the high cost of conversion to LNG, and lack of LNG fueling stations. According to a survey of seven Chinese cities in 2010, 2,800 LNG vehicles were in operation, of which 56% were city buses; a significant proportion of the remainder were accounted for by heavy trucks. For example, 400 LNG heavy trucks were used by the Guanghui Industry Investment Group (Xinjiang, China) for LNG delivery, and point-to-point transport between coal plants. In 2010, several cities (e.g., Erdos, Zhuhai) undertook pilot programs to promote the replacement of heavy diesel trucks by LNG alternatives.

Thus far, LNG vehicles are primarily being operated at the demonstration stage. Although LNG terminals began operation in Jiangsu and Dalian in 2011, they mainly targeted specific industrial users.

In recent 2 years, LNG vehicle market has developed very quickly and the ownership of LNG vehicles in China exceeded 70 thousand million; the numbers of LNG refueling stations reached to about 500 as the end of October of 2012 (CLNGVN, 2012) [5].

Consequently, LNG has played a more and more important role as an alternative to diesel as a fuel for vehicles.

1.1.3. Impact of Alternative Vehicle Fuels on Gasoline Demand in China. In 2010, the above-mentioned alternative fuels served as a replacement for approximately 3.3% (i.e., 7.1 million tons) of the total annual consumption of regular vehicle fuels. Specifically, gasoline alternatives (CNG, methanol, ethanol, and electricity) reduced annual vehicular gasoline consumption by approximately 6.6% (i.e., 60,000 tons). Diesel alternatives (LNG, coal-derived fuel, and biodiesel) were used to replace approximately 1.8% of annual vehicular diesel consumption. Of the various alternative vehicle fuels, CNG appears to be the most successful, accounting for 83% of gasoline substitution and 71% of that for diesel.

1.2. Life-Cycle Studies on Vehicle Fuels. Life-cycle analyses (LCAs) of energy consumption and greenhouse gas (GHG) emissions are an important component in assessing vehicle fuel pathways. Many studies have produced results that pertain to specific geographic locations (Zhang et al., 2008) [2].

In the last two decades, there have been intensive studies of alternative fuels and related vehicle technologies. Models have been used to analyze fuel use and carbon emissions of various alternative-fuel routes, such as the life-cycle emission model (LEM) and the greenhouse gas, regulated emissions, and energy use of transportation energy (GREET) model [6–9]. Many organizations and research groups have used the two models in their LCAs of alternative-fuel vehicles in various regions, such as Europe and North America [10, 11]. Their findings are highly region specific, and thus cannot be directly applied to other locations. By employing different models, researchers have investigated the LCA results of different energy pathways, including the natural gas pathway globally [12–21].

In China, earlier LCA studies focused on single-route analyses of passenger cars, new-energy vehicles, and vehicle operation (e.g., engine bench tests). Recent studies have increasingly focused on two- and multiple-route comparisons. However, these recent studies have lacked statistical information or basic data reflecting actual vehicle operations; many conclusions were drawn from experiments or predictions. Consequently, it is difficult to make a direct comparison between the findings from different studies.

Recently, Tsinghua University collaborated with Ford, General Motor, and the China Automotive Technology and Research Center (CATARC) to perform comparative LCAs for multifuel and vehicle pathways of vehicles using the GREET model. Because the original GREET model and default parameters were designed based on the energy production chain in the United States, the Chinese research team maximally incorporated localized data into its analyses. Thus, the results generally reflect the actual operation conditions in China [2, 3, 22].

2. Tsinghua Life-Cycle Analysis Model and Key Emission Intensities

The present study investigated the energy use and GHG emissions of various vehicle fuel pathways using the well-to-wheels (WTW) method and related LCA tools. This study focused only on LCA and included no subsequent evaluations.

The group has developed an LCA model termed the Tsinghua life-cycle analysis model (TLCAM), which is specifically designed to analyze vehicle fuels in China. The model is analogous to the GREET model (Wang, 1999)—a transportation energy model specific to the United States—and is implemented using Microsoft Excel (Microsoft, Redmond, WA, USA). TLCAM has been applied to WTW analyses of vehicle fuels under localized conditions in China [23–28].

The basic platform in this model was adapted from the GREET model, though it incorporates Chinese characteristics. The platform allows the user to determine the life-cycle fossil fuel intensities and GHG emission intensities of major end-use energies by means of iterative calculations.

In the present study, three primary fossil fuels were considered: coal, petroleum, and natural gas (NG). Nine forms of end-use energy were analyzed, including coal, petroleum, NG-based fuels, and electricity. The life cycle was divided into four phases for analysis: raw material production, raw material transport, fuel production, and fuel transport.

In our analyses, the system boundary was extended to cover the direct use of fuels (i.e., for processing and transport purposes), LCA energy consumption, and GHG emissions. However, indirect energy consumption (e.g., plant infrastructure construction and vehicle manufacture) was ignored.

This model clearly involves circular references. For example, petroleum mining and transportation require diesel, which is itself produced from petroleum. We solved this problem by using the circular reference function (i.e., iterative calculation) offered in Microsoft Excel. Finally, the platform outputs datasheets that detail the LCAs of the energy and carbon-emission intensities of various end-use energies.

In the current study, this model was used to generate the life-cycle primary fossil fuel intensities and major end-use energies in addition to their carbon emission intensities. The resulting datasheets were used as the basis (see Section 3) for subsequent analysis.

3. Basic Data

3.1. Energy Data

3.1.1. Petroleum and Refined Products. Table 1 summarizes data related to petroleum mining, transportation, and refining. Table 2 presents details of the transportation and distribution of refined petroleum products. It should be noted that petroleum-refining processes require the use of dry gas, which is not, however, one of the nine end-use energies considered in this study. This situation thus required special consideration: in the present study, dry gas was regarded as a raw material that involves the consumption of no additional primary fossil fuels, and it has a GHG intensity of 65 g/MJ.

Table 1: Energy-efficiency data related to petroleum mining and refining (%).[†]

Item	Value
Proportion of crude oil import	55.4
Extraction efficiency of crude oil	93.0
Energy efficiency of gasoline production	89.1
Energy efficiency of diesel production	89.7
Energy efficiency of fuel oil production	94.0

[†]Data assembled from publications [2, 3] and expert opinions.

3.1.2. Natural Gas. Tables 3 and 4 summarize the data related to the extraction, processing, transport, and distribution of natural gas.

3.2. Carbon Emission-Related Estimations. For our model, the carbon content and oxidation ratios of various types of energy were collected from an authoritative work containing basic data on GHG emissions in China. Methane (CH_4) emission factors for the combustion of oil and gas fuels under different facility conditions were obtained from IPCC report [29] and a report published by the 3E-THU (2003) [30]. The report was supported by the Ministry of Science and Technology of China and completed in collaboration with experts in the petroleum and petrochemical industries. The report estimated the methane emissions associated with various processes involved in the extraction, processing, transport, and consumption of petroleum and natural gas in China. Additionally, methane emission factors for coal combustion under different facility conditions were also collected. Nitrous oxide (N_2O) emission factors under various conditions used in the model were the default values published by IPCC (2006) [29].

It should be noted that natural gas extraction operations involve a small fraction (approximately 0.34%) of methane loss into the atmosphere.

4. WTW Classification of Phases of NG Fuels

After extraction and purification, NG is delivered to plants, where it is processed into liquid fuels or CNG for vehicle use. Figure 1 summarizes the life-cycle phases of NG fuels.

This section is primarily concerned with the energy use and GHG emissions of fuels. Specifically, our analyses were focused on two phases: well to pump (WTP) and pump to wheels (PTW). WTP refers to the upstream processes in vehicle fuel production, including resource extraction, resource transport, and the production, transport, distribution, storage, and tank-filling of fuel products. PTW refers to the downstream processes, including primarily the use (combustion) of fuels by vehicles and exhaust emissions.

In our analyses, the system boundary covered the direct use (i.e., process and traffic purposes) of fuels, as well as the LCA energy consumption and GHG emissions. Indirect

Table 2: Data related to petroleum transport and the transport and distribution of refined petroleum products.[†,‡]

Transport mode	Crude oil		Gasoline, diesel, and fuel oil	
	Percentage (%)	Average transport distance (km)	Percentage (%)	Average transport distance (km)
Ocean shipping	50	11000	0	0
Train	30	942	50	900
Pipeline	78	440	15	160
Ship	10	250	10	1200
Short-distance vehicle transport	0	0	10	50

[†]Data assembled from publications [2, 3] and expert opinions.
[‡]Owing to relay during transport, the sum of data for all modes may exceed 100%.

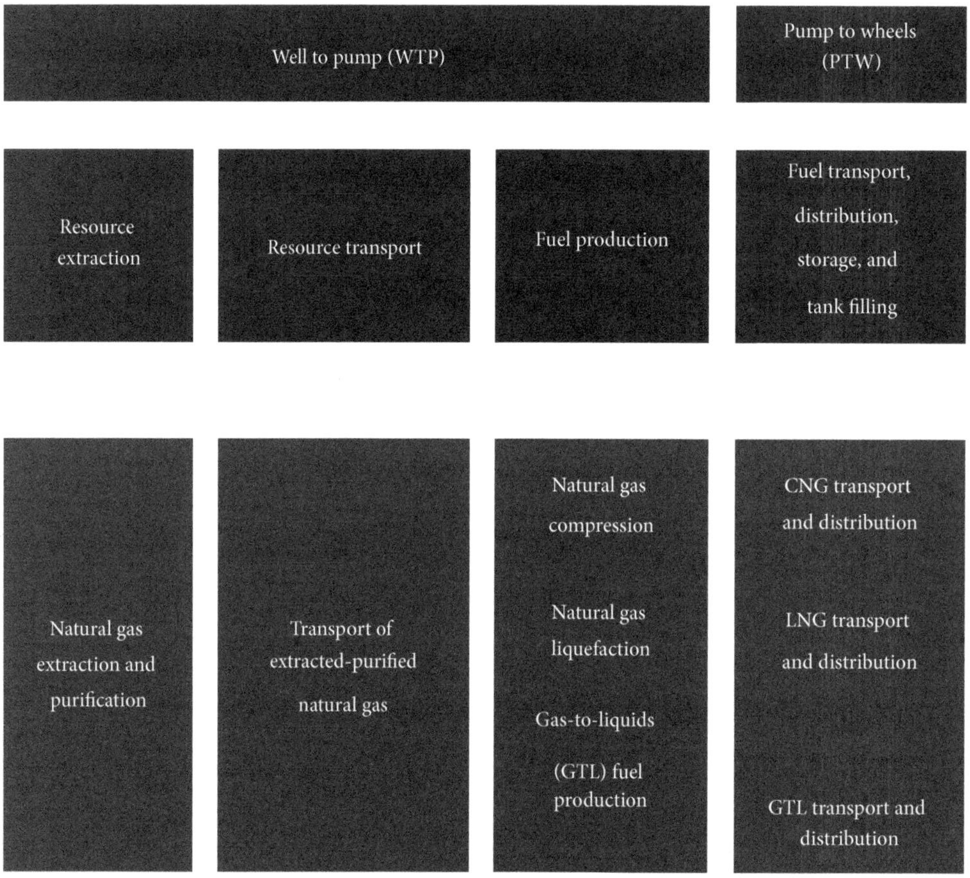

Figure 1: WTW classification of phases of NG fuels.

Table 3: Energy-efficiency data related to natural gas extraction and processing (%).

Item	Value
Extraction efficiency of natural gas	96.00
Natural gas processing efficiency	94.00

energy consumption (e.g., plant infrastructure construction and vehicle production) was ignored. However, our analyses considered methane loss (i.e., leakage to the atmosphere) during the production of coal, raw natural gas, and crude oil.

Table 4: Data related to pipeline transport of natural gas.

Application	Average transport distance (km)
Compressed natural gas	625
Process fuel	1500

5. Energy Consumption and Carbon Emissions during Various Phases

5.1. NG Extraction and Processing. In a previous study, CATARC (2007) [3] analyzed four major NG fields of

PetroChina (the fields account for 70% of all natural gas productivity in China) and predicted that the energy efficiency for NG extraction would be approximately 96.4% in 2015. Given the limited variation in this energy efficiency, we set it at 96% in our analyses. Additionally, Shen et al. (2012) [22] reported that the energy consumption for NG extraction and processing (i.e., purification) in China was approximately 10%. Accordingly, the energy efficiency of natural processing (i.e., purification) was estimated to be 94%; that is, (100%–10%)/96% = 94%. Meanwhile, NG extraction and processing primarily consume NG and electricity; the latter is usually provided by gas-fired power stations operated by the same NG fields and processing plants.

NG contains mainly methane, a major GHG. Consequently, methane leakage during NG extraction and processing contributes substantially to GHG emissions. The level of this leakage affects the energy efficiency and emission-reduction performance of NG fuels as alternative vehicle fuels.

5.2. CNG Production. CNG fueling stations mainly use electricity, water, and other resources. Additionally, they involve NG loss in the emptying of containers or pipelines. According to CATARC (2007) [3], the energy efficiency for NG compression is 96.9%, and this process involves the loss of NG (0.038 MJ/m^3 NG) and other energy sources (0.56 MJ/m^3 NG).

5.3. LNG Production. LNG-related processes primarily include liquefaction plants and LNG reception stations. The energy consumption involved in these processes includes electricity, water, and other resources. For imported LNG, the related energy consumption and efficiency were obtained from the GREET model. The comprehensive energy consumption was determined to be 90.2%, which included predominantly NG (98%) and a small fraction of electricity (2%). For LNG produced in China, there were two technical possibilities: (1) liquefaction near the gas field followed by short-distance transport and distribution for vehicle use; (2) long-distance pipeline transport of NG followed by liquefaction and short-distance transport and distribution for vehicle use. After consultation with the China National Offshore Oil Corp., we set the comprehensive energy efficiency for LNG production in China at 95%. LNG production consumes primarily electricity.

5.4. Gas-to-Liquid (GTL) Fuel Production. Currently, there is no GTL fuel production in China, and data related to the energy use of this process were obtained from the GREET model. The comprehensive energy consumption for this process was set at 54.2%. This process consumes predominantly NG (99.7%) and a minor fraction (0.3%) of n-butane.

5.5. NG Transport and Product Fuel Transport and Distribution. After purification, NG is generally transported by a long-distance pipeline to reach industrial users or city hubs. After delivery to city hubs, NG is further transported and distributed to final users. In recent years, the average distance of NG pipeline transport in China has rapidly increased—from 217 km in 1998 and 496 km in 2004 to 800 km in 2005. Moreover, several long-distance pipelines have been constructed, such as the west-east natural gas transmission engineering (3,900 km) and the China-Kazakhstan pipeline (3,000 km). In light of this trend, the average distance of NG pipeline transport was set at 1,500 km.

Because WTW analysis of NG-based vehicle fuels is sensitive to the mode and distance of NG transport, different transport distances were set for CNG, LNG, and GTL. The transport distance for CNG production is usually the average distance for NG transport within the country. Since CNG vehicles are predominantly used currently in regions with rich NG resources, the typical pipeline transport distance is approximately 300 km (e.g., the distance between the Kaixian gas field and Chongqing), and subsequent delivery is accomplished by vehicle transport.

Similar to earlier analyses (Section 5.3), three routes of NG supply application were considered: overseas import; local liquefaction followed by vehicle transport; pipeline transport followed by liquefaction and subsequent transport and distribution for vehicle use. For the first route, the average transport distance was set at 6,700 km (Shen et al., 2012) [22]. For the second route, the average vehicle transport distance was assumed to be 100 km. For the third route, the distance for transport and distribution was also set at 100 km.

For GTL fuels, an appropriate means of supply and application consists of production near gas fields followed by transport and distribution to final users. In this study, the typical distance between the gas field and the production plant was assumed to be 100 km, and the subsequent transport and distribution were assumed to be similar to those for diesel.

Moreover, the following assumption was made for estimating the energy consumption for NG transport: pipeline transport is driven by gas turbines that consume mainly NG (90%) and external electricity (10%) from the power grid. Data related to the energy consumption for overseas shipment of LNG were obtained from the GREET model.

5.6. Carbon Emissions during Various Phases of NG-Based Fuel Production and Use. In this section, carbon-emission patterns associated with the production and use of various phases of NG-based fuels (i.e., NG, CNG, LNG, and GTL) are analyzed.

5.6.1. Natural Gas. According to our calculations, the upstream processes account for 14.1% of the life-cycle GHG emissions of NG production and use: this comprises 12.6% emitted during NG extraction and processing and 1.5% by subsequent transport. Combining the GHG emissions during upstream processes and final applications, the NG energy chain produces total carbon emissions of 67.1 g $CO_{2,e}$/MJ.

5.6.2. CNG. For CNG, the upstream processes account for 21.2% of the life-cycle GHG emissions. Of these, NG extraction, NG transport, NG compression, and CNG transport amount to 11.6%, 0.3%, 9.3%, and 0.0%, respectively, of the total GHG emissions. In combination with GHG emissions from upstream processes and final application, this energy chain totally emits 73.2 g $CO_{2,e}$/MJ.

5.6.3. LNG Fuel. As described earlier, three supply routes were considered (Section 5.5). Our calculations suggest that for the first route (overseas import followed by supply to local cities for vehicle use), the upstream processes amount to 24.0% of the life-cycle GHG emissions: this comprises 11.2% from natural extraction, 9.4% from NG compression, 2.6% from LNG transport, and 0.8% from LNG distribution. Under this route, the LNG energy chain totally emits 75.7 g $CO_{2,e}$/MJ.

For the second route (liquefaction near gas fields followed by truck transport for vehicle use), the upstream processes account for 25.7% of the life-cycle GHG emissions: this comprises 10.9% from NG extraction, 14.0% from liquefaction, and 0.8% from LNG transport and distribution. Combining upstream processes and final application, the energy chain under this route totally emits 77.5 g $CO_{2,e}$/MJ.

For the third route (NG pipeline transport followed by liquefaction and further transport and distribution for vehicle use), the upstream processes account for 26.7% of the life-cycle GHG emissions: this comprises 10.8% from NG extraction and processing, 1.3% from NG transport and distribution, 13.8% from liquefaction, and 0.8% from LNG transport and distribution. Combining upstream processes and final application, the energy chain under this route totally emits 78.5 g $CO_{2,e}$/MJ.

5.6.4. GTL Fuel. For the application of GTL as an alternative vehicle fuel, the upstream processes amount to 48.0% of the life-cycle GHG emissions: this comprises 7.7% from NG extraction, 0.1% from NG transport, 40.1% from GTL production, and 0.2% from GTL transport and distribution. Combining upstream processes and final use, the GTL energy chain totally emits 146.8 g $CO_{2,e}$/MJ.

6. Life-Cycle GHG Emission: Analysis and Summary

Using a medium-size passenger car with an energy efficiency of 8 liters of gasoline consumed per 100 km as the baseline model, we can calculate the WTW fossil energy input and GHG emissions of such pathways as those for gas-based fuels. For different vehicle and fuel technology pathways, the fuel economy situation is presented in ratios using gasoline spark ignition (SI) vehicles as the baseline: diesel is 110%, CNG is 95.0%, LNG is 99.1%, and GTL is 110%. It should be noted that the vehicles are gauged under hypothetical conditions with heating and air-conditioning in use. The fuel consumption in real operating conditions is about 15% higher than in laboratory tests for inner combustion engine (ICE) vehicles.

Figures 2 and 3 summarize the GHG emission behavior of various NG-based fuels for vehicles. CNG- and LNG-powered vehicles have similar WTW fossil energy uses to conventional gasoline- and diesel-fueled vehicles, but differences emerge with the distance of NG transportation. Additionally, thanks to NG having a lower carbon content

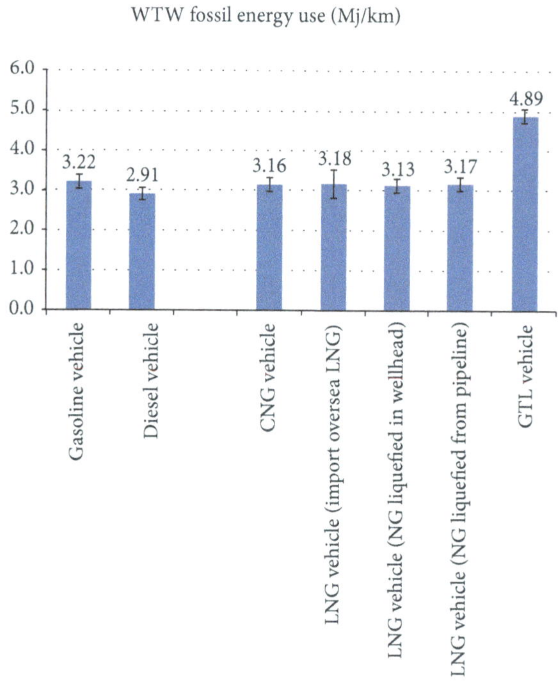

FIGURE 2: WTW fossil energy use of vehicles powered by NG-based fuels (MJ/km).

than petroleum, CNG- and LNG-powered vehicles emit 10–20% and 5–10% less GHGs than gasoline- and diesel-fueled vehicles, respectively. We assumed that errors in our modeling of CNG and LNG fuels would mainly arise with respect to the actual distance of transport.

However, GTL-powered vehicles involve approximately 50% more WTW fossil energy use than conventional gasoline- and diesel-fueled vehicles, primarily because of the low efficiency of GTL production. Nevertheless, since NG has a lower carbon content than petroleum, GTL-powered vehicles emit approximately 30% more GHGs than conventional-fuel vehicles. We considered uncertainty over the efficiency of future GTL production technology as representing a significant area of error in our analyses.

Since LNG is primarily targeted at commercial vehicles, we made efforts to compare the application of this alternative fuel for heavy trucks and buses. However, because of the lack of consistent data, it was difficult to make an accurate comparison. According to a study by Tang et al. (2011), the replacement ratio for diesel to NG is 90 : 100 (in terms of net energy). However, according to another source (http://wenku.baidu.com/view/75a1357e168884868762d6ba.html###), the diesel-to-NG replacement ratio should be 78 : 100 (in net energy). Because of this inconsistency, we performed calculations using these two values as the upper and lower limits. We found that, depending on the replacement ratio, LNG-powered vehicles may emit either 5% less or 12% more GHGs than conventional-fuel vehicles.

7. Sensitivity of Carbon Footprint of LNG

This section analyzes the sensitivity of the carbon footprint of LNG as an alternative vehicle fuel.

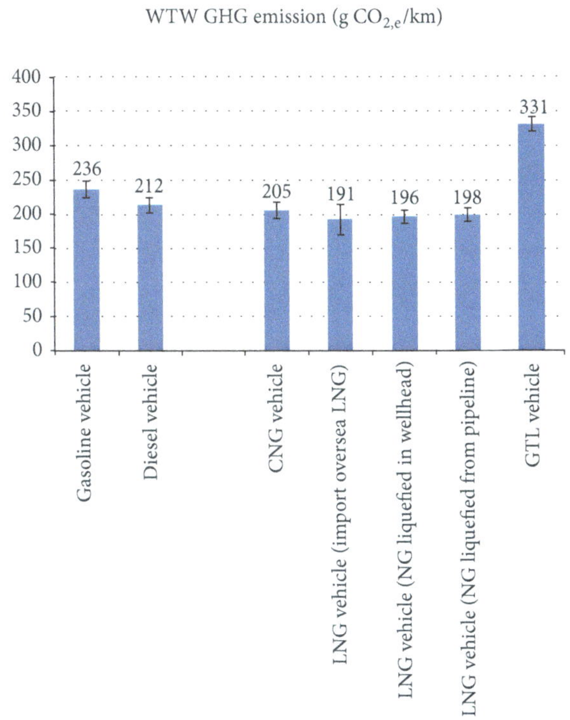

FIGURE 3: WTW GHG emission levels of vehicles powered by NG-based fuels (g $CO_{2,e}$/km).

7.1. Effect of LNG Supply Routes. As mentioned earlier (Section 5.5), three LNG supply routes were considered in this study: (1) overseas import (average shipment distance, 6,700 km (Shen et al., 2012) followed by supply to local cities (average transport distance, 100 km); (2) liquefaction near gas fields followed by truck transport (average distance, 100 km) for final vehicle use; (3) NG pipeline transport followed by liquefaction, transport, and distribution (distance, 100 km) for vehicle use. Our analyses suggest that the first route involves the lowest carbon intensity (75.7 g $CO_{2,e}$/MJ) and the second route the highest carbon intensity (78.5 g $CO_{2,e}$/MJ). The third route results in intermediate carbon intensity of 77.5 g $CO_{2,e}$/MJ.

7.2. Impact of Energy Efficiency and Energy Sources in NG Liquefaction. The carbon intensity of the LNG energy chain varied substantially with the type of energy used for NG liquefaction. Our calculations suggest that under the first supply route, assuming that the liquefaction plant uses electricity at a comprehensive energy efficiency of 95.2%, the upstream processes contribute 27.5% of the life-cycle GHG emissions. Combining these upstream processes and the subsequent LNG application, the LNG energy chain totally emits 79.3 g $CO_{2,e}$/MJ. This value amounts to a 4.8% higher carbon intensity than with another assumed situation (liquefaction primarily using NG).

Under the second route, assuming that the plant uses primarily NG at a comprehensive energy efficiency of 90.2%, the upstream processes contribute 21.6% of the life-cycle GHG emissions. Combining these upstream processes and final application, the LNG energy chain is expected to emit a total 73.5 g $CO_{2,e}$/MJ. This value amounts to a 5.2% lower carbon intensity than with another situation (liquefaction using electricity).

Moreover, with the third route, assuming that the liquefaction plant uses primarily NG at a comprehensive energy efficiency of 90.2%, the upstream processes contribute 22.7% of the life-cycle GHG emissions. Considering the upstream processes and final application, this energy chain totally emits 74.4 g $CO_{2,e}$/MJ. This emission level results in a 5.2% lower carbon intensity than with another assumed operation situation (liquefaction using electricity).

7.3. Effects of NG Transport. Under the three supply modes, reducing the transport distance for NG and LNG by 50% decreased the carbon intensity as follows: from 75.7 g $CO_{2,e}$/MJ to 74.5 g $CO_{2,e}$/MJ (first mode, 1.7% decrease); from 77.5 g $CO_{2,e}$/MJ to 77.2 g $CO_{2,e}$/MJ (second mode, 0.4% decrease); from 78.5 g $CO_{2,e}$/MJ to 77.7 g $CO_{2,e}$/MJ (third mode, 1.1% decrease).

7.4. Summary for the Sensitivity Analysis. Our calculations suggest that the carbon emission intensity of the LNG energy chain is highly sensitive to the efficiency of NG liquefaction and the form of energy used in that process. Moreover, this carbon emission intensity is moderately sensitive to different feedstock supply and fuel-production pathways but relatively insensitive to the distance of NG transport and LNG transport and distribution.

8. Concluding Remarks

(1) CNG- and LNG-powered vehicles have similar WTW fossil energy uses to conventional gasoline- and diesel-fueled vehicles, but differences emerge with the distance of NG transportation. Additionally, thanks to NG having a lower carbon content than petroleum, CNG- and LNG-powered vehicles emit 10–20% and 5–10% less GHGs than gasoline- and diesel-fueled vehicles, respectively.

(2) However, GTL-powered vehicles involve approximately 50% more WTW fossil energy uses than conventional gasoline- and diesel-fueled vehicles, primarily because of the low efficiency of GTL production. Nevertheless, since NG has a lower carbon content than petroleum, GTL-powered vehicles emit approximately 30% more GHGs than conventional-fuel vehicles.

(3) The carbon emission intensity of the LNG energy chain is highly sensitive to the efficiency of NG liquefaction and the form of energy used in that process.

Acknowledgments

The project is cosupported by the China National Natural Science Foundation (Grant no. 71041028, 71103109, and 71073095), China National Social Science Foundation (09 & ZD029), MOE Project of Key Research Institute of Humanities and Social Sciences at universities in China (2009JJD790029), and the CAERC program (Tsinghua/GM/SAIC-China).

References

[1] CAERC (China Automotive Energy Research Center, Tsinghua University), *China Automotive Energy Outlook 2012*, Scientific Press, Beijing, China, 2012.

[2] A. L. Zhang, W. Shen, W. J. Han, and Q. H. Chai, *Life Cycle Analysis of Automotive Alternative Fuel*, Tsinghua University Press, Beijing, China, 2008.

[3] CATARC (China Automotive Technology and Research Center), *Well-to-Wheels Analysis of Energy Consumption and GHG Emissions of Multi Vehicle Fuel in Future China*, CATARC, Beijing, China, 2007.

[4] NGV Global (natural gas vehicle global), "China's NGV Growth Accelerating," 2012, http://www.iangv.org/category/country/china/.

[5] CLNGVN (China LNG Vehicle Net), "LNG buses soar up in China," 2012, http://www.lngche.com/article-1396-1.html.

[6] M. A. Delluchi, *Emissions of GHG from the Use of Transportation Fuels and Electricity-Volume1: Main Text*, Center for Transportation Research, Argonne National Laboratory, Lemont, Ill, USA, 1991.

[7] M. A. Delluchi, *A Lifecycle Emissions Model (LEM): Main Texts*, Institute of Transportation Studies. University of California, Davis, Calif, USA, 2003.

[8] M. Wang, *GREET 1. 5-Transportaion Fuel-Cycle Model-Volume 1: Methodology, Development, Uses, and Results*, Center for Transportation Research, Argonne National Laboratory, Lemont, Ill, USA, 1999.

[9] M. Wang, H. Lee, and J. Molburg, "Allocation of energy use in petrol refineries to petrol products," *International Journal of Life Cycle Assessment*, vol. 9, no. 1, pp. 34–44, 2004.

[10] Concawe, EUCAR and EC Joint Research Centre, "Well-to-wheels analysis of future automotive fuels and powertrains in the European context," 2007, http://iet.jrc.ec.europa.eu/about-jec/sites/iet.jrc.ec.europa.eu.about-jec/files/documents/wtw3_wtw_report_eurformat.pdf.

[11] J. Wallace, M. Wang, T. Weber, and A. Finizza, "GM study: well-to-wheel energy use and greenhouse gas emissions of advanced fuel/vehicle systems—North American analysis," Tech. Rep., 2001, http://greet.es.anl.gov/publication-3plz9fyi.

[12] M. A. Weiss, J. B. Heywood, E. M. Drake, A. Schafer, and F. F. AuYeung, "On the road in 2020," Tech. Rep. MIT EL 00-003, Laboratory for Energy and the Environment (LFEE), Cambridge, Mass, USA, 2000.

[13] M. A. Weiss, J. B. Heywood, A. Schafer, and V. K. Natarajan, "Comparative assessment of fuel cell cars," Tech. Rep. MIT LFEE, 2003-001 RP, Laboratory for Energy and the Environment (LFEE), Cambridge, Mass, USA, 2003.

[14] J. Ally and T. Pryor, "Life-cycle assessment of diesel, natural gas and hydrogen fuel cell bus transportation systems," *Journal of Power Sources*, vol. 170, no. 2, pp. 401–411, 2007.

[15] D. Karman, "Life-cycle analysis of GHG emissions for CNG and diesel buses in Beijing," in *Proceedings of IEEE EIC Climate Change Technology Conference (EICCCC '06)*, pp. 1-6, May 2006.

[16] L. Kliucininkas, J. Matulevicius, and D. Martuzevicius, "The life cycle assessment of alternative fuel chains for urban buses and trolleybuses," *Journal of Environmental Management*, vol. 99, pp. 98–103, 2012.

[17] F. Ryan and B. Caulfield, "Examining the benefits of using bio-CNG in urban bus operations," *Transportation Research D*, vol. 15, no. 6, pp. 362–365, 2010.

[18] L. Rose, M. Hussain, S. Ahmed, K. Malek, R. Costanza, and E. Kjeang, "A comparative life cycle assessment of diesel and compressed natural gas powered refuse collection vehicles in a Canadian city," *Energy Policy*, vol. 52, pp. 453–461, 2013.

[19] A. Arteconi, C. Brandoni, D. Evangelista, and F. Polonara, "Life-cycle greenhouse gas analysis of LNG as a heavy vehicle fuel in Europe," *Applied Energy*, vol. 87, no. 6, pp. 2005–2013, 2010.

[20] CARB (California Air Resource Board), "Comparison of greenhouse gas emissions from natural gas and diesel vehicles," Tech. Rep. 08/10/2008.CARB, 2008.

[21] O. P. R. van Vliet, A. P. C. Faaij, and W. C. Turkenburg, "Fischer-Tropsch diesel production in a well-to-wheel perspective: a carbon, energy flow and cost analysis," *Energy Conversion and Management*, vol. 50, no. 4, pp. 855–876, 2009.

[22] W. Shen, W. J. Han, D. Chock, Q. H. Chai, and A. L. Zhang, "Well-to-wheels life-cycle analysis of alternative fuels and vehicle technologies in China," *Energy Policy*, vol. 49, pp. 296–307, 2012.

[23] X. Li, X. Ou, X. Zhang, Q. Zhang, and X. L. Zhang, "Life-cycle fossil energy consumption and greenhouse gas emission intensity of dominant secondary energy pathways of China in 2010," *Energy*, vol. 50, pp. 15–23, 2013.

[24] X. Ou, Y. Xiaoyu, and X. Zhang, "Life-cycle energy consumption and greenhouse gas emissions for electricity generation and supply in China," *Applied Energy*, vol. 88, no. 1, pp. 289–297, 2011.

[25] X. Ou, X. Zhang, S. Chang, and Q. Guo, "Energy consumption and GHG emissions of six biofuel pathways by LCA in China," *Applied Energy*, vol. 86, no. 1, pp. S197–S208, 2009.

[26] X. Ou, X. Zhang, and S. Chang, "Alternative fuel buses currently in use in China: life-cycle fossil energy use, GHG emissions and policy recommendations," *Energy Policy*, vol. 38, no. 1, pp. 406–418, 2010.

[27] X. Ou, X. Zhang, and S. Chang, "Scenario analysis on alternative fuel/vehicle for China's future road transport: life-cycle energy demand and GHG emissions," *Energy Policy*, vol. 38, no. 8, pp. 3943–3956, 2010.

[28] X. M. Ou and X. L. Zhang, *Life-Cycle Analysis of the Automotive Energy Pathways in China*, Tsinghua University Press, Beijing, China, 2011.

[29] IPCC (Intergovernmental Panel on Climate Change), "IPCC Guidelines for National GHG Inventories," 2006.

[30] 3E-THU (Institute of Energy, Environment and Economy at Tsinghua University), *Report of Inventory of CH4 Emission Sources and Sinks of China's Petroleum and Natural Gas Sector in 2000*, Institute of Energy, Environment and Economy, Tsinghua University, Beijing, China, 2003.

4

Determination of Free Fatty Acid by FT-NIR Spectroscopy in Esterification Reaction for Biodiesel Production

Djéssica Tatiana Raspe and Camila da Silva

Department of Technology, Maringá State University (UEM), 87506-370 Umuarama, PR, Brazil

Correspondence should be addressed to Camila da Silva; camiladasilva.eq@gmail.com

Academic Editor: Johan E. Hustad

This work reports the use of FT-NIR spectroscopy coupled with multivariate calibration to determine the percentage of free fatty acids (FFA) in samples obtained by the esterification of FFA in vegetable oils. The analytical method used as calibration matrix samples of the reaction medium of esterification of oleic acid in soybean oil in proportions of 0.3 to 40 wt% (by weight) of oleic acid obtained under different experimental conditions and utilized the partial least squares (PLS) regression. The efficiency of the method was tested to predict the content of FFA in reactions of esterification of oleic acid in soybean oil catalysed by KSF clay and Amberlyst 15 commercial resin, both in a batch mode. Good Correlations were observed between the FT-NIR/PLS method and the reference method (AOCS). The results confirm that FT-NIR spectroscopy, in combination with multivariate calibration, is a promising technique for monitoring esterification reaction for biodiesel production.

1. Introduction

The merits of biodiesel (fatty acid ethyl or methyl esters) obtained from vegetable oils or animal fats as an alternative to mineral diesel comprise a nontoxic, biodegradable, and a domestically produced and renewable source which is well documented in the literature [1, 2]. Because of the well-known environmental and economical benefits, biodiesel fuel may be expected to be a good alternative to petroleum-based fuel [3]. The biodiesel production is generally accomplished in homogeneous [4–6] and heterogeneous [7–9] reaction systems at low pressures. Recently, the noncatalytic reaction, using alcohol under supercritical conditions at high temperatures and pressures, has also been investigated as an alternative method for fatty acid esters production [10–12].

Fatty acid, methyl or ethyl, esters can be usually obtained from free fatty acid (FFA) esterification reaction, through vegetable oils hydrolysis followed by the fatty acid esterification or from direct vegetable oils transesterification [13]. The importance of examining the esterification reaction in a more detailed manner is justified by the huge amount and variety of vegetable oils worldwide available for biodiesel production [14–16], which may have a high percentage of FFA making the conventional alkali-catalyzed transesterification impracticable, since for this process, the percentage of FFA is to be less than 0.5% [1, 17, 18]. Recent studies propose to obtain esters in two reaction steps of substrates with high acidity, consisting of two approaches: (1) esterification of FFA and subsequent transesterification of triglycerides [19–22] or (2) hydrolysis of triglycerides, followed by esterification of fatty acids obtained [23–25].

According to AOCS [26], the acidity in vegetable oils can be determined by acid-base titration, using sodium hydroxide titrant and phenolphthalein as an indicator, such a method is commonly used because of simplicity of operation however, the method is slow, laborious; and lacks precision. In determining, large quantities of samples and reagents/solvents are consumed during operation [27–29]. The content of FFA can be determined by gas chromatography that involves complex steps of sample preparation and derivatization as the costs of internal standardization. Other methods involve techniques of high performance liquid chromatography [13] which eliminates the steps of derivatization; however, there is the high consumption of solvents used as mobile phase and the long time for analysis. A method that has been able to overcome such drawbacks is the FT-NIR spectroscopy, with the analytical procedure which provides rapid, direct (no pretreatment), reproducible, and moreover, the cost per analysis that is relatively low when compared to classical techniques [30, 31] and may be used in online monitoring of chemical reactions [32].

In this sense, the main objective of the present work is to investigate the application of spectroscopy FT-NIR for the quantitative determination of FFA in esterification reaction for biodiesel production.

2. Materials and Methods

2.1. Materials. Commercial refined soybean oil (Soya) and oleic acid (Sigma-Aldrich 99.9%) were used as substrates without further treatment. In esterification reactions, ethanol (Merck 99.9%) and as catalyst: KSF clay (Sigma-Aldrich) and Amberlyst 15 commercial resin (Supelco) were utilized. The characteristics of the catalysts are presented in the work of Zanette et al. [33]. In step titration of samples: solution of ethyl alcohol/ethyl ether 1 : 2 (v : v) (Vetec/Nuclear), 0.1 N sodium hydroxide (Nuclear), and phenolphthalein indicator (Nuclear) were used. All other reagents and solvents were of analytical grade.

2.2. Esterification Reactions of Oleic Acid in Soybean Oil. The samples obtained by esterification reaction were used for calibration and validation of the methodology. The esterification of oleic acid in soybean oil with ethanol in the presence of a heterogeneous catalyst (KSF clay and Amberlyst resin) was studied, and a synthetic sample of soybean oil with 40 wt% of oleic acid was utilized. For calibration and validation steps, different reactions were carried using different molar ratios of oleic acid to ethanol (from 1 : 10 to 1 : 40), temperature (from 50°C to 80°C), and amounts of catalyst (from 2 to 20 wt%) in order to obtain samples with different oleic acid contents.

The esterification reaction was carried out with magnetic stirrers in a batch reactor equipped with condenser and immersed in a temperature-controlled water bath. The oil was heated until the desired temperature was reached. At this point, ethanol and the catalyst were added to the oil, and the esterification reaction began. At the end of reaction, the catalyst was separated by centrifuged (3000 rpm for 10 minutes), and the ethanol and water were removed from the reaction mixture using a rotary evaporator.

2.3. FT-NIR Method. For the calibration of the FT-NIR method, FFA concentrations in samples from the reaction medium of esterification were determined by both FT-NIR and the conventional titration method by acid-base titration with 0.1 M NaOH in the presence of phenolphthalein. FFA content (FT-NIR and titration) was determined twice for each sample, and the average values are reported. The FT-NIR absorbance spectra were acquired in the range from 700 to 2500 nm in spectrophotometer FT-NIR model N-200 (Büchi).

Sixty spectra were used in these procedures, which were divided into two sets of data by the NIRCal software. The first set contained 40 spectra which were used to create the calibration protocol by extracting from them the relevant information. The second set contained 20 spectra which were used in the validation of calibration. Spectral treatments were applied, and calibration models were constructed using the software NIRCal 4.01 (Büchi). Multivariate calibration was performed by partial least squares (PLS) regression in order to study the relationship between the reference data and the FT-NIR data. This was assessed by a leave-one-out cross-validation procedure and optimized in terms of the appropriate number of factors and pretreatment by minimizing the root-mean-square error (RMSE) as in the following:

$$\mathrm{RMSE} = \sqrt{\sum_{i=1}^{n} \frac{\left(y_{\mathrm{FT\text{-}NIR}} - y_{\mathrm{REF}}\right)^2}{N}}, \quad (1)$$

where the terms "FT-NIR" and "REF" denote the predicted value and reference value, respectively. The relative standard error is designated as RMSEC (%) for calibration and RMSEV (%) for validation. The model that was finally adopted was that which led to the smallest possible RMSE value for a set of samples not previously used to construct it.

The calibration curve obtained is a mathematical algorithm that contains the necessary information to carry out the pretreatment of the obtained spectrum, more a linear function of Absorbance versus FFA content that can be used to predict the concentration of the chemical component of interest in samples of unknown composition using the instrumental response of them.

The predictive capability was tested against fourteen samples of reaction medium of esterification. The device outputs were compared with those obtained by the reference method [26] by means of linear correlation, using the software Statistica 6.0 (StatSoft Inc.). A prediction model based on the same test set was generated and evaluated via the root-mean-square error of prediction (RMSEP).

3. Results and Discussion

3.1. Calibration Model for FFA Content. Forty samples of esterification of oleic acid in soybean oil were used for the development of the calibration model for the FFA content. The FFA content in the samples ranged from 0.3 wt% to 45 wt%. The calibration of the method adopted as criterion the acidity values reported in the literature for the vegetable oils used for the biodiesel production [13]. The pretreated FT-NIR spectra utilized in the construction of the calibration curve as well the respective wavelength range analyzed are shown in Figure 1.

The calibration of the method was evaluated considering the correlation between the percentage of free fatty acids (FFA) predicted by FT-NIR and determined by the reference method [26]. In the chemometric treatment of data, the method of partial least squares (PLS) regression evaluated to the data for the spectrum obtained in the spectral region between 1000 and 2500 nm. Spectra pretreatment was necessary to avoid baseline fluctuations several methods of pretreatment have been tested, and the results are summarized in Table 1.

TABLE 1: Cross-validation and external validation results for the prediction of FFA content (%) using different pretreatment methods.

Parameter	Baseline correction[a]	[a]+SG smoothing[b]	[b]+ Derivative[c]
PLS factors	17	17	17
R^2 calibration	0.819	0.8124	0.9487
R^2 validation	0.966	0.9698	0.9407
RMSEC (%)	2.022	1.968	1.35
RMSEV (%)	3.706	2.685	1.68

[a]Obtained using as reference the spectra without pretreatment.
[b]Savitzky-Golay smoothing.
[c]First derivative obtained by the Savitzky-Golay method.

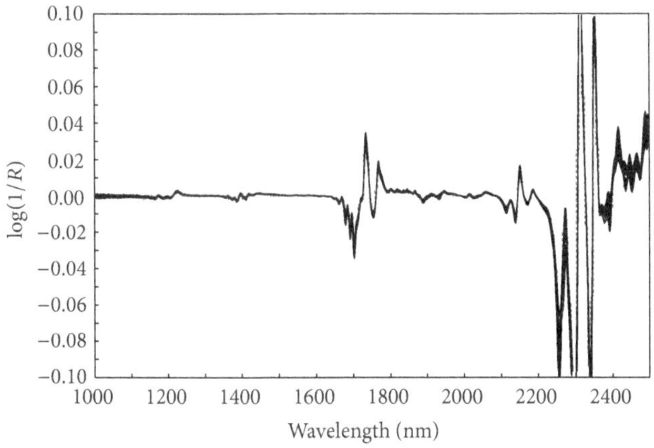

FIGURE 1: Pretreated FT-NIR spectra of samples of esterification medium.

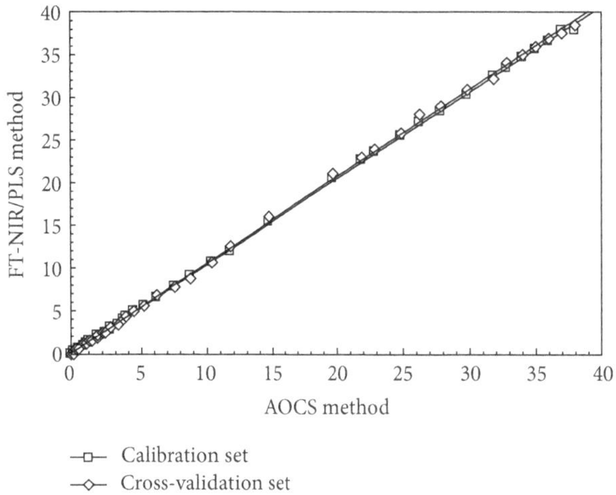

FIGURE 2: Correlation between the method of reference (AOCS, 1998) and FT-NIR/PLS for FFA content (wt%) in samples of reaction medium of esterification (□), Calibration set; (◊) cross-validation set.

The lowest RMSEV obtained was 1.35 and 1.68 using 17 (seventeen) PLS factors for calibration and validation of the method, respectively. The linear regression made between the points that represent the predicted content of FFA against the measured points was 0.9487. In cross-validation test, this regression was 0.9407.

Chang et al. [34] reported the application of PLS method and FT-NIR spectroscopy to monitor the enzymatic inter-esterification. The correlation found by these authors was 0.903 at 20°C, and values of RMSEC and RMSEP was 1.096 and 1.827, respectively, using 5 (five) factors for the PLS method. Multivariate curve resolution-alternating least squares (MCR-ALS) methodology applied to near-infrared spectroscopy (NIR) data for the esterification reaction between glycerol and a mixture of caprylic and capric acids was reported by Blanco et al. [35]. The authors reported values of 3.7 and 6.1 for steps of calibration and prediction of methodology in concentration interval of 0.4 to 162 (mg KOH/g sample) for acid value. Zhang et al. [36] showed the application of FT-NIR spectroscopy in order to monitor the enzymatic interesterification process for butterfat modification using PLS method and values of 0.173 and 0.675 were reported for RMSEC and RMSEV, respectively. Blanco et al. [37] show the potential of near-infrared (NIR) spectroscopy for monitoring the esterification of stearic acid by glycerol constructed PLS models and reported relative standard error (RSEC and RSEV) of 2.1 and 6.7, respectively.

Figure 2 shows the performance of the model developed using PLS method to the experimental data. The results presented in Table 1 and Figure 2 demonstrated that the calibration models developed perform very well in the prediction of the FFA content.

3.2. Prediction Efficiency of FT-NIR/PLS Method for Monitoring Esterification Reactions. After determining the mathematical algorithm of prediction (linear function of the calibration curve more the data pretreatment condition), it was inserted into the prediction software of the FT-NIR equipment. Then, its predictive capability was tested in an independent dataset of reaction medium of esterification, and the FFA content was determined as the previously described methodology.

The output data were then inserted into the Statistica 6.0 software. In this independent dataset, the linear regression between the points that represents the reference methodology (AOCS, 1998) against the predicted by FT-NIR/PLS methodology was 0.99 ± 0.05 (Figure 3). The RMSEPs obtained for the results reported in Figure 3 were 1.15 and 1.07 for reaction with KSF and Amberlyst as catalysts, respectively.

According to our results obtained with the prediction of FFA content in the reaction medium of esterification from the independent set of samples, it was observed that both methodologies are well correlated and now, this methodology can be used routinely to predict the yields of esterification reaction for biodiesel production.

The use of ^{1}H NMR method for the quantification of FFA content during the esterification of oleic acid in methanol was reported by Satyarthi et al. [29] with high correlation (0.99) with this method and the titration method. Aryee et al. [28] evaluated the suitability of FT-IR method for the determination of FFA in fish oils for the production of

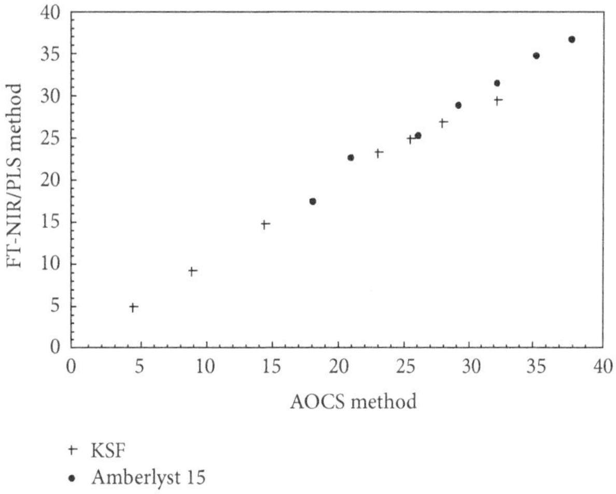

FIGURE 3: Comparison of FFA content determined by the reference method (AOCS, 1998) and predicted by the FT-NIR/PLS for samples of reaction medium of esterification with different catalysts.

biodiesel and reported a high correlation to the proposed methodology for FFA concentration from 0 to 6.5% by weight.

For the quantification of methyl esters in samples of transesterification reactions by NIR spectroscopy and PLS method, Baptista et al. [38] reported an R^2 prediction of 0.930 and RMSEP of 0.9%, and Killners et al. [39] for on-line monitoring of the biodiesel production reported RMSEPs of 0.74 and 1.27 for the transesterification reaction carried out at 20°C and 55°C, respectively.

4. Conclusion

This work reported experimental data for FT-NIR calibration models for prediction of the esterification yields for biodiesel production. Results demonstrate that FT-NIR spectroscopy is a valuable tool to predict the FFA content in samples of reaction medium of esterification, with errors similar to the maximum errors admitted by the reference method. The FT-NIR/PLS method shows RMSEP of about 1.1% and high correlation ($R > 0.99$) between the methodology developed and the reference method. The FT-NIR/PLS is nondestructive and requires a small amount of the sample. It can be used to monitor the esterification reaction of FFA to biodiesel.

Acknowledgments

The authors thank CNPq (472962/2009-4), CAPES (PROCAD and ProEngenharias), and Maringá State University (UEM).

References

[1] H. Fukuda, A. Kondo, and H. Noda, "Biodiesel fuel production by transesterification of oils," *Journal of Bioscience and Bioengineering*, vol. 92, no. 5, pp. 405–416, 2001.

[2] R. Altin, S. Çetinkaya, and H. S. Yücesu, "Potential of using vegetable oil fuels as fuel for diesel engines," *Energy Conversion and Management*, vol. 42, no. 5, pp. 529–538, 2001.

[3] D. C. Grotta, A. Lopes, C. E. A. Furlani, R. P. Silva, G. N. Reis, and J. W. Cortez, "Biodiesel etílico filtrado de óleo residual de soja: desempenho de um trator agrícola na operação de gradagem," *Acta Scientiarum Tecnhology*, vol. 30, no. 2, pp. 135–138, 2008.

[4] D. De Oliveira, M. Di Luccio, C. Faccio et al., "Optimization of alkaline transesterification of soybean oil and castor oil for biodiesel production," *Applied Biochemistry and Biotechnology Part A*, vol. 122, no. 1-3, pp. 553–560, 2005.

[5] G. Vicente, M. Martínez, and J. Aracil, "Kinetics of Brassica carinata oil methanolysis," *Energy and Fuels*, vol. 20, no. 4, pp. 1722–1726, 2006.

[6] S. M. P. Meneghetti, M. R. Meneghetti, C. R. Wolf et al., "Biodiesel from castor oil: a comparison of ethanolysis versus methanolysis," *Energy and Fuels*, vol. 20, no. 5, pp. 2262–2265, 2006.

[7] G. S. Silva, D. Y. Inoue, G. Dors, A. Furigo Jr., and H. F. Castro, "Desempenho de diferentes lipases imobilizadas na síntese de biodiesel de óleo de palma," *Acta Scientiarum Tecnhology*, vol. 33, no. 2, pp. 197–203, 2011.

[8] S. Semwal, A. K. Arora, R. O. Badoni, and D. K. Tuli, "Biodiesel production using heterogeneous catalysts," *Bioresource Technology*, vol. 102, no. 3, pp. 2151–2161, 2011.

[9] A. P. S. Chouhan and A. K. Sarma, "Modern heterogeneous catalysts for biodiesel production: a comprehensive review," *Renewable and Sustainable Energy Reviews*, vol. 15, no. 9, pp. 4378–4399, 2011.

[10] T. Pinnarat and P. E. Savage, "Assessment of noncatalytic biodiesel synthesis using supercritical reaction conditions," *Industrial and Engineering Chemistry Research*, vol. 47, no. 18, pp. 6801–6808, 2008.

[11] D. Wen, H. Jiang, and K. Zhang, "Supercritical fluids technology for clean biofuel production," *Progress in Natural Science*, vol. 19, no. 3, pp. 273–284, 2009.

[12] C. Silva, G. Borges, F. Castilhos, J. V. Oliveira, and L. Cardozo Filho, "Continuous production of fatty acid ethyl esters from soybean oil at supercritical conditions," *Acta Scientiarum Tecnhology*, vol. 34, no. 2, pp. 185–192, 2012.

[13] G. Knothe, J. V. Gerpen, and J. Krahl, *The Biodiesel Handbook*, AOCS Press, Champaign, Ill, USA, 2005.

[14] M. K. Lam, K. T. Lee, and A. R. Mohamed, "Homogeneous, heterogeneous and enzymatic catalysis for transesterification of high free fatty acid oil (waste cooking oil) to biodiesel: a review," *Biotechnology Advances*, vol. 28, no. 4, pp. 500–518, 2010.

[15] P. Nakpong and S. Wootthikanokkhan, "High free fatty acid coconut oil as a potential feedstock for biodiesel production in Thailand," *Renewable Energy*, vol. 35, no. 8, pp. 1682–1687, 2010.

[16] A. Hayyan, M. Z. Alam, M. E. S. Mirghani et al., "Reduction of high content of free fatty acid in sludge palm oil via acid catalyst for biodiesel production," *Fuel Processing Technology*, vol. 92, no. 5, pp. 920–924, 2011.

[17] F. Ma and M. Hanna, "Biodiesel production: a review," *Bire-source Technology*, vol. 70, no. 1, pp. 1–15, 1999.

[18] A. P. Vyas, J. L. Verma, and N. Subrahmanyam, "A review on FAME production processes," *Fuel*, vol. 89, no. 1, pp. 1–9, 2010.

[19] M. Berrios, M. A. Martín, A. F. Chica, and A. Martín, "Study of esterification and transesterification in biodiesel production from used frying oils in a closed system," *Chemical Engineering Journal*, vol. 160, no. 2, pp. 473–479, 2010.

[20] K. Srilatha, T. Issariyakul, N. Lingaiah, P. S. Sai Prasad, J. Kozinski, and A. K. Dalai, "Efficient esterification and transesterification of used cooking oil using 12-tungstophosphoric acid (TPA)/Nb_2O_5 catalyst," *Energy and Fuels*, vol. 24, no. 9, pp. 4748–4755, 2010.

[21] L. T. Thanh, K. Okitsu, Y. Sadanaga, N. Takenaka, Y. Maeda, and H. Bandow, "A two-step continuous ultrasound assisted production of biodiesel fuel from waste cooking oils: a practical and economical approach to produce high quality biodiesel fuel," *Bioresource Technology*, vol. 101, no. 14, pp. 5394–5401, 2010.

[22] G. Corro, N. Tellez, T. Jimenez, A. Tapia, F. Banuelos, and O. Vazquez-Cuchillo, "Biodiesel from waste frying oil. Two step process using acidified SiO_2 for esterification step," *Catalysis Today*, vol. 166, no. 1, pp. 116–122, 2011.

[23] S. Gan, H. K. Ng, C. W. Ooi, N. O. Motala, and M. A. F. Ismail, "Ferric sulphate catalysed esterification of free fatty acids in waste cooking oil," *Bioresource Technology*, vol. 101, no. 19, pp. 7338–7343, 2010.

[24] L. L. Rocha, A. L. D. Ramos, N. R. A. Filho, N. C. Furtado, C. A. Taft, and D. A. G. Aranda, "Production of biodiesel by a two-step niobium oxide catalyzed hydrolysis and esterification," *Letters in Organic Chemistry*, vol. 7, no. 7, pp. 571–578, 2010.

[25] J. S. de Sousa, E. D. Cavalcanti-Oliveira, D. A. G. Aranda, and D. M. G. Freire, "Application of lipase from the physic nut (*Jatropha curcas* L.) to a new hybrid (enzyme/chemical) hydroesterification process for biodiesel production," *Journal of Molecular Catalysis B*, vol. 65, no. 1–4, pp. 133–137, 2010.

[26] R. E. Walker, Ed., *Official Methods and Recommended Practices of the American Oil Chemists' Society (Method AOCS Cd 3d-63)*, 5th edition, 1998.

[27] A. F. C. Pereira, M. J. C. Pontes, F. F. Gambarra Neto, S. R. B. Santos, R. K. H. Galvão, and M. C. U. Araújo, "NIR spectrometric determination of quality parameters in vegetable oils using iPLS and variable selection," *Food Research International*, vol. 41, no. 4, pp. 341–348, 2008.

[28] A. N. A. Aryee, F. R. van de Voort, and B. K. Simpson, "FTIR determination of free fatty acids in fish oils intended for biodiesel production," *Process Biochemistry*, vol. 44, no. 4, pp. 401–405, 2009.

[29] J. K. Satyarthi, D. Srinivas, and P. Ratnasamy, "Estimation of free fatty acid content in oils, fats, and biodiesel by 1H NMR spectroscopy," *Energy and Fuels*, vol. 23, no. 4, pp. 2273–2277, 2009.

[30] H. Cen and Y. He, "Theory and application of near infrared reflectance spectroscopy in determination of food quality," *Trends in Food Science and Technology*, vol. 18, no. 2, pp. 72–83, 2007.

[31] A. Muller and H. Steinhart, "Recent developments in instrumental analysis for food quality," *Food Chemistry*, vol. 101, no. 3, pp. 1136–1144, 2007.

[32] M. Garrido, F. X. Rius, and M. S. Larrechi, "Multivariate curve resolution-alternating least squares (MCR-ALS) applied to spectroscopic data from monitoring chemical reactions processes," *Analytical and Bioanalytical Chemistry*, vol. 390, no. 8, pp. 2059–2066, 2008.

[33] A. F. Zanette, R. A. Barella, S. B. C. Pergher et al., "Screening, optimization and kinetics of *Jatropha curcas* oil transesterification with heterogeneous catalysts," *Renewable Energy*, vol. 36, no. 2, pp. 726–731, 2011.

[34] T. Chang, X. Lai, H. Zhang, I. Søndergaard, and X. Xu, "Monitoring lipase-catalyzed interesterification for bulky fat modification with FT-IR/NIR spectroscopy," *Journal of Agricultural and Food Chemistry*, vol. 53, no. 26, pp. 9841–9847, 2005.

[35] M. Blanco, M. Castillo, A. Peinado, and R. Beneyto, "Application of multivariate curve resolution to chemical process control of an esterification reaction monitored by near-infrared spectroscopy," *Applied Spectroscopy*, vol. 60, no. 6, pp. 641–647, 2006.

[36] H. Zhang, H. Mu, and X. Xu, "Monitoring lipase-catalyzed butterfat interesterification with rapeseed oil by Fourier transform near-infrared spectroscopy," *Analytical and Bioanalytical Chemistry*, vol. 386, no. 6, pp. 1889–1897, 2006.

[37] M. Blanco, D. Zamora, M. Mir, and R. Mulero, "Study of the lipase-catalyzed esterification of stearic acid by glycerol using in-line near-infrared spectroscopy," *Industrial and Engineering Chemistry Research*, vol. 48, no. 15, pp. 6957–6960, 2009.

[38] P. Baptista, P. Felizardo, J. C. Menezes, and M. J. N. Correia, "Multivariate near infrared spectroscopy models for predicting the methyl esters content in biodiesel," *Analytica Chimica Acta*, vol. 607, no. 2, pp. 153–159, 2008.

[39] M. H. N. Killners, J. J. R. Rohwedder, and C. Pasquini, "A PLS regression model using NIR spectroscopy for on-line monitoring of the biodiesel production reaction," *Fuel*, vol. 90, no. 11, pp. 3268–3273, 2011.

5

Solar Radiation: Models and Measurement Techniques

C. K. Pandey and A. K. Katiyar

Department of Applied Sciences, Institute of Engineering & Technology, Gautam Buddh Technical University, Lucknow 226021, India

Correspondence should be addressed to C. K. Pandey; chanchalpandey44@yahoo.com

Academic Editor: Mattheos Santamouris

In order to grasp the significance of the work accomplished by the author, it is necessary to keep abreast of the present developments in this field. The research work reported in the paper is an attempt to get knowledge to assess the solar energy potential for practical and efficient utilization in India. Our work is centered on estimating realistic values of solar (global and diffuse) radiation on horizontal and tilted surfaces using measured meteorological data and geographical and geometrical parameters for India.

1. Introduction

The sun is the driving force for all atmospheric processes. Solar radiant intensity is the expression of that input of energy upon the planet. Therefore, the ability to understand and quantify its value and distribution accurately is important in the initial understanding and modeling of any other thermodynamic or dynamic process in the earth-ocean-atmosphere system. Unfortunately, however, too little is known about the spatial and temporal distribution of incoming solar radiation. A more complete and precise description of that distribution will prove usefulness to many fields of study that rely on atmospheric energy input, such as agricultural [1], architectural [2], and engineering [3] planning. For these reasons, analysis of the solar radiation distribution in India—a state with a relatively high loading of input radiation and relatively high spatial and temporal variability—is both important and relevant.

2. Renewable Energy

From the beginning of 19th century, dwelling of fossil fuels is increasing continually toward the development of industrialization and modern life style. The fossil fuels are being used from domestic to industrial applications on the cost of pollution, health hazards, and ecology of earth. Excessive exploitation of conventional fuels directly and indirectly assist in global warming and many more factors which drive the planet towards dark future.

To overcome the dependency on conventional fuels, researchers and many organizations are working on alternative fuels, which should be commercially viable, easy to use, less pollutant, and must be abundant in nature. In this direction, renewable energies, like solar energy, tidal energy, wind energy, biofuels, and so forth, are more suitable than conventional sources of energy. These nonconventional forms are not only renewable but also maintain ecology and environment as they are ecofriendly and do not contribute to global warming and production of green house gases and so forth.

In context of India and developing countries, it is awful condition, because most of the Indian families exploit forest and fossil fuels for their domestic uses, which directly and indirectly not only affect growth of country but also health of user. Quest for energy security and sustainable development depend on the ability to get energy from renewable sources, and to use it optimistically to meet growing and diverse needs of India.

India is blessed with an abundance of nondepleting and environment friendly renewable energy resources such as solar, wind, biomass and hydro, and so forth. The Uttar-Pradesh, where solar energy and wind energy are profusely available throughout the year in comparison to the rest of India, has higher potential to exploit renewable energies for domestic and industrial applications.

3. Need for Renewable Energy

Increasing rate of energy consumption is essential for progress of our civilization and therefore main problem is how we produce energy. Extensive use of fossil fuels and nuclear energy has created bad impact on environmental, social, and sustainability problems. So we need such energy sources that will forever and can be used without pollution. Furthermore, conventional energy systems using fossil resources, especially old ones in numerous and small scale, are found to be major contributors to atmosphere pollution and greenhouse effect. In this sense the earth is already showing many signs of worldwide climate change (http://www.geosunnrg.com/) as follows.

(i) Average temperatures have climbed 1.4 degrees Fahrenheit (0.8 degree Celsius) around the world since 1880, much of this in recent decades, according to NASA's Goddard Institute for Space Studies.

(ii) The rate of warming is increasing. According to a number of climate studies, the last two decades of 20th century were the hottest in 400 years and possibly the warmest for several millennia. The UN'S Intergovernmental Panel on Climate Change (IPCC) reports that 11 of the past 12 years are among the dozen warmest since 1850.

(iii) The Arctic is feeling atmosphere pollution and greenhouse effect the most. Average temperatures in Alaska, western Canada, and eastern Russia have risen twice the global average, according to the multinational Arctic Climate Impact Assessment report compiled between 2000 and 2004.

(iv) Arctic ice is rapidly disappearing, and the region may have its first completely ice-free summer by 2040 or earlier. Polar bears and indigenous cultures are already suffering from the sea-ice loss.

(v) Glaciers and mountain snows are rapidly melting; for example, Montana's Glacier National Park now has only 27 glaciers versus 150 in 1910. In the Northern Hemisphere, thaws also come a week earlier in spring and freezes begin a week later.

(vi) Coral reefs, which are highly sensitive to small changes in water temperature, suffered the worst bleaching—or die off in response to stress—ever recorded in 1998, with some areas seeing bleach rates of 70 percent. Experts expect these sorts of events to increase in frequency and intensity in the next 50 years as sea temperatures rise.

(vii) The IPPC, in a 2007 report, claims that humans are "very likely" behind global warming. The report, based on the work of some 2,500 scientists in more than 130 countries, concluded that humans have caused all or most of the current planetary warming. Human-caused global warming is often called anthropogenic climate change.

(viii) Industrialization, deforestation, and pollution have greatly increased atmospheric concentrations of water vapor, carbon dioxide, methane, and nitrous oxide, all greenhouse gases that help to trap heat near Earth's surface.

(ix) Humans are pouring carbon dioxide into the atmosphere much faster rate than absorbing rate of plants and oceans.

(x) These gases persist in the atmosphere for years, meaning that even if such emissions were eliminated today, it would not immediately stop the global warming.

(xi) Some experts point out that the natural cycles in Earth's orbit can alter the planet's exposure to sunlight, which may explain the current trend. Earth has indeed experienced warming and cooling cycles roughly every hundred thousand years due to these orbital shifts, but such changes have occurred over the span of several centuries. Today's changes have taken place over the past 100 years or less.

4. Renewable Energy Future

Clearly, human kind has to set a different course in its need for energy, one that involves less intrusive sources such as solar, wind, and geothermal energy. There are energy sources that do not harm the planet and will never run out. The clock is ticking down but there is still time. Humankind has proven to be resourceful and prudent in the past. It needs to be again in the crucial areas of energy and the environment in order to assure sustainability for future generations (http://www.geosunnrg.com/).

5. Renewable Energy in India

India in 2006 had a population of 1.1 billion with a Gross Domestic Product of Rupees 33 trillion (728 billion US$) (http://indiabudget.nic.in/es2006-07/esmain.htm; 2006-2007). A breakup of India's primary commercial energy shows that more than 80% is supplied from fossil fuels. If we also consider traditional fuels and biomass, India's total primary energy consumption was about 20 EJ in 2004-2005 (an average of 18 GJ/capita/year). Figure 1 shows the share of different kind of energy sources in total India's primary energy supply (Government of India, Planning Commission; 2006). Fossil fuels account for about 64% of the total primary energy while traditional biomass accounts for about 33%. The India's population accounts for 17% of the world's total population; however, energy consumption is only 4% of the total world's primary energy consumption. The modern renewable energy accounts for only small portion of the total energy mix. India is the only country in the world that has a separate Ministry of New and Renewable Energy (MNRE), earlier known as the Ministry of Non-Conventional Energy Sources.

In view of the scarce fossil fuel reserves, energy security, and climate change concerns, it is expected that renewable energy will play a significant role in India's future energy mix. Figure 2 provides an overview of the different renewable energy sources. Renewable energy can be used for the entire spectrum of end-uses as given in Figure 3 [4].

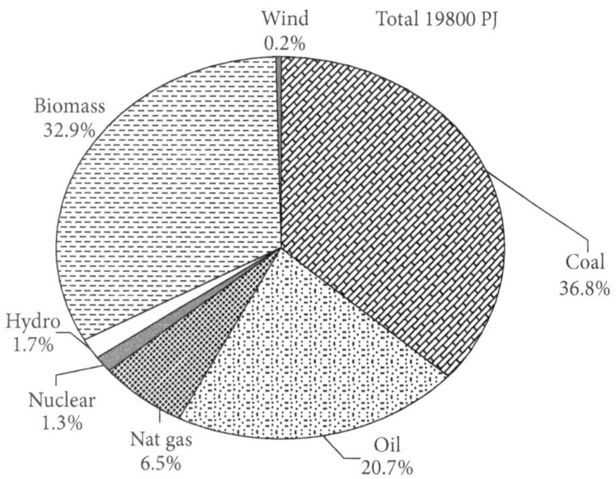

FIGURE 1: Primary energy production in India.

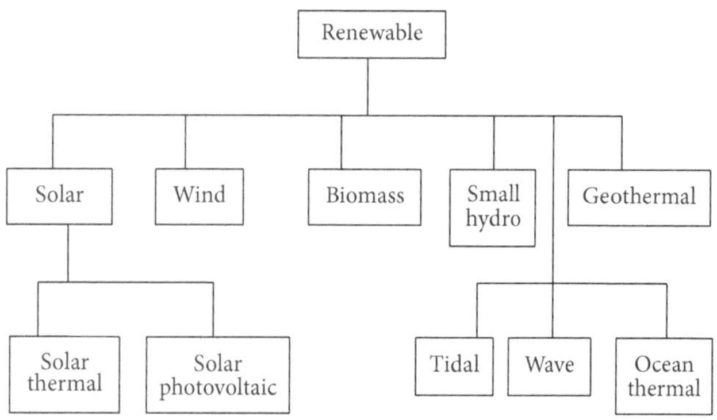

FIGURE 2: Schematic of renewable energy options.

6. Modeling Techniques

In the literature, there exist several methods for modeling solar radiation components (global, beam, and diffuse) on the ground of parametric models and decomposition models. Parametric models like Iqbal, Gueymard and ASHRAE models [5–7], require detailed information of atmospheric conditions. Meteorological parameters frequently used as predictors include the type, amount, and distribution of clouds or other observations, such as the fractional sunshine, atmospheric turbidity, and perceptible water content [8]. On other hand, decomposition models usually use information only on global radiation to predict the beam and sky components. These relationships are usually expressed in terms of the irradiations which are the time integrals of the radiant flux or irradiance. Decomposition models which are based on the correlations between the clearness index k_t (it is the ratio of global to the extraterrestrial solar radiation) and the diffuse fraction k_d (it is the ratio of diffuse to the global solar radiation), diffuse coefficient k_D (it is the ratio of diffuse to the extraterrestrial solar radiation) or the direct transmittance k_b (it is the ratio of beam to the extraterrestrial solar radiation) as an example Orgill and Hollands, Erbs et al., Reindle et al. and Liu and Jordan models [9–12] developed to estimate direct and diffuse radiations from global radiation. Instead

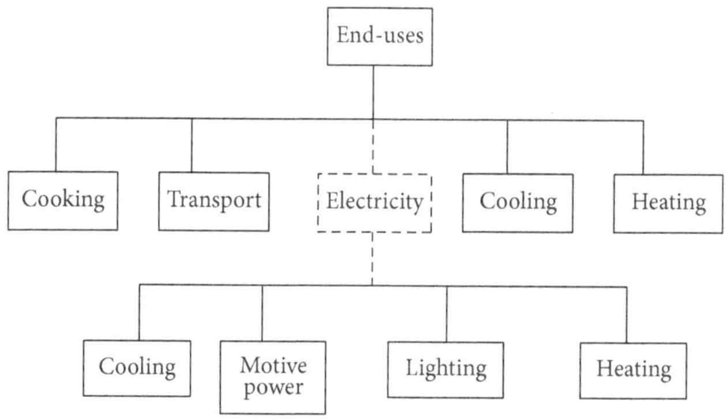

FIGURE 3: Energy end.

of these two methods nowadays empirically method [13–15] and the artificial neural network method (which constitutes a widely accepted novel approach offering an alternative way to synthesize complex problems) can also be used for the determination of global and diffuse components of solar radiation [16, 17].

7. Measurement Techniques of Radiation

To asses the availability of solar radiation at different locations, measurements of global radiation, diffuse radiation, beam radiation, sun shine hours, bright sun shine hours, maximum and minimum temperature, humidity, pressure, visibility, wind speed and direction, gust speed, water precipitation, and air mass are very important parameters.

To measure the abovementioned parameters we require a big laboratory and group of skilled fellows. Although it is a difficult task to maintain and run such a laboratory but the quality and reliability of data of the site then can only be ensured. However there is a wide network of Indian meteorological department which provides wide variety of data that includes radiation and meteorological and pollution data. But the radiation data are scared. World Radiation Centre also provides data of global, beam, and diffuse radiation of cities of the world.

The direct measurement of solar radiation and its components (direct and diffuse) is done in two basic ways as well. The values are measured either by using ground-based instrumentation as pyranometers, or remotely with satellites. These methods are often used in combination to validate one another [18–20]. In general, pyranometer data from adequately maintained instruments provide an accurate description of the solar radiation values in the immediate area. It has been suggested that extrapolation of daily values beyond the discrete point represented by the location of the pyranometer can result in the misrepresentation of the extrapolated areas. Suckling [21] found that, for areas in the Tennessee Valley Authority region, permissible extrapolation distances of daily solar radiation values were ~200 km, but that these distances may vary by season. However, Younes and Muneer [22] claimed that "...for a given location that is, farther than 50 km from the measurement station the use of the respective measurement station's data is obsolete in the assessment of solar energy applications." In his study of solar radiation variability in San Diego County, California, Aguado [23] suggested that the relative proximity of two points to the coast further complicates the abilities of researchers to extrapolate beyond the discrete points at which solar radiation was measured.

Global and diffuse solar radiation can be measured with the help of a thermoelectric pyranometer. Basically pyranometer is consisting of a thin-blackened surface supported inside a relatively massive well-polished case. When solar radiation falls on this surface, the temperature of the surface rises until its rate of loss of the heat by all causes is equal to the rate of gain of heat by radiation. This rise in temperature sets up a thermal e.m.f. which is measured on a recording millivoltmeter or recorder. Each pyranometer is calibrated and a certificate is provided by the manufacturer. For quality data of solar radiation pyranometer have to be calibrated at regular intervals and required proper maintenance.

Instrument used for measuring the intensity of direct solar radiation, that is, beam radiation is called pyrheliometer. The Ångström compensation pyrheliometer is a standard instrument for the measurement of direct solar radiation, in which the sensor is fixed at the lower end of a tube provided with a diaphragm so that when the tube is directed towards the sun, the sensing surface is normal to the line joining the sun to the receiver, and only radiation from the sun and a narrow annulus of the sky is received by the sensor. The alignment is done by means of a sighting device called the diopter.

In pyrheliometer the absorption of the radiant energy by a blackened metal strip, exposed to the sun's ray, is determined by measuring the electric current necessary to heat an identical shielded strip to the same temperature. Since both strips are mounted similarly and are at the same temperature and the heat exchange of the strips with surroundings is identical, the rate of generation of heats in shielded strip due to electric current is equal to the rate of absorption of radiant energy by the exposed strip. The equivalence or otherwise of the temperature of the two strips is determined by two fine thermocouples attached to the back of the strips, connected in series with a sensitive galvanometer. The current through the shielded strip is determined with an accurate digital milliammeter. The advantage of the instrument is that the equilibrium current through the shielded strip is not affected by change in the rate of heat loss from the strips, provided that the changes affect both the strips equally.

Such an instrument in theory is an absolute instrument as all the relevant factors required for the calculation of the radiation intensity can be measured:

let I = the intensity of direct solar radiation in watts/cm^2

A = area of the strip

α = absorption of the strip

i = heating current in amperes.

Then we have $I \cdot A \cdot \alpha \;=\; R \cdot i^2$

$$I \;=\; (R \cdot i^2)/(A \cdot \propto) \;=K \cdot i^2 \text{ watts/cm}^2,$$

where $K \;=\; R/(A \cdot \propto)$ is a constant for the instrument and depends upon resistance, length, breath, and absorptance of the strip. The constant of the instrument is actually determined by comparison with other standard instrument with constant traceable to the group of standards maintained at the World Radiation Centre in Switzerland.

For a study of the intensity of solar radiation in different regions, Scott glass filter OG1, RG2, and RG8 are used. The transmission of the filters is given below:

OG1 transmits from 0.525 μ to 2.800 μ

RG2 transmits from 0.630 μ to 2.800 μ

RG8 transmits from 0.700 μ to 2.800 μ.

8. Estimation of Radiation

It is generally accepted that models for solar radiation prediction are necessary, because in most cases the density and number of solar radiation measuring stations cannot describe the necessary variability [24]. It is then understandable that new models and improvements to existing modeling techniques are continually proposed which intend to improve estimates of solar radiation values with the use of more readily available meteorological variables [22, 25, 26].

8.1. Estimation of Solar Radiation on Horizontal Surface. Ångström [27] proposed first theoretical model for estimating global solar radiation based on sunshine duration. Page [28] and Prescott [29] reconsidered this model in order to make it possible to calculate monthly average of the daily global radiation $\overline{H}$ (MJ/m^2 day) on a horizontal surface from monthly average daily total insolation on an extraterrestrial horizontal surface as per the following relation:

$$\frac{\overline{H}}{\overline{H}_0} = a + b\left(\frac{\overline{s}}{\overline{s}_0}\right), \tag{1}$$

where $\overline{H}_0$ is the monthly average daily extraterrestrial radiation (MJ/m^2 day), $\overline{s}$ is the monthly average daily bright sunshine hours, $\overline{s}_0$ is the maximum possible monthly average daily sunshine hours or the day length, and a and b are constants.

Although a number of correlations included more parameters and have been developed by different workers [12, 30–32], one has been found to be very convenient, applicable to a large number of locations, and is the most widely used correlation [15, 33–37].

Solar radiation coming through the atmosphere is partially absorbed or reflected by its constituents (aerosol particles cause diffuse radiation) which actually reduces the beam component and affects the performance of energy systems. Therefore, knowledge of diffuse irradiation on a horizontal surface is also important in designing the various energy utilization systems. Many researchers have presented empirical correlations to estimate daily diffuse radiation. Two of the most widely used correlations are due to Liu and Jordan [12] and Page [28] to which the irradiation data fitted are given as

$$\frac{H_d}{H} = a + b\left(\frac{H}{H_0}\right), \tag{2}$$

where H_d is daily mean diffuse radiation and $H/H_0 = K_t$, the clearness index.

A third type approach has been proposed by Iqbal [38] to find a correlation between H_d and the ratio of bright sunshine (S) to the day length (S_0) given as

$$\frac{H_d}{H} = c + d\left(\frac{S}{S_0}\right), \quad (3)$$

where c and d are the constants.

Gopinathan [30] used the above models, estimated the radiation, and suggested that the applicability of the correlations to locations in various parts of the word is to be tested. Therefore, an extensive work has to be done in this area [25, 39–41]. Recently some others researchers [13, 14, 42–51] have also reported new developments in solar energy.

Orgill and Holland [9], Erbs et al. [10], and Collars-Pareira and Rabl [31] have all related the diffuse fraction of global radiation to an index of the atmospheric clarity on hourly basis. Elminir [47] examined the correlations of Orgill and Hollands [9] and Erbs et al. [10] and recommended then later for computing hourly diffuse fraction. Ahmad and Tiwari [52] presented modified version of ASHRAE [7] model for New Delhi station. Katiyar [53], Gueymard [54], Jacovides et al. [55], Katiyar et al. 2010 [36, 37], and Chen et al. [56] have also reported new developments in solar energy, which could be used to estimate long-term global, direct, and diffuse radiation.

8.2. Estimation of Solar Radiation on Tilted Surfaces. Most of the solar energy systems of interest redesigned with tilted collected surfaces. Therefore, it is necessary to have knowledge about the availability of solar radiation on tilted surfaces.

The total amount of radiation incident on an inclined plane (H_t) is composed of beam (H_b), sky-diffuse (H_s), and ground reflected (H_r) components:

$$H_t = H_b + H_s + H_r. \quad (4)$$

The daily beam radiation received on the tilted surface can be expressed as

$$H_b = (H - H_d) r_b. \quad (5)$$

Under isotropic conditions, the global radiation on inclined surface can be written as

$$H_t = (H - H_d) r_b + \frac{1}{2} H_d (1 + \cos\beta) + \frac{1}{2} H\rho (1 - \cos\beta), \quad (6)$$

where ρ is the ground albedo, β is the surface slope from the horizontal (degree), and H and H_d are monthly mean daily total and diffuse radiation on horizontal surface, respectively.

The estimation of solar radiation on tilted surfaces has been made by different authors. Recently, Mefti et al. [57] gave the hourly solar radiation model for inclined surfaces using sunshine duration data. An extensive work has to done in this area [13, 14, 58–63].

9. Present Trends in Correlation

Ångström-Prescott correlation has served as a basic approach to estimate global radiation for long time. Simplicity of Ångström-Prescott equation has dominated over its several demerits. Coefficients in Ångström-Prescott equation are site dependent. Yeboah-Amankwah and Agyeman [64] believed that "a_1" and "b_1" are time dependent and so developed a differential Ångström model with a set of coefficients which vary with time. Sahin and Sen [65] proposed a method to dynamically estimate the coefficients. These works do not include radiation damping process when solar rays pass through the atmosphere. Some researchers [66, 67] are employing a damping structure to calculate global solar radiation in clear sky. Their models consider physical process in detail, so the effect of latitude, elevation, and other factors are taken into account automatically. However damping spectrums are very irregular and hence numerical integration is indispensable. To overcome all these difficulties Yang et al. [68] proposed a "Hybrid model" which considers the physical process but still maintains the simplicity of the Ångström equation. The author tested their hybrid model with data of Japan and found that it needs greater turbidity when applied to urban areas due to air pollution; otherwise, global radiation may be estimated. Also, the estimation under completely cloudy sky is still difficult. Hybrid model assumed that global radiation has linear relationship with effective beam radiation and diffuse radiation as well as fractional sunshine time. Yang and Koike [69] also proposed a model to calculate solar radiation from upper air humidity. This could be used to predict numerical weather prediction (NWP).

Lingamgunta and Veziroglu [70] proposed a universal relationship for estimating clear sky insolation. This relation predicts annual mean daily clear sky insolation and is a function of latitude and altitude only. The author claimed that relation (the author called it as Lingamgunta-Veziroglu relation) predicts radiation that fairly accurate over conventional methods.

Suehrcke [71] proposed an entirely new relation between relative sunshine duration and global radiation. This relation, unlikely Ångström-Prescott equation, is nonlinear and it does not require any empirical constants. The local atmospheric conditions are considered through the value of the average daily clear index. Further fractional sunshine can be calculated from Suehrcke relation, and Suehrcke's sunshine radiation relationship has been verified by Driesse and Thevenard [72] using a global data sheet.

Recently there have been emphases [73–76] on empirical correlation using satellite images. Most estimation of daily global radiation values from satellite images requires the useof models which allow us to calculate these values from a small number of images per day (usually three). Different methodologies have been developed in this regard.

Atmospheric pollutants and aerosols absorb and scatter shortwave solar radiations. The interactions have resultant impacts on atmospheric radiative energy transfer and balance. If the pollution increases, then diffuse component of global solar radiation will also increase. Spectral and diffuse radiation study of global solar radiation has now gained importance as reflected in recent literature [77–79]. There has been another area of modeling gaining importance which is estimation of solar radiation potential using artificial neural network [16, 60, 80–82].

10. Conclusion

This paper presents a brief account of the general introduction, principle, experimental technique, measurements of solar radiation data, and review of literature of solar radiation models and describes present trend of solar energy modeling which is of major interest to solar energy engineers, architects, designing building, and thermal devices for optimum and efficient utilization of this nonconventional energy resource.

References

[1] S. A. Changnon and D. Changnon, "Importance of sky conditions on the record 2004 Midwestern crop yields," *Physical Geography*, vol. 26, no. 2, pp. 99–111, 2005.

[2] Z. Yang, X. H. Li, and Y. F. Hu, "Study on solar radiation and energy efficiency of building glass system," *Applied Thermal Engineering*, vol. 26, no. 8-9, pp. 956–961, 2006.

[3] E. H. Amer and M. A. Younes, "Estimating the monthly discharge of a photovoltaic water pumping system: model verification," *Energy Conversion and Management*, vol. 47, no. 15-16, pp. 2092–2102, 2006.

[4] I. R. Pillai and R. Banerjee, "Renewable energy in India: status and potential," *Energy*, vol. 34, no. 8, pp. 970–980, 2009.

[5] M. Iqbal, *An Introduction to Solar Radiation*, Academic Press, Toronto, Canada, 1983.

[6] C. Gueymard, "Critical analysis and performance assessment of clear sky solar irradiance models using theoretical and measured data," *Solar Energy*, vol. 51, no. 2, pp. 121–138, 1993.

[7] American Society of Heating Refrigeration and Air-Conditioning Engineers, *ASHRAE Applications Handbook (SI)*, ASHRAE, Atlanta, Ga, USA, 1999.

[8] L. T. Wong and W. K. Chow, "Solar radiation model," *Applied Energy*, vol. 69, no. 3, pp. 191–224, 2001.

[9] J. F. Orgill and K. G. T. Hollands, "Correlation equation for hourly diffuse radiation on a horizontal surface," *Solar Energy*, vol. 19, no. 4, pp. 357–359, 1977.

[10] D. G. Erbs, S. A. Klein, and J. A. Duffie, "Estimation of the diffuse radiation fraction for hourly, daily and monthly-average global radiation," *Solar Energy*, vol. 28, no. 4, pp. 293–302, 1982.

[11] D. T. Reindl, W. A. Beckman, and J. A. Duffie, "Diffuse fraction correlations," *Solar Energy*, vol. 45, no. 1, pp. 1–7, 1990.

[12] B. Y. H. Liu and R. C. Jordan, "The inter-relationship and characteristic distribution of direct, diffuse and total solar radiation," *Solar Energy*, vol. 4, no. 3, pp. 1–19, 1960.

[13] C. K. Pandey and A. K. Katiyar, "A note on diffuse solar radiation on a tilted surface," *Energy*, vol. 34, no. 11, pp. 1764–1769, 2009.

[14] C. K. Pandey and A. K. Katiyar, "A comparative study to estimate daily diffuse solar radiation over India," *Energy*, vol. 34, no. 11, pp. 1792–1796, 2009.

[15] O. P. Singh, S. K. Srivastava, and A. Gaur, "Empirical relationship to estimate global radiation from hours of sunshine," *Energy Conversion and Management*, vol. 37, no. 4, pp. 501–504, 1996.

[16] A. Sözen, E. Arcaklioglu, and M. Özalp, "Estimation of solar potential in Turkey by artificial neural networks using meteorological and geographical data," *Energy Conversion and Management*, vol. 45, no. 18-19, pp. 3033–3052, 2004.

[17] J. Soares, A. P. Oliveira, M. Z. Božnar, P. Mlakar, J. F. Escobedo, and A. J. Machado, "Modeling hourly diffuse solar-radiation in the city of São Paulo using a neural-network technique," *Applied Energy*, vol. 79, no. 2, pp. 201–214, 2004.

[18] J. A. Otkin, M. C. Anderson, J. R. Mecikalski, and G. R. Diak, "Validation of GOES-based insolation estimates using data from the U.S. Climate Reference Network," *Journal of Hydrometeorology*, vol. 6, no. 4, pp. 460–475, 2005.

[19] H. Deneke, A. Feijt, A. van Lammeren, and C. Simmer, "Validation of a physical retrieval scheme of solar surface irradiances from narrowband satellite radiances," *Journal of Applied Meteorology*, vol. 44, no. 9, pp. 1453–1466, 2005.

[20] S. Kimothi, B. K. Bhattacharya, P. D. Semalty, V. K. Pandey, and V. K. Dadhwal, "Estimation of ground insolation using METEOSAT data over India," *Current Science*, vol. 86, no. 9, pp. 1308–1312, 2004.

[21] P. W. Suckling, "Extrapolation of solar radiation measurements: mesoscale analyses from Arizona and Tennessee Valley Authority regions," *Journal of Climate & Applied Meteorology*, vol. 22, no. 3, pp. 488–494, 1983.

[22] S. Younes and T. Muneer, "Improvements in solar radiation models based on cloud data," *Building Services Engineering Research and Technology*, vol. 27, no. 1, pp. 41–54, 2006.

[23] E. Aguado, "Local-scale variability of daily solar radiation—San Diego County, California," *Journal of Climate & Applied Meteorology*, vol. 25, no. 5, pp. 672–678, 1986.

[24] T. Muneer, S. Younes, and S. Munawwar, "Discourses on solar radiation modeling," *Renewable and Sustainable Energy Reviews*, vol. 11, no. 4, pp. 551–602, 2007.

[25] M. Donatelli, G. Bellocchi, and F. Fontana, "RadEst3.00: software to estimate daily radiation data from commonly available meteorological variables," *European Journal of Agronomy*, vol. 18, no. 3-4, pp. 363–367, 2003.

[26] S. Safi, A. Zeroual, and M. Hassani, "Prediction of global daily solar radiation using higher order statistics," *Renewable Energy*, vol. 27, no. 4, pp. 647–666, 2002.

[27] A. Ångström, "Solar and terrestrial radiation," *Quarterly Journal of the Royal Meteorological Society*, vol. 50, no. 210, pp. 121–125, 1924.

[28] J. K. Page, "The estimation of monthly mean values of daily total short wave radiation on vertical and inclined surfaces from sun shine records for latitudes 40°N-40°S," *Proceedings of the United Nations Conference on New Sources of Energy*, vol. 98, no. 4, p. 378, 1961.

[29] J. A. Prescott, "Evaporation from water surface in relation to solar radiation," *Transactions of The Royal Society of South Australia*, vol. 64, pp. 114–118, 1940.

[30] K. K. Gopinathan, "Empirical correlations for diffuse solar irradiation," *Solar Energy*, vol. 40, no. 4, pp. 369–370, 1988.

[31] M. Collares-Pereira and A. Rabl, "The average distribution of solar radiation-correlations between diffuse and hemispherical and between daily and hourly insolation values," *Solar Energy*, vol. 22, no. 2, pp. 155–164, 1979.

[32] J. Glover and J. S. G. Mc Culloch, "The empirical relation between solar radiation and hours of sunshine," *Quarterly Journal of the Royal Meteorological Society*, vol. 84, pp. 172–175, 1958.

[33] V. Bahel, H. Bakhsh, and R. Srinivasan, "A correlation for estimation of global solar radiation," *Energy*, vol. 12, no. 2, pp. 131–135, 1987.

[34] N. A. Elagib and M. G. Mansell, "New approaches for estimating global solar radiation across Sudan," *Energy Conversion and Management*, vol. 41, no. 5, pp. 419–434, 2000.

[35] I. T. Togrul, H. Togrul, and D. Evin, "Estimation of global solar radiation under clear sky radiation in Turkey," *Renewable Energy*, vol. 21, no. 2, pp. 271–287, 2000.

[36] A. K. Katiyar, A. Kumar, C. K. Pandey, V. K. Katiyar, and S. H. Abdi, "Correlations for the estimation of monthly mean hourly diffuse solar radiation: a time dependent approach," *The International Journal of Energy and Environment*, vol. 1, no. 5, pp. 833–840, 2010.

[37] A. K. Katiyar, A. Kumar, Akhilesh, C. K. Pandey, and B. Das, "A comparative study of monthly mean daily clear sky radiation over India," *The International Journal of Energy and Environment*, vol. 1, no. 1, pp. 177–182, 2010.

[38] M. Iqbal, "Correlation of average diffuse and beam radiation with hours of bright sunshine," *Solar Energy*, vol. 23, no. 2, pp. 169–173, 1979.

[39] D. Y. Goswami, S. Vijayaraghavan, S. Lu, and G. Tamm, "New and emerging developments in solar energy," *Solar Energy*, vol. 76, no. 1–3, pp. 33–43, 2004.

[40] A. Madhlopa, "Solar radiation climate in Malawi," *Solar Energy*, vol. 80, no. 8, pp. 1055–1057, 2006.

[41] S. Tarhan and A. Sari, "Model selection for global and diffuse radiation over the Central Black Sea (CBS) region of Turkey," *Energy Conversion and Management*, vol. 46, no. 4, pp. 605–613, 2005.

[42] E. O. Falayi, J. O. Adepitan, and A. B. Rabiu, "Empirical models for the correlation of global solar radiation with meteorological data for Iseyin, Nigeria," *International Journal of Physical Sciences*, vol. 3, no. 9, pp. 210–216, 2008.

[43] K. Skeiker, "Correlation of global solar radiation with common geographical and meteorological parameters for Damascus province, Syria," *Energy Conversion and Management*, vol. 47, no. 4, pp. 331–345, 2006.

[44] M. M. Kvalevåg and G. Myhre, "Human impact on direct and diffuse solar radiation during the industrial era," *Journal of Climate*, vol. 20, no. 19, pp. 4874–4883, 2007.

[45] E. O. Ogolo, "Evaluating the performance of some predictive models for estimating global solar radiation across varying climatic conditions in Nigeria," *Indian Journal of Radio and Space Physics*, vol. 39, pp. 121–131, 2010.

[46] C. K. Pandey and A. K. Katiyar, "Temperature base correlation for the estimation of global solar radiation on horizontal surface," *The International Journal of Energy and Environment*, vol. 1, no. 4, pp. 737–744, 2010.

[47] H. K. Elminir, "Experimental and theoretical investigation of diffuse solar radiation: data and models quality tested for Egyptian sites," *Energy*, vol. 32, no. 1, pp. 73–82, 2007.

[48] K. Bakirci, "Correlations for estimation of daily global solar radiation with hours of bright sunshine in Turkey," *Energy*, vol. 34, no. 4, pp. 485–501, 2009.

[49] A. M. Al-Salihi, M. M. Kadum, and A. J. Mohammed, "Estimation of global solar radiation on horizontal surface using routine meteorological measurements for different cities in Iraq," *Asian Journal of Scientific Research*, vol. 3, pp. 240–248, 2010.

[50] Y. Jiang, "Estimation of monthly mean daily diffuse radiation in China," *Applied Energy*, vol. 86, no. 9, pp. 1458–1464, 2009.

[51] R. F. Mechlouch and A. B. Brahim, "A global solar radiation model for the design of solar energy systems," *Asian Journal of Scientific Research*, vol. 1, pp. 231–238, 2008.

[52] M. J. Ahmad and G. N. Tiwari, "Evaluation and comparison of hourly solar radiation models," *International Journal of Energy Research*, vol. 33, no. 5, pp. 538–552, 2009.

[53] A. K. Katiyar, C. K. Pandey, and V. K. Katiyar, "Correlation model of hourly diffuse solar radiation based on ASHRAE model: a study case in India," *International Journal of Renewable Energy Technology*, vol. 3, no. 4, pp. 341–355, 2012.

[54] C. Gueymard, "Prediction and performance assessment of mean hourly global radiation," *Solar Energy*, vol. 68, no. 3, pp. 285–303, 2000.

[55] C. P. Jacovides, F. S. Tymvios, V. D. Assimakopoulos, and N. A. Kaltsounides, "Comparative study of various correlations in estimating hourly diffuse fraction of global solar radiation," *Renewable Energy*, vol. 31, no. 15, pp. 2492–2504, 2006.

[56] R. Chen, E. Kang, X. Ji, J. Yang, and J. Wang, "An hourly solar radiation model under actual weather and terrain conditions: a case study in Heihe river basin," *Energy*, vol. 32, no. 7, pp. 1148–1157, 2007.

[57] A. Mefti, M. Y. Bouroubi, and A. Adane, "Generation of hourly solar radiation for inclined surfaces using monthly mean sunshine duration in Algeria," *Energy Conversion and Management*, vol. 44, no. 19, pp. 3125–3141, 2003.

[58] P. G. Loutzenhiser, H. Manz, C. Felsmann, P. A. Strachan, T. Frank, and G. M. Maxwell, "Empirical validation of models to compute solar irradiance on inclined surfaces for building energy simulation," *Solar Energy*, vol. 81, no. 2, pp. 254–267, 2007.

[59] C. K. Pandey and A. K. Katiyar, "A comparative study of solar irradiation models on various inclined surfaces for India," *Applied Energy*, vol. 88, no. 4, pp. 1455–1459, 2011.

[60] N. L. Forero, L. M. Caicedo, and G. Gordillo, "Correlation of global solar radiation values estimated and measured on an inclined surface for clear days in Bogotá," *Renewable Energy*, vol. 32, no. 15, pp. 2590–2602, 2007.

[61] G. A. Kamali, I. Moradi, and A. Khalili, "Estimating solar radiation on tilted surfaces with various orientations: a study case in Karaj (Iran)," *Theoretical and Applied Climatology*, vol. 84, no. 4, pp. 235–241, 2006.

[62] A. De Miguel, J. Bilbao, and M. Diez, "Solar radiation incident on tilted surfaces in Burgos, Spain: isotropic models," *Energy Conversion and Management*, vol. 36, no. 10, pp. 945–951, 1995.

[63] A. Skartveit and J. A. Olseth, "Modelling slope irradiance at high latitudes," *Solar Energy*, vol. 36, no. 4, pp. 333–344, 1986.

[64] D. Yeboah-Amankwah and K. Agyeman, "Differential Ångstrom model for predicting insolation from hours of sunshine," *Solar Energy*, vol. 45, no. 6, pp. 371–377, 1990.

[65] A. D. Şahin and Z. Şen, "Statistical analysis of the Angstrom formula coefficients and application for Turkey," *Solar Energy*, vol. 62, no. 1, pp. 29–38, 1998.

[66] R. E. Bird, "A simple, solar spectral model for direct-normal and diffuse horizontal irradiance," *Solar Energy*, vol. 32, no. 4, pp. 461–471, 1984.

[67] B. Leckner, "The spectral distribution of solar radiation at the earth's surface-elements of a model," *Solar Energy*, vol. 20, no. 2, pp. 143–150, 1978.

[68] K. Yang, G. W. Huang, and N. Tamai, "Hybrid model for estimating global solar radiation," *Solar energy*, vol. 70, no. 1, pp. 13–22, 2001.

[69] K. Yang and T. Koike, "Estimating surface solar radiation from upper-air humidity," *Solar Energy*, vol. 72, no. 2, pp. 177–186, 2002.

[70] C. Lingamgunta and T. N. Veziroglu, "A universal relationship for estimating clear sky insolation," *Energy Conversion and Management*, vol. 45, no. 1, pp. 27–52, 2004.

[71] H. Suehrcke, "On the relationship between duration of sunshine and solar radiation on the earth's surface: Ångstrom's equation revisited," *Solar Energy*, vol. 68, no. 5, pp. 417–425, 2000.

[72] A. Driesse and D. Thevenard, "A test of Suehrcke's sunshine-radiation relationship using a global data set," *Solar Energy*, vol. 72, no. 2, pp. 167–175, 2002.

[73] C. Rigollier, M. Lefèvre, and L. Wald, "The method Heliosat-2 for deriving shortwave solar radiation from satellite images," *Solar Energy*, vol. 77, no. 2, pp. 159–169, 2004.

[74] R. Perez, P. Ineichen, K. Moore et al., "A new operational model for satellite-derived irradiances: description and validation," *Solar Energy*, vol. 73, no. 5, pp. 307–317, 2002.

[75] S. Janjai, T. Jantarach, and J. Laksanaboonsong, "A model for calculating global illuminance from satellite data," *Renewable Energy*, vol. 28, no. 15, pp. 2355–2365, 2003.

[76] L. Ramirez, L. Mora-López, and M. Sidrach-de-Cardona, "A multivariate qualitative model for the prediction of daily global radiation from three hourly global radiation values," *Energy*, vol. 26, no. 2, pp. 205–215, 2001.

[77] C. P. Jacovides, M. D. Steven, and D. N. Asimakopoulos, "Spectral solar irradiance and some optical properties for various polluted atmospheres," *Solar Energy*, vol. 69, no. 3, pp. 215–227, 2000.

[78] A. S. Rapti, "Atmospheric transparency, atmospheric turbidity and climatic parameters," *Solar Energy*, vol. 69, no. 2, pp. 99–111, 2000.

[79] Z. Jin, W. Yezheng, and Y. Gang, "Estimation of daily diffuse solar radiation in China," *Renewable Energy*, vol. 29, no. 9, pp. 1537–1548, 2004.

[80] C. A. Roulet, "Solar energy and global heat balance of a city," *Solar Energy*, vol. 70, no. 3, pp. 255–261, 2001.

[81] K. S. Reddy and M. Ranjan, "Solar resource estimation using artificial neural networks and comparison with other correlation models," *Energy Conversion and Management*, vol. 44, no. 15, pp. 2519–2530, 2003.

[82] A. S. S. Dorvlo, J. A. Jervase, and A. Al-Lawati, "Solar radiation estimation using aritificial neural networks," *Applied Energy*, vol. 71, no. 4, pp. 307–319, 2002.

A New Strategy for Accurately Predicting *I-V* Electrical Characteristics of PV Modules Using a Nonlinear Five-Point Model

Sakaros Bogning Dongue, Donatien Njomo, and Lessly Ebengai

Environmental Energy Technologies Laboratory (EETL), Faculty of Sciences, University of Yaoundé 1, P. O. Box 812, Yaoundé, Cameroon

Correspondence should be addressed to Sakaros Bogning Dongue; sakarosb@gmail.com

Academic Editor: Mohamed Benghanem

This paper presents the modelling of electrical *I-V* response of illuminated photovoltaic crystalline modules. As an alternative method to the linear five-parameter model, our strategy uses advantages of a nonlinear analytical five-point model to take into account the effects of nonlinear variations of current with respect to solar irradiance and of voltage with respect to cells temperature. We succeeded in this work to predict with great accuracy the *I-V* characteristics of monocrystalline shell SP75 and polycrystalline GESOLAR GE-P70 photovoltaic modules. The good comparison of our calculated results to experimental data provided by the modules manufacturers makes it possible to appreciate the contribution of taking into account the nonlinear effect of operating conditions data on *I-V* characteristics of photovoltaic modules.

1. Introduction

The performance of photovoltaic modules is highly dependent on both the availability of solar radiation in the locality and the temperature of photovoltaic cells [1]. A real knowledge and mastery of photovoltaic modules are important as are the design and technology choice for a project operating photovoltaic electricity.

Important research has been carried out throughout the world in view of mastering electrical performance of photovoltaic modules [2, 3]. Numerical models are developed for this purpose [4, 5], mainly to reduce the costs of experimental studies. Unfortunately, these studies are very useful, particularly to obtain accurate results [6, 7]. It is obvious that the application of the best model will combine in a better way simplicity and accuracy [8, 9].

This work consists in doing comparative study between a five-parameter three-point model hereafter referred to as five-parameter model characterized by both the linear variation of the current with respect to the irradiance and the linear variation of the voltage with respect to the temperature of the cells and an adjusted five-parameter five-point model hereafter referred to as five-point model which takes into account the nonlinear effects of both the variation of the current with respect to the irradiance and the variation of the voltage with temperature.

2. Single-Diode Equivalent Circuit of a PV Module

A photovoltaic module under illumination can be represented either by two-diode [10] or by one-diode equivalent circuit. In normal operating conditions, the one-diode model is the most common [11], and the photovoltaic module is equivalent to the following electrical circuit (see Figure 1).

The current-voltage (*I-V*) characteristic under a solar radiation and temperature set is expressed in

$$I = I_{ph} - I_d - I_{sh}, \tag{1}$$

where I_d is the current through the diode and I_{sh} is the current through the shunt resistor both given by (2) as follows:

$$I_d = I_0 \left[\exp\left(\frac{V + IR_S}{a} \right) - 1 \right], \qquad I_{sh} = \frac{V + IR_S}{R_{sh}}. \tag{2}$$

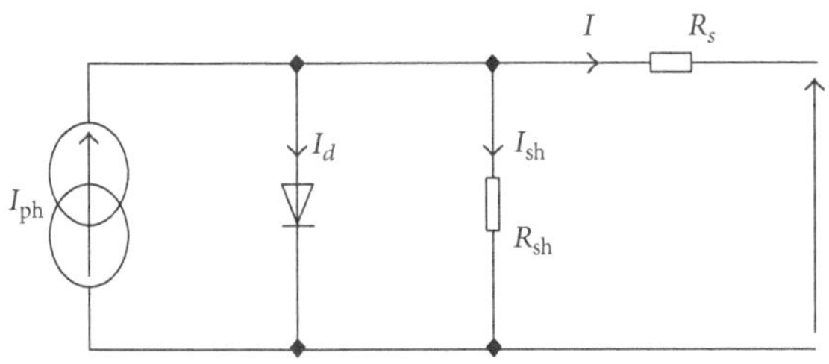

FIGURE 1: Equivalent electrical circuit for the one-diode model.

Therefore,

$$I = I_{ph} - I_0 \left[\exp\left(\frac{V + IR_S}{a} \right) - 1 \right] - \frac{V + IR_S}{R_{sh}}, \quad (3)$$

where I_{ph}, I_0, R_s, and R_{sh} are the photocurrent, the reverse saturation current of the diode, the series resistance and shunt resistance, respectively. The modified ideality factor of the junction is given as

$$a = \frac{N_s A_0 K T_c}{q}, \quad (4)$$

where A_0 is the normal ideality factor; N_s, k, T_c, and q are the number of cells in series, the Boltzmann constant, the temperature of the cells, and elementary charge, respectively.

The determination of the electrical behaviour of a photovoltaic module in given operating conditions is based on the current-voltage (*I-V*) curve. However, (1) that gives this characteristic is implicit and cannot be easily solved. Many numerical algorithms to model the *I-V* curve from one-diode representation of photovoltaic module have been proposed. Some of these models are known as "resistive companion" [12], the nonlinear least squares optimization, and other iterative solutions described in [8, 9, 13]. Although many improvements have been done, these approaches still require a great effort of numerical calculations. Recently, some authors have used artificial intelligence methods (AI) such as fuzzy logic [14] and artificial neural network (ANN) [15, 16] for modelling (*I-V*) curves.

Another interesting approach is based on the evaluation of unknown parameters of (1) based on information related to the field operation of the module in order to obtain an explicit *I-V* relationship. To this end, an iterative process may be necessary to evaluate the different parameters [17, 18]. One may also perform a direct analytical calculation of each parameter [19, 20]. The latter approach requires information often available in the data sheets provided by photovoltaic modules manufacturers or experimental data collected on site. Both five-parameter and five-point models described in this study use only data provided by module manufacturers, together with semiempirical correlation equations and weather data, to predict *I-V* characteristics of photovoltaic crystalline modules.

3. Five-Parameter Model

The five parameters that give name to the model are the photocurrent, the reverse saturation current, series resistance, shunt resistance, and modified ideality factor. These parameters are firstly evaluated at so-called reference operating conditions or standard test conditions (STC) for which the irradiance on the modules is 1000 W/m^2 and the cells temperature is 25°C with an air mass number AM = 1.5. This environmental situation is used by manufacturers of the modules to measure the data laid out on the data sheet.

The five parameters defined above corresponding to STC are designated by $I_{ph_{ref}}$, I_{0ref}, $R_{s_{ref}}$, $R_{sh_{ref}}$, and a_{ref} (the index is relative to reference conditions). We obtain the (*I-V*) characteristic equation at STC as

$$I = I_{ph_{ref}} - I_{0_{ref}} \left[\exp\left(\frac{V + IR_{S_{ref}}}{a_{ref}} \right) - 1 \right] - \frac{V + IR_{S_{ref}}}{R_{sh_{ref}}}. \quad (5)$$

3.1. Calculation of Reference Parameters. Data provided by module manufacturers for STC are used when writing the following five equations [21].

At short-circuit point: $V = 0$, $I = I_{sc}$,

$$I_{sc_0} = I_{ph_{ref}} - I_{0_{ref}} \left[\exp\left(\frac{I_{sc_0} R_{S_{ref}}}{a_{ref}} \right) - 1 \right] - \frac{I_{sc_0} R_{S_{ref}}}{R_{sh_{ref}}}. \quad (6)$$

At the open-circuit point: $I = 0$, $V = V_{oc}$,

$$0 = I_{ph_{ref}} - I_{0_{ref}} \left[\exp\left(\frac{V_{oc_{ref}}}{a_{ref}} \right) - 1 \right] - \frac{V_{oc_{ref}}}{R_{sh_{ref}}}. \quad (7)$$

At maximum power point (MPP): $I = I_{mp}$, $V = V_{mp}$,

$$I_{mp_{ref}} = I_{ph_{ref}} - I_{0_{ref}} \left[\exp\left(\frac{V_{mp_{ref}} + I_{mp_{ref}} R_{S_{ref}}}{R_{sh_{ref}}} \right) - 1 \right] - \frac{V_{mp_{ref}} + I_{mp_{ref}} R_{S_{ref}}}{R_{sh_{ref}}}. \quad (8)$$

At MPP, the derivative of the power with respect to the voltage is equal to zero, that is,

$$\left[\frac{d(IV)}{dV} \right]_{mpp,ref} = 0. \quad (9)$$

At the short-circuit point, the slope of the *I-V* curve at short circuit is given by

$$\left(\frac{dI}{dV} \right)_{I_{sc}} = -\frac{1}{R_{sh_{ref}}}. \quad (10)$$

A simultaneous resolution of (6)–(10) yields the five reference parameters, and then the *I-V* curve for STC is built using (5). As one can notice, in this strategy we made use of only three critical points of the *I-V* curve, namely, the short-circuit point, the maximum power point, and the open-circuit point.

3.2. Dependence of Operating Conditions. The determination of the *I-V* characteristic in all other operating conditions can be made possible by evaluating the various parameters of the *I-V* equation in the new weather environment using the semiempirical relationships that follow as shown in [21]. One may consider the following:

$$\frac{a}{a_{\text{ref}}} = \frac{T_c}{T_{c_{\text{ref}}}},$$

$$\frac{I_0}{I_{0_{\text{ref}}}} = \left[\frac{T_c}{T_{c_{\text{ref}}}}\right]^3 \exp\left[E_G \frac{N_s}{a_{\text{ref}}}\left(1 - \frac{T_c}{T_{c_{\text{ref}}}}\right)\right],$$

$$I_{\text{ph}} = \frac{G}{G_{\text{ref}}}\left[I_{\text{ph,ref}} + \alpha_{cc}\left(T_c - T_{c_{\text{ref}}}\right)\right],$$

$$\frac{G}{G_{\text{ref}}} = \left[\frac{-(I_0/a)\exp\left[\left(V_{\text{mp}} + I_{\text{mp}}R_s\right)/a\right] - 1/R_{\text{sh}}}{1 + (I_0 R_s/a)\exp\left[\left(V_{\text{mp}} + I_{\text{mp}}R_s\right)/a\right] + R_s/R_{\text{sh}}}\right]$$
$$\times\left[-\frac{I_{0_{\text{ref}}}}{a_{\text{ref}}}\exp\left[\frac{\left(V_{\text{mp}_{\text{ref}}} + I_{\text{mp}_{\text{ref}}}R_{s_{\text{ref}}}\right)}{a_{\text{ref}}}\right] - \frac{1}{R_{\text{sh}_{\text{ref}}}}\right.$$
$$\left.\times\left(1+\left(\frac{I_{0_{\text{ref}}}R_{s_{\text{ref}}}}{a_{\text{ref}}}\right)\exp\left[\frac{\left(V_{\text{mp}_{\text{ref}}}+I_{\text{mp}_{\text{ref}}}R_{s_{\text{ref}}}\right)}{a_{\text{ref}}}\right]+\frac{R_{s_{\text{ref}}}}{R_{\text{sh}_{\text{ref}}}}\right)^{-1}\right]^{-1},$$

$$R_{\text{sh}} = R_{\text{sh,ref}}. \tag{11}$$

Finally at the MPP, we write the following two equations:

$$I_{\text{mp}} = I_{\text{mp}_{\text{ref}}}\frac{G}{G_{\text{ref}}},$$
$$V_{\text{mp}} = V_{\text{mp}_{\text{ref}}} + \beta_{V_{\text{oc}}}\left(T_c - T_{c_{\text{ref}}}\right). \tag{12}$$

4. Five-Point Model

The five-point model is an improved method to extract the five unknown parameters a, I_{ph}, I_0, R_s, and R_{sh} of (3) using five key physical quantities of the *I-V* characteristic: the short-circuit current, the open-circuit voltage, current and voltage at the maximum power point, and the nominal series and shunt resistances R_{s_0} and R_{sh_0}, respectively. This method was first introduced in [22] using experimental values of key physical quantities. Data issued by manufacturers of modules do not contain the values of the nominal series and shunt resistances. Indeed, the exact five-point model and the approximate five-point model that use approximate values of the nominal series and shunt resistances yield comparable results [22], provided that R_{s_0} and R_{sh_0} are well defined.

4.1. Calculation of Key Quantities. The nominal series and shunt resistances are determined using the slope method. The quantities R_{s_0} and R_{sh_0} are defined at STC as the slope of *I-V* curves at $I = 0$ and at $V = 0$, respectively, as it can be seen in following equations [22]:

$$R_{s_0} = -\left(\frac{dV}{dI}\right)_{V=V_{\text{oc}}}, \tag{13}$$

$$R_{\text{sh}_0} = -\left(\frac{dV}{dI}\right)_{I=I_{\text{sc}}}. \tag{14}$$

In this paper, the concept of nonlinearity is introduced for a better prediction of the behaviour of the tested photovoltaic modules. Thus, the dependency of the others key quantities on temperature and irradiance is given by the following equations:

$$I_{\text{sc}} = \left[I_{\text{sc}_{\text{ref}}} + \mu_{I\text{sc}}\left(T_c - T_{c_{\text{ref}}}\right)\right]\left(\frac{G}{G_{\text{ref}}}\right)^{\alpha},$$
$$V_{\text{oc}} = \frac{V_{\text{oc}_{\text{ref}}}}{1 + \beta\ln\left(G_{\text{ref}}/G\right)}\left(\frac{T_{c_{\text{ref}}}}{T_c}\right)^{\gamma},$$
$$I_{\text{mp}} = \left[I_{\text{mp}_{\text{ref}}} + \mu_{I\text{mp}}\left(T_c - T_{c_{\text{ref}}}\right)\right]\left(\frac{G}{G_{\text{ref}}}\right)^{\alpha},$$
$$V_{\text{mp}} = \frac{V_{\text{mp}_{\text{ref}}}}{1 + \beta\ln\left(G_{\text{ref}}/G\right)}\left(\frac{T_{c_{\text{ref}}}}{T_c}\right)^{\gamma}. \tag{15}$$

Exponents α and γ are constants describing all effects related to the nonlinear dependence of current on the intensity of solar radiation and considering all the nonlinear effects of voltage variation with the cells temperature, respectively [23]. β is a dimensionless coefficient related to the specific technology of photovoltaic module [24, 25].

4.2. Calculation of Parameters. After expressing the above essential quantities as functions of operating conditions, the five parameters required are evaluated using the following equations [26, 27]:

$$I_{\text{ph}} = I_{\text{sc}}\left(1 + \frac{R_s}{R_{\text{sh}}}\right) + I_0\left[\exp\left(\frac{I_{\text{sc}}R_s}{a}\right) - 1\right],$$
$$I_0 = \left(I_{\text{sc}} + \frac{V_{\text{oc}}}{R_{\text{sh}}}\right)\exp\left(-\frac{V_{\text{oc}}}{a}\right),$$
$$a = \left(V_{\text{mp}} + I_{\text{mp}}R_{s_0} - V_{\text{oc}}\right)$$
$$\times\left[\ln\left(I_{\text{sc}} - \frac{V_{\text{mp}}}{R_{\text{sh}}} - I_{\text{mp}}\right) - \ln\left(I_{\text{sc}} - \frac{V_{\text{oc}}}{R_{\text{sh}}}\right)\right.$$
$$\left.+\frac{I_{\text{mp}}}{I_{\text{sc}} - \left(V_{\text{oc}}/R_{\text{sh}}\right)}\right]^{-1},$$
$$R_s = R_{s_0} - \left[\frac{a}{I_0}\exp\left(-\frac{V_{\text{oc}}}{a}\right)\right],$$
$$R_{\text{sh}} = R_{\text{sh}_0}\left(\frac{G_{\text{ref}}}{G}\right). \tag{16}$$

4.3. Calculation of the Constants. The constants α, β, and γ are related to important effects in the photovoltaic module. They are empirical, and the precision of model depends on the acuity of their calculated results. Accurate determination of these constants requires significant experimentation to find the values of current and voltage at different points for different operating conditions in view to calculate them using (17)–(19):

$$\alpha = \frac{\ln\left(\left(I_{sc_1} - \mu_{Isc}\left(T_1 - T_0\right)\right)/I_{sc_0}\right)}{\ln\left(G_1/G_0\right)}, \tag{17}$$

where I_{sc}, G, and T are values of short-circuit current, irradiance, and temperature, respectively; and the indices 0 and 1 are related to the fact that these quantities have to be measured for two different operating conditions.

The variation of the open-circuit voltage is related to the variations of the solar radiation intensity and cells temperature. To calculate β, temperature is kept constant while irradiance varies from G_0 to G_1. β is obtained as

$$\beta = \frac{\left(V_{oc_0}/V_{oc_1}\right) - 1}{\ln\left(G_0/G_1\right)}. \tag{18}$$

For the same value of irradiance, the temperature of the cells varies from T_0 to T_1, and γ is obtained as follows:

$$\gamma = \frac{\ln\left(V_{oc_0}/V_{oc_1}\right)}{\ln\left(T_1/T_0\right)}. \tag{19}$$

5. Results and Discussion

The calculations were made for the Shell SP75 module from Shell Solar and the GES-P70 module from GESOLAR. Shell SP75 is a commercial module with an output of 75 W_p, consisting of 36 single crystal silicon cells connected in series. GES-P70 is also a commercial module of 36 cells connected in series with an output of 70 W_p but in polycrystalline silicon material. Both panels have similar high fill factor values (≈0.75) and at first glance, their curves seem to be quite similar, but if more accurately examined, they show some differences, probably due to their different processing technology. The data of the modules are presented in Table 1.

Manufacturers usually provide only limited operational data for PV modules. These data are available at STC and sometimes under nominal operating cell temperature (NOCT) conditions, that is, irradiance of 800 W/m^2, temperature of 45 or 47°C (depending on the manufacturer), ambient temperature of 20°C, wind speed of 1 m/s, and AM 1.5 spectrum.

The constant α, β, and γ can be calculated following the constant estimation procedures given above, and the results are shown in Table 2.

Utilization of available data aforementioned in the equations above presented for the five-parameter and the five-point models yields computed parameters of the *I-V* characteristic at given operating conditions for both modules used. Following Figures 2 and 3 shows that *I-V* curves fitted

Table 1: Data of the photovoltaic modules used.

Parameters of modules at STC	Shell SP75	GES-P70
Short-circuit current	4,8 A	4,28 A
Open-circuit voltage	21,7 V	21,8 V
Maximum power point current	4,4 A	3,98 A
Maximum power point voltage	17 V	17,6 V
Temperature coefficient of I_{sc}	0,002 A/°C	0,0017 A/°C
Temperature coefficient of V_{oc}	−0,076 V/°C	−0,069 V/°C
Number of cells	36	36
Maximum power	75 W	70 W

Table 2: Constants estimation for PV modules.

Constant	α	β	γ
Shell SP75	1,022	0,058	1,116
GES-P70	0,974	0,060	1,076

from computed parameters of five-parameter and five-point models and experimental data issue by the manufacturers of both the photovoltaic modules used.

Figure 2 shows the *I-V* curves for the modules Shell SP75 Figure 2(a) and GES-P70 Figure 2(b) for different levels of irradiance. It can be seen that the five-point model strongly agrees with experimental data than the five-parameter model for both types of modules. Another significant observation is the very accurate prediction in low irradiance of the five-point model.

Besides verifying the accuracy of the models to fit with the solar irradiance, it is also very important to analyze their capabilities to reproduce the way photovoltaic panel performance is affected by the silicon temperature. The *I-V* curves of both modules have been realized by both mathematical models used in this work when subjected to the variation of temperature as it is shown in Figure 3. The accuracy of both models was tested for different levels of temperature. All measurements were performed at 1000 W/m^2.

As it has been noticed previously with the variation of irradiance, Figure 3 shows that five-point model better fits *I-V* curves of both modules than five-parameter model when subjected to the temperature variation. It is well known that irradiance and temperature of the cells strongly affect the performance of operating photovoltaic devices.

However, latest observations made on modelled *I-V* characteristics when subjected to irradiance and temperature of the cells variations are insufficient to make a true judgement about the capability of both the five-point and the five-parameter models to reproduce the behaviour of operating photovoltaic modules. To further investigate the performance of both the models in various operating conditions of both analyzed photovoltaic modules, inaccuracies on *I-V* curves prediction may be quantified.

Root mean square error (RMSE) is used as the statistical tool for assessing the performance of the models to predict the current. The %RMSE compares the predicted (*I-V*) characteristic at different operating conditions to the measured one, thus providing a clear picture of the precision of the

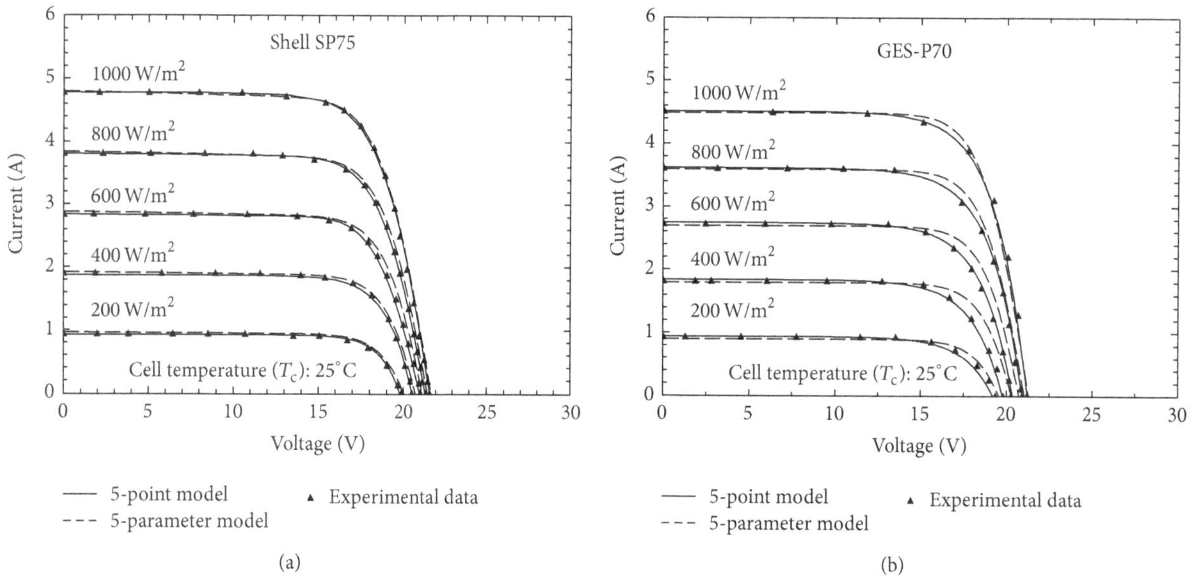

FIGURE 2: Calculated *I-V* curves and experimental data of Shell SP75 (a) and GES-P70 (b) at different irradiance, 25°C.

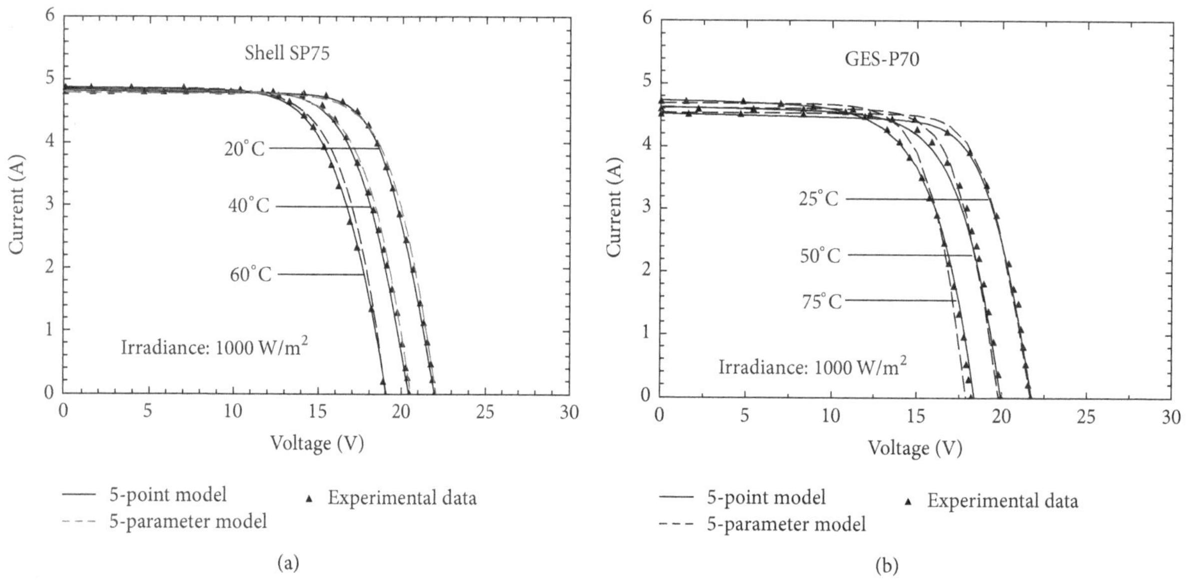

FIGURE 3: Calculated *I-V* curves and experimental data of Shell SP75 (a) and GES-P70 (b) at different temperatures, 1000 W/m^2.

model. Figures 4 and 5 show %RMSE for five-point and five-parameter models for each operating condition analyzed.

Figures 4 and 5 show the measuring deviation of computed *I-V* curves from experimental *I-V* data issue by manufacturers of modules for the whole of operating conditions analyzed. In general, five-point model exhibits lower %RMSEs than five-parameter model for both the studied photovoltaic modules for all environmental conditions. These observations are as expected because it has been above observed that computed *I-V* curves using five-parameter model are more unfasten from experimental *I-V* data than those computed using five-point model.

In this work, it has been noticed that the five-parameter model for which nonlinearities are neglected either overestimates or underestimates the values of current while the five-point model that takes into account the effects of nonlinear shows a remarkable fitting around the key points and a sufficiently accurate prediction at the other points of the *I-V* characteristic for the whole set of the weather conditions analyzed. This is to be foreseeable because the formulation of the equations of the five-parameter model presents an obsolete dependence on the environmental variations. The five-parameter model like many photovoltaic models used up to now, in addition to considering that the variations of

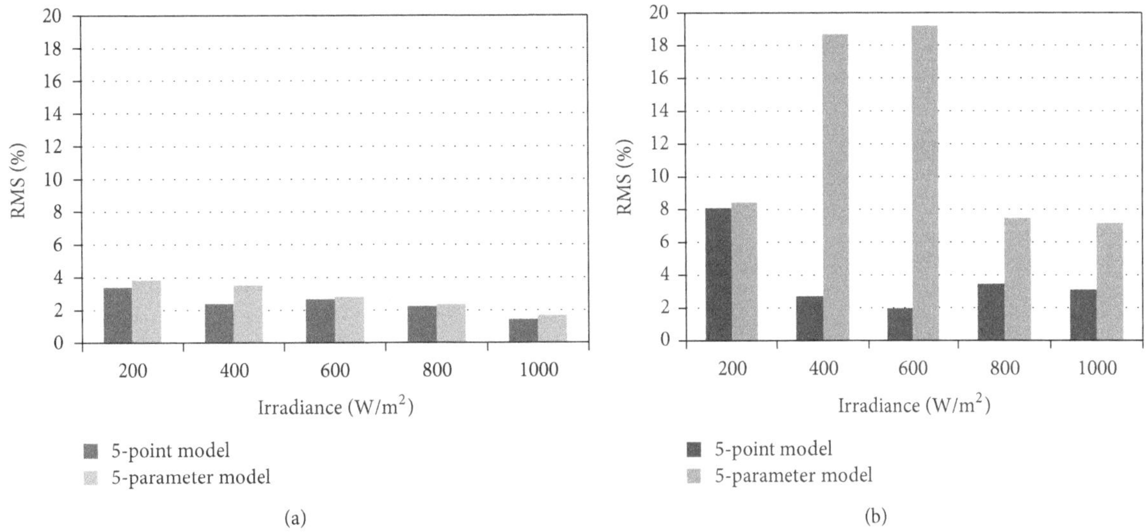

FIGURE 4: %RMS for Shell SP75 (a) and GES-P70 (b) at different irradiations, 25°C.

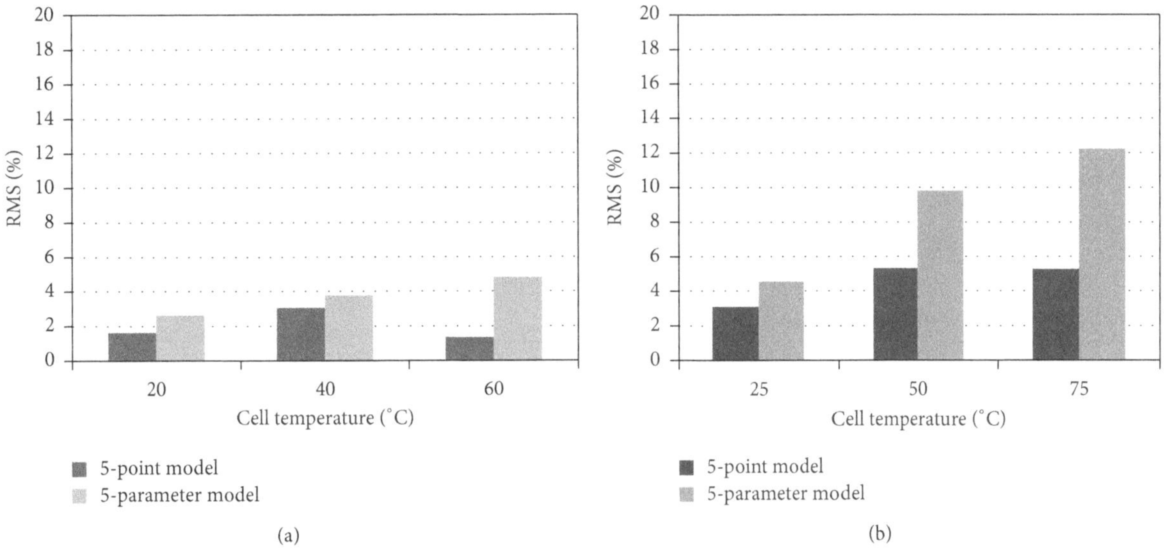

FIGURE 5: %RMS for Shell SP75 (a) and GES-P70 (b) at different temperatures, 1000 W/m^2.

the current and voltage with irradiance and temperature of cells, respectively, at the key operating points are made in linear ways or deviate slightly from their linearities, does not well define the dependence on irradiance and temperature of the cells of certain parameters. The shunt resistance and the ideality factor are considered independent from irradiance and temperature of the cells; the variations of the parameter a (modified ideality factor) with respect to temperature are not related to ideality factor but to its definition such as the above mentioned in (4). It is obvious that the use of constant parameters determined under STC conditions must bring about inaccuracies in other operating conditions [8, 28, 29].

In conventional one-diode representation of photovoltaic module, physical phenomena are considered globally, so that parameters (which are related to these phenomena) lose their meaning [30]. In the five-parameter model, variations of parameters with environmental data (irradiance and temperature) are from their physical conception. Contrary to this technique, five-point model uses only experimental data and information of key operating points at STC and at any other operating conditions. Furthermore, the effects of nonlinearities considered in key operating points are transported in any other point of the characteristic through the calculation of parameters using key physical quantities. Methods that use a minimum simplification assumption from (3) such as five-point model are expected to give more realistic values for parameters under investigation [22, 30].

6. Conclusion

In this work five-parameter and five-point models were used to predict electrical *I-V* characteristics of illuminated commercial photovoltaic modules Shell SP75 and GES-P70 of monocrystalline and polycrystalline silicon material, respectively. Validation of these models was carried out with comparison to the data provided by module manufacturers. The following conclusions were drawn from our analysis.

(i) Most parameters depend on both the cell temperature and the solar irradiance. Thus, the mastery of their behavior is crucial to correctly predict the performance of the photovoltaic cells and arrays.

(ii) For both models used in this work, the precision varies on one hand from one module to another and on the other hand from one operating condition to another. Furthermore, the five-point model is more precise than the five-parameter model as well for the two photovoltaic module types studied for the whole set of weather conditions analyzed.

(iii) Taking into account the effects of nonlinearities has made it possible to achieve better fitting of the *I-V* characteristics. This study therefore shows that the five-point model is an excellent tool for the prediction of the electrical response of the illuminated crystalline photovoltaic modules.

Nomenclature

Symbols

A_0: Ideality factor
a: Modified ideality factor
E_G: Band gap of the material
G: Irradiance
I: Current
MPP: Maximum power point
R: Resistance
STC: Standard test conditions
V: Voltage
I_0: Saturation current
I_{ph}: Photocurrent
N_s: Number of cells in series
V_{oc}: Open-circuit voltage
I_{sc}: Short-circuit current.

Subscript

ref: Reference conditions
mpp: Maximum power point
oc: Open circuit
s: Series
sc: Short circuit
sh: Shunt.

Greek Letters

μ: Temperature coefficient
α: Constant
β: Constant
γ: Constant.

Conflict of Interests

There is no competing interest in the validity of this paper due to the choice of the Shell SP75 and GES-P70 modules. However, the choice of the Shell SP75 and GES-P70 module was made with reference to other research carried on these modules.

Acknowledgments

The authors are grateful to Dr. Obounou and Dr. Akana for their efforts in the realization of this work.

References

[1] D. L. King, W. E. Boyson, and J. A. Kratochvil, "Analysis of factors influencing the annual energy production of photovoltaic systems," in *Proceedings of the 29th IEEE Photovoltaic Specialists Conference*, pp. 1356–1361, May 2002.

[2] B. Kroposki, W. Marion, D. King et al., *Comparison of Module Performance Characterization Methods for Energy Production*, National Renewable Energy Laboratory, Golden, Colo, USA, 2000.

[3] B. Marion, B. Kroposki, K. Emery, J. del Cueto, D. Myers, and C. Osterwald, *Validation of a Photovoltaic Module Energy Ratings Procedure at NREL*, National Renewable Energy Laboratory, Golden, Colo, USA, 1999.

[4] D. F. Menicucci, "Photovoltaic array performance simulation models," *Solar Cells*, vol. 18, no. 3-4, pp. 383–392, 1986.

[5] A. S. Joshi, I. Dincer, and B. V. Reddy, "Performance analysis of photovoltaic systems: a review," *Renewable and Sustainable Energy Reviews*, vol. 13, no. 8, pp. 1884–1897, 2009.

[6] D. L. King, "Photovoltaic module and array performance characterization methods for all system operating conditions," in *Proceeding of NREL/SNL Photovoltaics Program Review Meeting*, AIP Press, New York, NY, USA, 1997.

[7] A. H. Fanney, M. W. Davis, B. P. Dougherty, D. L. King, W. E. Boyson, and J. A. Kratochvil, "Comparison of photovoltaic module performance measurements," *Journal of Solar Energy Engineering, Transactions of the ASME*, vol. 128, no. 2, pp. 152–159, 2006.

[8] E. Saloux, A. Teyssedou, and M. Sorin, "Explicit model of photovoltaic panels to determine voltages and currents at the maximum power point," *Solar Energy*, vol. 85, no. 5, pp. 713–722, 2011.

[9] M. G. Villalva, J. R. Gazoli, and E. R. Filho, "Comprehensive approach to modeling and simulation of photovoltaic arrays," *IEEE Transactions on Power Electronics*, vol. 24, no. 5, pp. 1198–1208, 2009.

[10] U. Eicker, *Solar Technologies for Buildings*, Wiley, New York, NY, USA, 2003.

[11] V. L. Brano, A. Orioli, G. Ciulla, and A. di Gangi, "An improved five-parameter model for photovoltaic modules," *Solar Energy Materials and Solar Cells*, vol. 94, no. 8, pp. 1358–1370, 2010.

[12] S. Liu and R. A. Dougal, "Dynamic multiphysics model for solar array," *IEEE Transactions on Energy Conversion*, vol. 17, no. 2, pp. 285–294, 2002.

[13] D. Sera, R. Teodorescu, and P. Rodriguez, "PV panel model based on datasheet values," in *Proceedings of the IEEE International Symposium on Industrial Electronics (ISIE '07)*, pp. 2392–2396, Vigo, Spain, June 2007.

[14] T. F. Elshatter, M. E. Elhagree, Aboueldahab, and A. A. Elkousry, "Fuzzy modeling and simulation of photovoltaic system," in *Proceedings of the 14th European Photovoltaic Solar Energy Conference*, 1999.

[15] A. Mellit, M. Benghanem, and S. A. Kalogirou, "Modeling and simulation of a stand-alone photovoltaic system using an adaptive artificial neural network: proposition for a new sizing procedure," *Renewable Energy*, vol. 32, no. 2, pp. 285–313, 2007.

[16] A. N. Celik, "Artificial neural network modelling and experimental verification of the operating current of mono-crystalline photovoltaic modules," *Solar Energy*, vol. 85, no. 10, pp. 2507–2517, 2011.

[17] T. Easwarakhanthan, J. Bottin, I. Bouhouch, and C. Boutrit, "Nonlinear minimization algorithm for determining the solar cell parameters with microcomputers," *International Journal of Solar Engeneering*, vol. 4, pp. 1–12, 1986.

[18] T. U. Townsend, *A method for estimating the long term performance of direct-coupled photovoltaic systems [M.S. thesis]*, Solar Energy Laboratory, University of Wisconsin, Madison, Wis, USA, 1989.

[19] M. Chegaar, Z. Ouennoughi, and F. Guechi, "Extracting dc parameters of solar cells under illumination," *Vacuum*, vol. 75, no. 4, pp. 367–372, 2004.

[20] M. Haouari-Merbah, M. Belhamel, I. Tobías, and J. M. Ruiz, "Extraction and analysis of solar cell parameters from the illuminated current-voltage curve," *Solar Energy Materials and Solar Cells*, vol. 87, no. 1–4, pp. 225–233, 2005.

[21] W. de Soto, S. A. Klein, and W. A. Beckman, "Improvement and validation of a model for photovoltaic array performance," *Solar Energy*, vol. 80, no. 1, pp. 78–88, 2006.

[22] D. K. Schroder, *Semiconductor Material and Device Characterization*, John Willey & Sons, NEW York, NY, USA, 1998.

[23] W. Zhou, H. Yang, and Z. Fang, "A novel model for photovoltaic array performance prediction," *Applied Energy*, vol. 84, no. 12, pp. 1187–1198, 2007.

[24] A. J. Anderson, "Photovoltaic translation equations: a new approach," NREL Final Subcontract Report No. DE-AC36-83CH10093, 1996.

[25] E. E. van Dyk, E. L. Meyer, F. J. Vorster, and A. W. R. Leitch, "Long-term monitoring of photovoltaic devices," *Renewable Energy*, vol. 25, no. 2, pp. 183–197, 2002.

[26] A. H. Arab, F. Chenlo, and M. Benghanem, "Loss-of-load probability of photovoltaic water pumping systems," *Solar Energy*, vol. 76, no. 6, pp. 713–723, 2004.

[27] A. N. Celik and N. Acikgoz, "Modelling and experimental verification of the operating current of mono-crystalline photovoltaic modules using four- and five-parameter models," *Applied Energy*, vol. 84, no. 1, pp. 1–15, 2007.

[28] J. A. Eikelboom and A. H. Reinders, "Determination of the irradiation dependent efficiency of multicrystalline Si PV modules on basis of IV curve fitting and its influence on the annual performance," in *Proceedings of the 14th European PV Solar Energy Conference*, pp. 293–296, Barcelona, Spain, 1997.

[29] K. Ishaque and Z. Salam, "An improved modeling method to determine the model parameters of photovoltaic (PV) modules using differential evolution (DE)," *Solar Energy*, vol. 85, no. 9, pp. 2349–2359, 2011.

[30] M. Bashahu and A. Habyarimana, "Review and test of methods for determination of the solar cell series resistance," *Renewable Energy*, vol. 6, no. 2, pp. 129–138, 1995.

Numerical Investigation of Forced Convection in a Channel with Solid Block inside a Square Porous Block

Neda Janzadeh and Mojtaba Aghajani Delavar

Islamic Azad University Science and Research Ayatollah Amoli Branch, Amol, Iran

Correspondence should be addressed to Mojtaba Aghajani Delavar; m.a.delavar@nit.ac.ir

Academic Editor: S. A. Kalogirou

The fluid flow and heat transfer in a porous medium have received considerable attention due to its importance in many engineering applications. In this study numerical investigation of fluid flow and heat transfer over a hot solid block inside a square porous block located in a channel was carried out. The lattice Boltzmann method with nine velocities, D2Q9, was used for numerical simulations. Brinkman-Forchheimer model was successfully used to simulate fluid flow in porous media. The effects of parameters such as porosity, Reynolds number on flow pattern, and heat transfer were studied. The different effects of mentioned parameters were discussed in the paper.

1. Introduction

The fluid flow and transport phenomena in a porous medium have been used in many fields of science and engineering, such as geothermal energy extraction, solar collector, solar absorbers food processing, fuel cells, petroleum processing, catalytic and chemical particle beds, transpiration cooling, electronic cooling, drying processes, porous bearing, nuclear reactors, and many others.

Thermal Insulation. Fluid flow and convective heat transfer in conduits fully and partially filled with porous medium have been investigated analytically, experimentally, and numerically by many researchers. Poulikakos and Kazmierczak [1] theoretically studied the convection heat transfer in the developed region of two parallel plates and in a circular pipe with porous media attached to the wall. Fu et al. [2] numerically simulated the convection heat transfer through a single porous block mounted on the heated wall, and they studied the effect of different parameters including the porous block size and length and porosity of porous block and Reynolds number on convection heat transfer enhancement. Li et al. [3] numerically investigated the effect of staggered porous blocks mounted on the wall in a laminar flow channel on the velocity field and local heat transfer. Huang et al. [4] numerically studied steady-state heat transfer to a fluid passing through multiple heated block mounted on the wall of a channel with porous covers.

Guo et al. [5] investigated numerically the heat transfer of pulsating flow through a pipe with a porous medium attached to the pipe wall. In their work they used conventional control volume method. Hamdan et al. [6] used an implicit finite-difference method to simulate forced convection in a porous substrate inserted in the core of a parallel plate channel. Jiang et al. [7] investigate the forced convection heat transfer in plate channels saturated with air and water which contain sintered bronze porous media.

The lattice Boltzmann method (LBM) is a powerful numerical technique based on kinetic theory for simulation of fluid flows and modeling the physics in fluids [8–10]. The problem of convective heat in porous medium was investigated numerically by Guo and Zhao [11]. In their work, they used the lattice Boltzmann method to simulate the temperature field. Seta et al. [12] applied the lattice Boltzmann method to simulation of natural convection in porous media. Shokouhmand et al. [13] simulated the laminar flow and convective heat transfer in a porous medium inside the core of two parallel plates of a conduit using lattice Boltzmann method.

Hot solid block inside a square porous block may be applied in some engineering application such as porous media as an active layer in reacting chemical flows, fuel cell, heat exchangers, and electronic devices cooling.

2. Numerical Model

The general form of lattice Boltzmann equation with nine velocities, D2Q9, with external force can be written as [9]

$$f_k(\vec{x}+\vec{c}_k\Delta t, t+\Delta t) = f_k(\vec{x},t) + \frac{\Delta t}{\tau}\left[f_k^{\mathrm{eq}}(\vec{x},t) - f_k(\vec{x},t)\right] + \Delta t \vec{F}_k,$$
$$f_k^{\mathrm{eq}} = \omega_k \cdot \rho \left[1 + \frac{\vec{c}_k\cdot\vec{u}}{c_s^2} + \frac{1}{2}\frac{(\vec{c}_k\cdot\vec{u})^2}{c_s^4} - \frac{1}{2}\frac{\vec{u}\cdot\vec{u}}{c_s^2}\right], \tag{1}$$

where $\vec{c}_k$ is the discrete lattice velocity in direction k, $\vec{F}_k$ is the external force, Δt denotes the lattice time step, τ is the lattice relaxation time, f_k^{eq} denotes the equilibrium distribution function, ω_k is weighting factor, and ρ is the lattice fluid density. To consider both the flow and the temperature fields, the thermal LBM utilizes two distribution functions, f and g, for flow and temperature fields, respectively. The f distribution function is as same as discussed previously; the g distribution function is as follows [9]:

$$g_k(\vec{x}+\vec{c}_k\Delta t, t+\Delta t) = g_k(\vec{x},t) + \frac{\Delta t}{\tau_g}\left[g_k^{\mathrm{eq}}(\vec{x},t) - g_k(\vec{x},t)\right],$$
$$g_k^{\mathrm{eq}} = \omega_k \cdot T \cdot \left[1 + \frac{\vec{c}_k\cdot\vec{u}}{c_s^2}\right]. \tag{2}$$

The flow properties are defined as (i denotes the component of the Cartesian coordinates)

$$\rho = \sum_k f_K, \qquad \rho u_i = \sum_k f_k c_{ki}, \qquad T = \sum_k g_k. \tag{3}$$

The Brinkman-Forchheimer equation was used for flow in porous regions that is written as [9, 12]

$$\frac{\partial \vec{u}}{\partial t} + (\vec{u}\cdot\nabla)\left(\frac{\vec{u}}{\varepsilon}\right) = -\frac{1}{\rho}\nabla(\varepsilon p) + \upsilon_{\mathrm{eff}}\nabla^2\vec{u} + \left(-\frac{\varepsilon\upsilon}{K}\vec{u} - \frac{1.75}{\sqrt{150\varepsilon K}}|\vec{u}|\,\vec{u} + \varepsilon\vec{G}\right), \tag{4}$$

where ε is the porosity, K is the permeability, υ_{eff} is the effective viscosity, υ is the kinematic viscosity, and G is the acceleration due to gravity. The last term in the right hand in the parenthesis is the total body force, F, which was written by using the widely used Ergun's relation [16]. For porous medium the corresponding distribution functions are as the same as (1). But the equilibrium distribution functions and the best choice for the forcing term are

$$f_k^{\mathrm{eq}} = \omega_k \cdot \rho \cdot \left[1 + \frac{\vec{c}_k\cdot\vec{u}}{c_s^2} + \frac{1}{2}\frac{(\vec{c}_k\cdot\vec{u})^2}{\varepsilon c_s^4} - \frac{1}{2}\frac{\vec{u}^2}{\varepsilon c_s^2}\right],$$
$$F_k = \omega_k \rho\left(1 - \frac{1}{2\tau_v}\right)\left[\frac{\vec{c}_k\cdot\vec{F}}{c_s^2} + \frac{(\vec{u}\vec{F} : \vec{c}_k\vec{c}_k)}{\varepsilon c_s^4} - \frac{\vec{u}\cdot\vec{F}}{\varepsilon c_s^2}\right]. \tag{5}$$

The forcing term F_k defines the fluid velocity $\vec{u}$ as

$$\vec{u} = \sum_k \frac{c_k F_k}{\rho} + \frac{\Delta t}{2}\vec{F}. \tag{6}$$

According the previous equations $\vec{F}$ is related to $\vec{u}$, so (6) is nonlinear for the velocity. A temporal velocity $\vec{v}$ is used to solve this nonlinear problem [13] as follows:

$$\vec{u} = \frac{\vec{v}}{c_0 + \sqrt{c_0^2 + c_1|\vec{v}|}}, \qquad \vec{v} = \sum_k \frac{c_k f_k}{\rho} + \frac{\Delta t}{2},$$
$$c_0 = \frac{1}{2}\left(1 + \varepsilon\frac{\Delta t}{2}\frac{\upsilon}{K}\right), \qquad c_1 = \varepsilon\frac{\Delta t}{2}\frac{1.75}{\sqrt{150\varepsilon^3 K}}. \tag{7}$$

The effective thermal conductivity, k_{eff} of the porous medium should be recognized for proper investigation of conjugate convection and conduction heat transfer in porous zone, which was calculated by [16]

$$k_{\mathrm{eff}} = k_f\left[\left(1-\sqrt{1-\varepsilon}\right) + \frac{2\sqrt{1-\varepsilon}}{1-\sigma B} \times \left(\frac{(1-\sigma)B}{(1-\sigma B)^2}\ln\left(\frac{1}{\sigma B}\right) - \frac{B+1}{2} - \frac{B-1}{1-\sigma B}\right)\right],$$
$$B = 1.25\left[\frac{1-\varepsilon}{\varepsilon}\right]^{10/9}, \qquad \sigma = \frac{k_f}{k_s}. \tag{8}$$

3. Boundary Conditions

From the streaming process the distribution functions out of the domain are known. The unknown distribution functions are those toward the domain. In Figure 1 the unknown distribution function which needs to be determined, are shown as dotted lines. Regarding the boundary conditions of the flow field, the solid walls are assumed to be no slip, and thus the bounce-back scheme is applied. This scheme specifies the outgoing directions of the distribution functions as the reverse of the incoming directions at the boundary sites. For example, for flow field in the north boundary the following conditions is used:

$$f_{4,n} = f_{2,n}, \qquad f_{7,n} = f_{5,n}, \qquad f_{8,n} = f_{6,n}. \tag{9}$$

In this channel the inlet velocity at west boundary is known $u = u_w = u_{\mathrm{inlet}}$. In LBM method the inward distribution functions at the boundaries have to be specified. So the

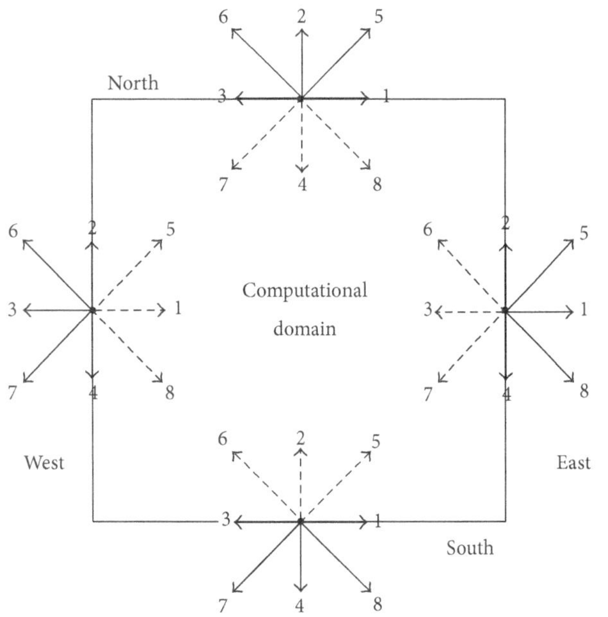

Figure 1: Domain boundaries and known (solid lines) and unknown (dotted lines) distribution functions.

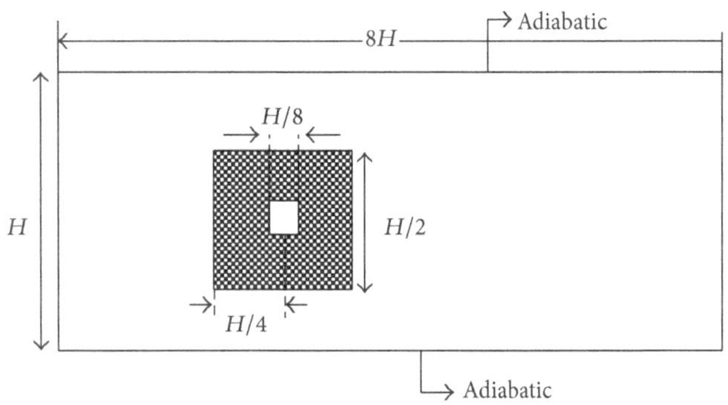

Figure 2: The computational domain.

f_1, f_5, and f_8 are unknowns for west boundary condition. The unknown distribution functions are calculated as

$$\rho_w = \frac{1}{1-u_w}\left[f_0 - f_2 + f_4 + 2\left(f_3 + f_6 + f_7\right)\right],$$

$$f_{1,n} = f_{3,n} + \frac{2}{3}\rho_w \qquad f_{5,n} = f_{7,n} - \frac{1}{2}\left(f_{2,n} - f_{4,n}\right) + \frac{1}{6}\rho_w u_w,$$

$$f_{8,n} = f_{6,n} + \frac{1}{2}\left(f_{2,n} - f_{4,n}\right) + \frac{1}{6}\rho_w u_w. \tag{10}$$

In this problem the outlet velocity is unknown. The east boundary condition in Figure 1 represents the outlet condition. Then f_3, f_6, and f_7 need to be calculated at the east boundary as

$$f_{3,n} = 2f_{3,n-1} - f_{3,n-2}, \qquad f_{6,n} = 2f_{6,n-1} - f_{6,n-2},$$
$$f_{7,n} = 2f_{7,n-1} - f_{7,n-2}. \tag{11}$$

Table 1: Simulation parameters.

L	8.0 cm
ε	0.3-0.5-0.7-0.9
Pr	0.7
H	1.0 cm
Re	25-50-75-100
T_{inlet}-$T_{\text{wall solid block}}$	20°C -40°C

Table 2: Comparison of averaged Nusselt number between LBM and Kays and Crawford [17].

	q_2''/q_1''		
	0.5	1.0	1.5
Nu_1			
Kays and Crawford [17]	17.48	8.23	11.19
LBM	17.25	8.16	11.10
Nu_2			
Kays and Crawford [17]	6.51	8.23	7.00
LBM	6.49	8.16	6.91

Furthermore, for the temperature field, the local temperature is defined by (3). For isothermal boundaries such as a bottom hot wall the unknown distribution functions are evaluated as

$$g_{2,n} = T_h\left(\omega_2 + \omega_4\right) - g_{4,n}, \qquad g_{5,n} = T_h\left(\omega_5 + \omega_7\right) - g_{7,n},$$
$$g_{6,n} = T_h\left(\omega_6 + \omega_8\right) - g_{8,n}, \tag{12}$$

where n is the lattice on the boundary. For adiabatic boundary condition such as a bottom wall the unknown distribution functions are evaluated as

$$g_{2,n} = g_{2,n-1}, \qquad g_{5,n} = g_{5,n-1}, \qquad g_{6,n} = g_{6,n-1}. \tag{13}$$

4. Computational Domain

Computational domain consists of a hot solid block inside a square porous block located in a channel (Figure 2). The channel height, length, and other simulation parameters are illustrated in the Figure 2 and Table 1.

5. Validation and Grid Independent Check

In this study, flow pattern and thermal field were simulated in a channel with a hot solid block inside a porous block by using lattice Boltzmann method. Figure 3 compares well the velocity profile for flow in a clear channel captured by LBM and results presented by Nield [14]. In this figure the result of simulation is compared well with Seta et al. [15]. Table 2 shows the computed Nusselt number by LBM and Kays and Crawford [17], and good agreement is obtained.

6. Results and Discussion

In the present work the thermal lattice Boltzmann model with nine velocities was used to solve the force convection

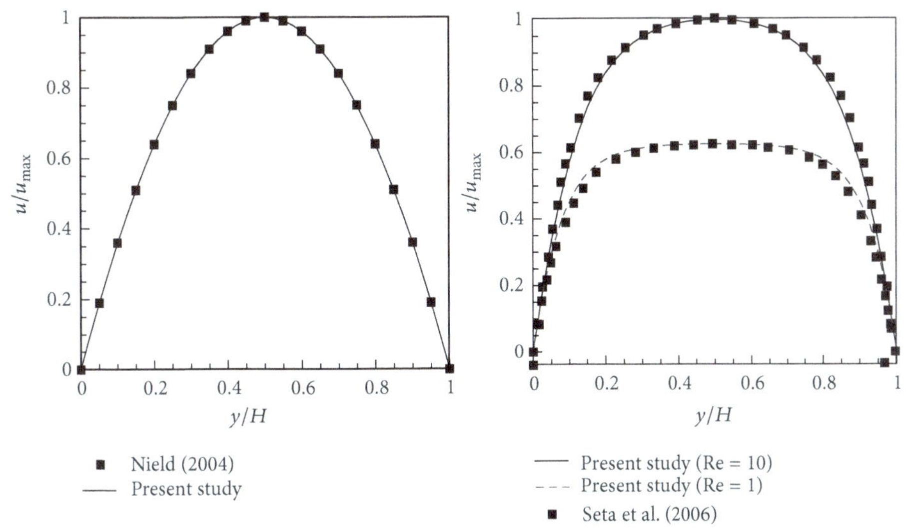

FIGURE 3: Comparison of velocity profile for clear channel between Nield [14] and LBM and porous channel flow between LBM and Seta et al. [15].

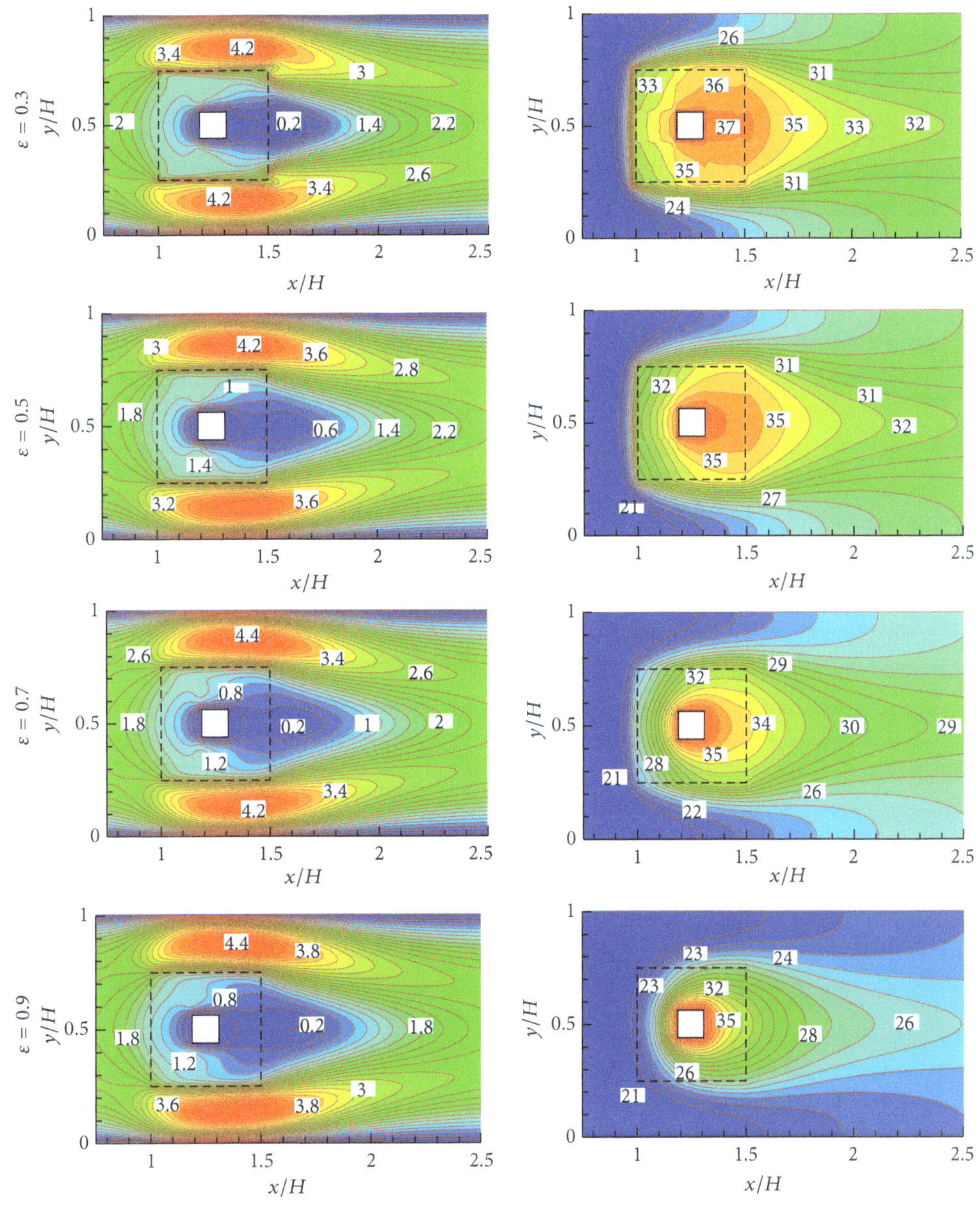

FIGURE 4: Velocity contours ($U/U_{inlet\ (Re=25)}$) temperature contours for different porosities.

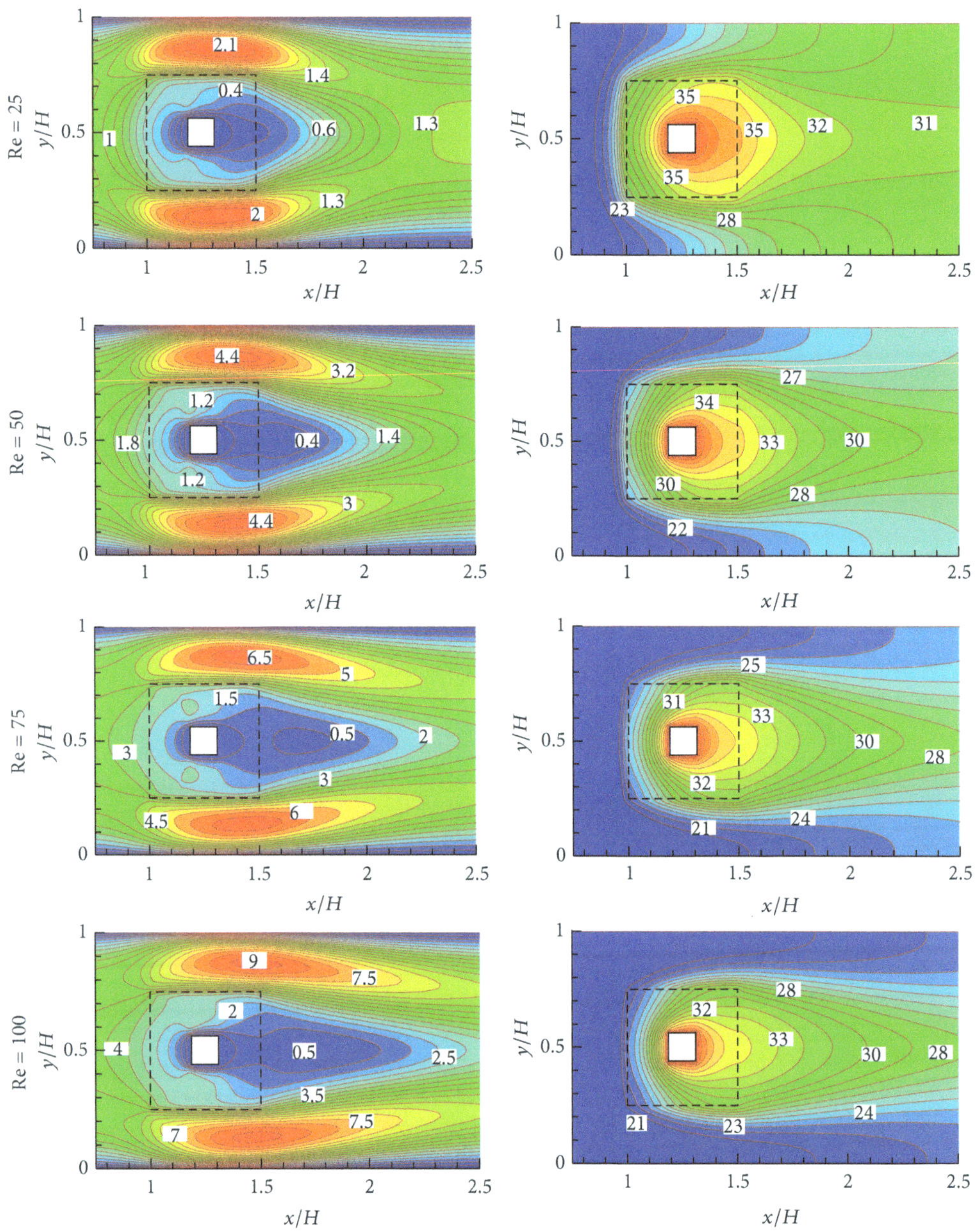

FIGURE 5: Velocity vectors and contours ($U/U_{inlet\ (Re=25)}$) for different Reynolds numbers.

heat transfer in a channel containing hot solid block inside a square porous block. The effects of porosity and Reynolds number, on the flow field and convective heat transfer, were investigated.

Figure 4 shows the velocity and temperature contours for different values of porosity at Re = 50. With increasing the porosity, velocity rises slightly in the top and bottom of porous block, because fluid tends to flow in clear passages with lower pressure drop. At higher values of porosity, fluid changes easier its path to clear passages, so velocity increases in these passages.

According to (8) with increasing the porosity, the effective thermal conductivity in the porous block reduces. At higher value of porosity, the heat transfer between fluid and solid block decreases. So the fluid temperature will increase.

The velocity and temperature contours for different Reynolds number have been drawn in Figure 5.

It can be seen that as Reynolds number increases the flow velocity increases. With increasing the velocity, the convective coefficient (h) increased according to ($Q = hA(T_{wall} - T_{fluid})$), and the magnitude of heat transfer increases. But According to ($Q = mc\Delta T = \rho uA(T_{out} - T_{in})$) with increasing the velocity, the temperature gradient of the fluid reduces. So the outlet fluid temperature will decrease.

The temperature and velocity profiles for various Reynolds numbers are represented in Figure 6. As expected the flow velocity increases with increasing the Reynolds number. The higher values of Reynolds number cause the temperature of zones to be decreased due to the decrease in heat transfer process. In Figure 7 the average fluid

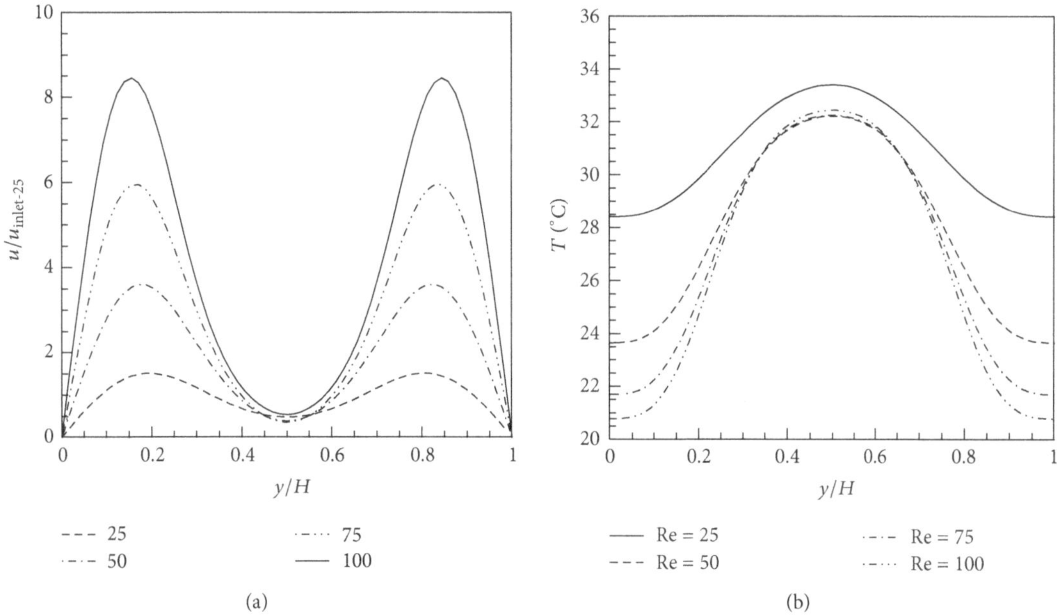

FIGURE 6: velocity and temperature profiles at different Reynolds numbers (porosity = 0.7, $x/H = 7/16$).

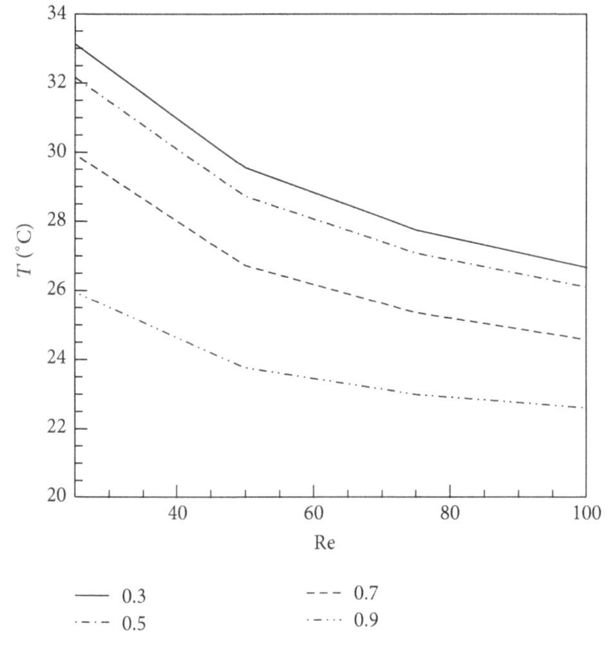

FIGURE 7: Averaged fluid temperature at different porosities.

temperature for various Reynolds numbers at different porosity is drown that shows the effect of Reynolds number on temperature more than porosity. As mentioned before at higher porosity obtains a lower average temperature for fluid.

7. Conclusion

Fluid flow and heat transfer in a channel with solid block inside a square porous block were carried out in this paper. The effect of parameters including porosity and Reynolds number on the velocity and thermal field was investigated numerically by using lattice Boltzmann method. Decreasing the porosity leads to a higher temperature of fluid due to increasing K_{eff} which causes the heat transfer enhancement in porous block. With increasing the Reynolds number the fluid velocity increased, and the average fluid temperature will be reduced.

References

[1] D. Poulikakos and M. Kazmierczak, "Forced convection in a duct partially filled with a porous material," *Journal of Heat Transfer*, vol. 109, no. 3, pp. 653–662, 1987.

[2] W. S. Fu, H. C. Huang, and W. Y. Liou, "Thermal enhancement in laminar channel flow with a porous block," *International Journal of Heat and Mass Transfer*, vol. 39, no. 10, pp. 2165–2175, 1996.

[3] H. Y. Li, K. C. Leong, L. W. Jin, and J. C. Chai, "Analysis of fluid flow and heat transfer in a channel with staggered porous blocks," *International Journal of Thermal Sciences*, vol. 49, no. 6, pp. 950–962, 2010.

[4] P. C. Huang, C. F. Yang, J. J. Hwang, and M. T. Chiu, "Enhancement of forced-convection cooling of multiple heated blocks in a channel using porous covers," *International Journal of Heat and Mass Transfer*, vol. 48, no. 3-4, pp. 647–664, 2005.

[5] Z. Guo, S. Y. Kim, and H. J. Sung, "Pulsating flow and heat transfer in a pipe partially filled with a porous medium," *International Journal of Heat and Mass Transfer*, vol. 40, no. 17, pp. 4209–4218, 1997.

[6] M. O. Hamdan, M. A. Al-Nimr, and M. K. Alkam, "Enhancing forced convection by inserting porous substrate in the core of a parallel-plate channel," *International Journal of Numerical Methods for Heat and Fluid Flow*, vol. 10, no. 5, pp. 502–517, 2000.

[7] P. X. Jiang, M. Li, T. J. Lu, L. Yu, and Z. P. Ren, "Experimental research on convection heat transfer in sintered porous plate

channels," *International Journal of Heat and Mass Transfer*, vol. 47, no. 10-11, pp. 2085–2096, 2004.

[8] S. Succi, *The Lattice Boltzmann Equation for Fluid Dynamics and Beyond*, Clarendon Press, Oxford, UK, 2001.

[9] A. A. Mohammad, *Lattice Boltzmann Method, Fundamentals and Engineering Applications with Computer Codes*, Springer, London, UK, 2011.

[10] P. H. Kao, Y. H. Chen, and R. J. Yang, "Simulations of the macroscopic and mesoscopic natural convection flows within rectangular cavities," *International Journal of Heat and Mass Transfer*, vol. 51, no. 15-16, pp. 3776–3793, 2008.

[11] Z. Guo and T. S. Zhao, "A lattice Boltzmann model for convection heat transfer in porous media," *Numerical Heat Transfer, Part B*, vol. 47, no. 2, pp. 157–177, 2005.

[12] T. Seta, E. Takegoshi, and K. Okui, "Lattice Boltzmann simulation of natural convection in porous media," *Mathematics and Computers in Simulation*, vol. 72, no. 2-6, pp. 195–200, 2006.

[13] H. Shokouhmand, F. Jam, and M. R. Salimpour, "Simulation of laminar flow and convective heat transfer in conduits filled with porous media using lattice Boltzmann Method," *International Communications in Heat and Mass Transfer*, vol. 36, no. 4, pp. 378–384, 2009.

[14] D. A. Nield, "Forced convection in a parallel plate channel with asymmetric heating," *International Journal of Heat and Mass Transfer*, vol. 47, no. 25, pp. 5609–5612, 2004.

[15] T. Seta, E. Takegoshi, and K. Okui, "Thermal lattice Boltzmann model for incompressible flows through porous media," *Journal of Thermal Science and Technology*, vol. 1, no. 2, pp. 90–100, 2006.

[16] S. Ergun, "Fluid flow through packed columns," *Chemical Engineering Progress*, vol. 48, no. 2, pp. 89–94, 1952.

[17] W. M. Kays and M. E. Crawford, *Solutions Manual, Convective Heat and Mass Transfered*, McGraw-Hill, New York, NY, USA, 3rd edition, 1993.

8

Biochemical Methane Potential of Agro Wastes

Vidhya Prabhudessai, Anasuya Ganguly, and Srikanth Mutnuri

Applied and Environmental Biotechnology Laboratory, Department of Biological Sciences, Birla Institute of Technology and Science, Pilani, K. K. Birla Goa Campus, Goa 403726, India

Correspondence should be addressed to Srikanth Mutnuri; srikanth.mutnuri@gmail.com

Academic Editor: Guohe Huang

The focus of our work is on anaerobic digestion of locally available agro wastes like coconut oil cake, cashew apple waste, and grass from lawn cuttings. The most productive agro waste, in terms of methane yield, was coconut oil cake and grass. The results showed that the initial volatile solids concentration significantly affected the biogas production. The methane yield from coconut oil cake was found to be 383 ml CH_4/g VS and 277 ml CH_4/g VS added at 4 and 4.5 g VS/l. In case of grass the biogas production increased with increasing VS concentrations with methane yield of 199, 250, 256, 284, and 332 ml CH_4/g VS at 3, 3.5, 4, 4.5, and 5.0 g VS/l. For cashew apple waste single-stage fermentation inhibited biogas production. However, phase separation showed methane yield of 60.7 ml CH_4/g VS and 64.6 ml CH_4/g VS at 3.5 and 4.0 g VS/l, respectively. The anaerobic biodegradability of coconut oil cake was evaluated in fed batch mode in a 5 L anaerobic reactor at 4 g VS/L per batch, and the maximum methane yield was found to be 320 ml CH_4/g VS.

1. Introduction

In the light of rapidly rising costs associated with energy supply, waste disposal, and increasing concerns with environmental quality degradation, conversion of wastes to energy is becoming an economically viable solution [1]. Anaerobic digestion (AD) is a biological process in which bacteria break down organic matter producing biogas as the end product. AD is a sequence of chemical reactions during which organic matter is decomposed through the metabolic pathways of naturally occurring microorganisms in an oxygen depleted environment. AD is a promising technology which could effectively address the problem of waste disposal yielding valuable outputs like biogas and fertilizer. AD without any pretreatment but with energy recovery is the most attractive method for treating solid wastes [2]. AD of organic wastes and energy crops to produce methane would benefit society by providing a clean fuel from renewable feed stocks. This could substitute fossil fuel-derived energy and reduce environmental impact including global warming and acid rain [3]. Agro wastes and energy crops represent an important source of biomass that can serve as a substrate in anaerobic digestion, resulting in the production of renewable energy. The wastes selected for our study are coconut oil cake (residue obtained after oil extraction), cashew apple waste (crushed apple waste obtained after extraction of "Feni"—"Feni" is a form of liquor i.e., distilled from the pure fermented juice of cashew apple without addition of spirit), and grass from lawn cuttings.

The direct comparison of biomethane production from different feedstocks is difficult as performance data for specific types are often produced under a wide variety of experimental conditions, for example, mixing regime, temperature, total solids, volatile, solids and hydraulic retention time. For this reason, it is better to compare feedstocks by their ultimate methane yield determined by the biochemical methane potential (BMP) assay [4–6]. Chynoweth et al. [7] studied the biochemical methane potential of a variety of feedstocks. Gunaseelan [8] reported the ultimate methane yields of several fractions of fruit and vegetable solid wastes, sorghum and Napier grass, using the BMP assay. Cavinato et al. [9] studied the codigestion of cattle manure with agro wastes and energy crops. Similar studies on biomethanization of energy crops were reported by Demirel and Scherer [10]. The purpose of this work was to study the anaerobic digestion of three different substrates, namely, coconut oil cake, cashew apple waste, and grass cuttings, in order to evaluate the potential of anaerobic digestion as an alternative to the conventional solution-like composting and incineration while reducing the energy consumption of fossils fuels.

2. Materials and Methods

2.1. Feedstocks. Coconut oil cake and cashew apple waste were collected from local industries in and around Goa. Grass cuttings were from the lawns of our institute campus (BITS, Pilani, K. K. Birla Goa Campus). Cashew apple waste was collected from a cashew plantation located in Sancoale, Goa. Coconut oil cake was collected from a factory extracting coconut oil located in Cansaulim, Goa. And grass clippings were from the lawns of our institute campus.

Prior to use, the substrates were ground in a blender to give a fraction with particle size less than 2 mm and stored at 4°C until use.

2.2. Inoculum. Cow dung was used as inoculum with total solid content of 20 g/L. The inoculum was preincubated in order to deplete the residual biodegradable organic material present in it.

2.3. Analytical Methods

2.3.1. Total Solids (TS) Estimation. For total solids, a known amount of sample was transferred into a previously weighed crucible and dried at 105°C for 24 h. The increase in weight over that of the empty crucible represents the total solids [11].

2.3.2. Volatile Solids (VS) Estimation. For volatile solids estimation, the dried sample obtained after TS estimation was ignited in a muffle furnace at 550°C for 2 h. The weight lost on ignition represents the volatile solids [11].

2.3.3. Chemical Oxygen Demand. The chemical oxygen demand measurement is performed on fresh waste. Prior to use, the substrates were ground in a blender to give a fraction with particle size less than 2 mm. The substrate (1 g) is then suspended in 1 L mL of distilled water, stirred on a magnetic stirrer for one hour, and the COD of the suspension is measured as described by Raposo et al. [12]. Briefly, 10 mL of suspension was digested with potassium dichromate and concentrated sulphuric acid at 150°C for 2 h in a COD Block Digestion Unit. After cooling, the digestate is titrated against ferrous ammonium sulphate (0.5 N) using 1, 10-phenantrholine as indicator.

2.3.4. Biochemical Methane Potential Assays. The biochemical methane potential (BMP) of the substrates was performed according to [5, 6]. The methane potential of the substrates was determined over the following range of VS, that is, 3.0, 3.5, 4.0, 4.5, and 5.0 g VS/L. The reactors were supplemented with nutrients, trace elements, and bicarbonate. Finally, the reactors were made up to the working volume 0.1 L with distilled water, and the headspace was flushed with nitrogen. A control without substrate was also set up to account for the endogenous biogas produced from the inoculum. All the experiments were carried out in triplicates. The bottles were shaken manually once a day. Biogas production was measured using water displacement technique. Gas samples were taken periodically for composition analysis by gas chromatography using helium as carrier gas. The calculated biogas production is also corrected for blank biogas production.

One mL macronutrients, 0.8 mL micronutrients, and 5 mL of buffer from the above-mentioned stock solutions were added (Table 1).

2.3.5. Biogas Measurement. The biogas production was measured by water displacement setup [13]. A tube was used to connect the reactor with an inverted 250 mL graduated measuring cylinder immersed in a 1000 mL beaker filled with water. Biogas produced was collected in the graduated cylinder connected with a water reservoir which allowed volumetric biogas measurements at atmospheric pressure.

2.3.6. Gas Composition. Gas samples were taken periodically for composition analysis by gas chromatography. The samples were analysed with a gas chromatograph (GC-7610, Chemito) equipped with thermal conductivity detector. The carrier gas was hydrogen. The oven, injector, and detector temperatures were 80, 150, and 250°C, respectively.

2.3.7. Batch Reactors. A double-walled reactor of 3 L effective volume was used in this work. Temperature was maintained at 35°C by a water recirculation. Mixing was done by magnetic stirring.

Series of experiments were carried out at the same time. The reactor was seeded at a volatile solid (VS) concentration of 3.0 g/L with anaerobic sludge, taken from the outlet of an anaerobic pilot reactor treating food waste.

3. Results and Discussion

3.1. Selection of Feedstocks

3.1.1. Coconut Oil Cake. India is the third largest coconut producing country in the world with a cultivation area of about 1.78 million hectares [14]. In a conventional edible oil manufacturing mill, oil is extracted from dried copra (matured coconut Kernel) leaving behind a protein and lignocellulosic rich oil cake.

The advantage of using oil cakes as a substrate for biogas production is their cheaper availability throughout the year. Moreover, with increasing emphasis on cost reduction of industrial processes and value addition to agro-industrial residues, utilization of oil cakes as an energy source seems to be promising because of their higher energetic value. Typically coconut oil cake was used as cattle feed, but the dominant factor that affected coconut oil cake market was the situation of other oil cakes (meal), especially, soybean meal and sunflower meal. Also the reason for looking for alternative use of coconut oil cake rather than as cattle feed is the introduction of attractive incentives for green fodder

Table 1: Composition of solutions.

		(g/L)
Macronutrients		
NH_4Cl	g/L	26.6
KH_2PO_4	g/L	10
$MgCl_2$, $6H_2O$	g/L	6
$CaCl_2$, $2H_2O$	g/L	3
Micronutrients		
$FeCl_2$, $4H_2O$	g/L	2
$CoCl_2$, $6H_2O$	g/L	0.5
$MnCl_2$, $4H_2O$	g/L	0.1
$NiCl_2$, $6H_2O$	g/L	0.1
$ZnCl_2$	g/L	0.05
H_3BO_3	g/L	0.05
Na_2SeO_3	g/L	0.05
$CuCl_2$, $2H_2O$	g/L	0.04
Na_2MoO_4, $2H_2O$	g/L	0.01
Buffer		
$NaHCO_3$	g/L	50

cultivation by Government of Goa, India. High incentives are given by the local government, that is, incentive under Perennial Fodder Cultivation at 370 USD per hectare area of land for the 1st year, 185 USD each per hectare area of land for the 2nd and 3rd years. This has resulted in decline in demand for coconut oil cake as cattle feed in Goa, India.

3.1.2. Cashew Apple Waste. Cashew is an important cash crop grown in India, on an area of 820,000 hectares which produced 539,000 tons of raw nuts in 2004-2005 [15]. The fruit consists of mainly the nuts containing an embryo (Kernel) and a false fruit commonly called as cashew apple. The nuts represent only 10% of the total fruit weight, and large amounts of cashew apples are lost in the field after nut removal [15]. Although cashew apples can be consumed as juice, jams, and other food stuffs, the cashew tree cultivation is an agricultural activity directed at the production of the cashew nuts. Due to its large availability and low cost, cashew apple has been studied as a substrate for fermentative and enzymatic processes for several applications [16–20]. As such, cashew apple is considered an agricultural residue rich in reducing sugars (fructose and glucose), vitamins, minerals, and some amino acids [21] and can be a suitable low cost substrate for anaerobic digestion studies.

3.1.3. Grass. Energy production from lignocellulosic biomass will be a major alternative to conventional energy sources. The efficient conversion of plant biomass to biogas remains a challenge because of the presence of recalcitrant and insoluble starting materials [22].

The characteristics of the different substrates selected are shown below in Table 2.

The TCOD/VS ratio close to unity indicates better biogas potential, and thus methanogenesis can be elucidated from COD/VS ratio [23].

Table 2: Characteristics of substrates.

Feedstock	TS %	VS %	VS/TS %	TCOD g/gVS	COD g/kg
Coconut oil cake	90.93	84	0.92	0.87	1.057
Cashew apple	21.8	21.2	0.97	0.9	1.077
Grass	56	53	0.95	0.4	2.3

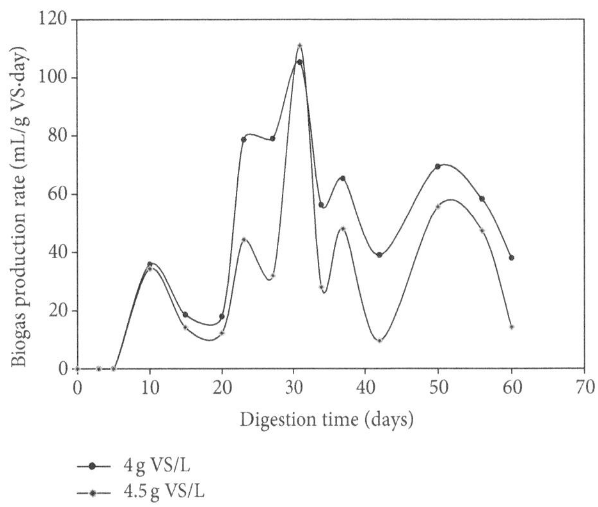

Figure 1: Daily biogas production rates at different volatile solid rates.

3.2. Digestion of Coconut Oil Cake. The daily biogas production rate showed multiple peak values over time (Figure 1). Multiple peaks indicate the presence of multiple substrates within coconut oil cake or solid matrix of the coconut oil cake, and as a result, additional cellulose becomes bioavailable [24]. Coconut oil cake is rich in lignocellulosic content and protein [14]. Biogas production increased until the 30th day, and then there was slight increase until the 60th day. There was an initial lag phase for 5 days. The average biogas production was found to be 451 and 662 mL/g VS at 4 and 4.5 g VS/L, respectively (Figure 2). The percentage of methane in the biogas produced from coconut oil cake was found to be 55% which corresponds to 383 mL CH_4/g VS and 277 mL CH_4/g VS added, respectively. At lower 3.0 and 3.5 g VS/L as well as higher concentration of volatile solids (5 g VS/L) the reactors did not recover from the initial lag phase, and no significant amount of biogas was produced. Increasing the volatile solid loading rate decreased biogas production; this could be due to accumulation of intermediates as reported by Hansen et al. [5] for waste rich in fats and lipids. Coconut oil cake contains about 5-6% of oil in it even after extraction as the extraction of oil is only by mechanical extraction.

3.3. Digestion of Cashew Apple. The biogas production started from the first day but declined rapidly with no biogas produced at the end of 14 days of digestion (Figures 3 and 4).

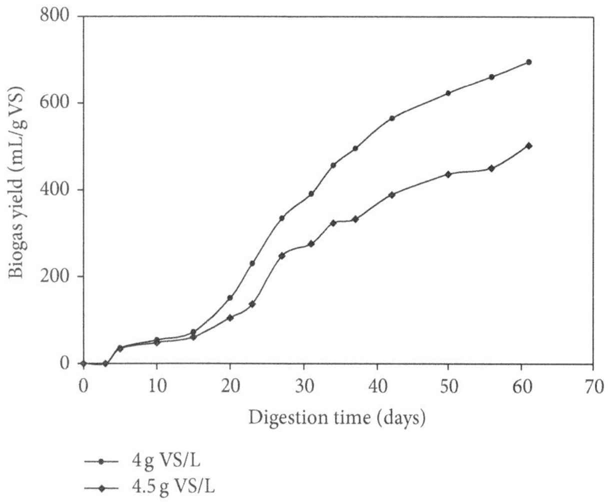

Figure 2: Biogas yield for coconut oil cake at different volatile solid rates.

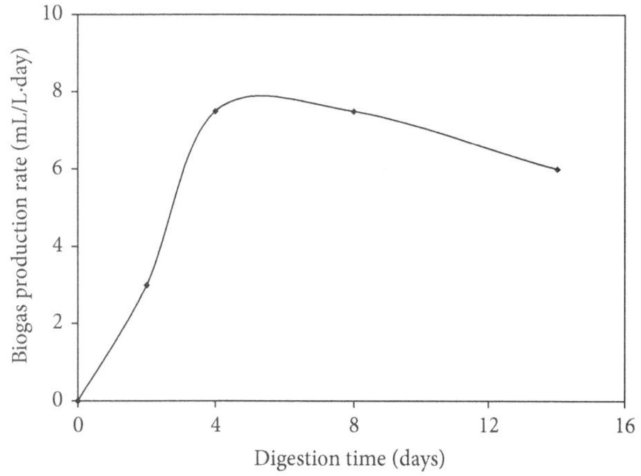

Figure 3: Biogas production rate for cashew apple waste.

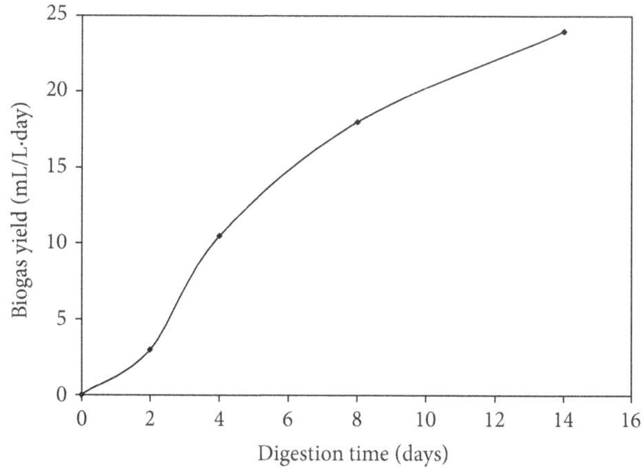

Figure 4: Biogas yield for cashew apple waste mL/g VS.

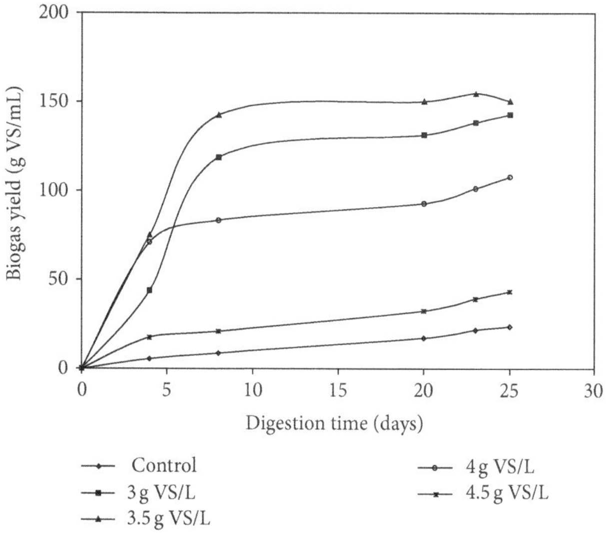

Figure 5: Biogas yield mL/g VS for cashew apple waste.

The pH values in the digesters at the end of digestion time were below 6.0. Fruit and vegetable wastes tend to have low total solids and high volatile solids. The VS/TS ratio for cashew apple was very high 0.97 g VS/g TS. The rapid hydrolysis of the feedstocks might have led to acidification and consequent inhibition of methanogenesis, which is a major limitation of one-stage anaerobic digestion system [25].

3.4. Two-Stage Studies for Cashew Apple Waste. The substrate, that is, cashew apple waste, was kept in the hydrolytic stage in reactors at different loading rates by following the pH which indicates acidification of the reactor. Hydrolysis stabilized at pH 5.5 for OLR of 3.5–4.0 g VS/L after 15 days. After 15 days, the pH was adjusted to 7.2 so as to start the methanogenesis.

Maximum biogas 132 and 140 mL/g VS was produced at OLR of 3.5 and 4 g VS/L (Figure 5). The methane content of the biogas was 46% and 46.2% which corresponds to 60.7 mL and 64.6 mL at 3.5 and 4.0 g VS/L, respectively, at the end of 25 days of digestion.

3.5. Digestion of Grass. The daily production rates and average biogas yield for digestion of grass waste are shown in Figures 6 and 7. The daily biogas production rate showed multiple peak value overtime. Multiple peaks are due to presence of multiple substrates within a single component as explained earlier. The daily biogas production rate was highly variable during the 25 days of digestion and then stabilized until the end of the experiment. As shown in Figure 7 the biogas yield increased with increasing volatile solid rates. Volatile solids analysis determines the total amount of organic matter in a substrate. With increasing VS, the amount of organic matter being added to the digester or the reactor increases, hence contributing to increasing methane yield per VS with increasing VS loading. The average biogas yields were 391, 490, 501, 557, and 651 mL/g VS, respectively. The percentage of methane in the biogas was found to be 51% which

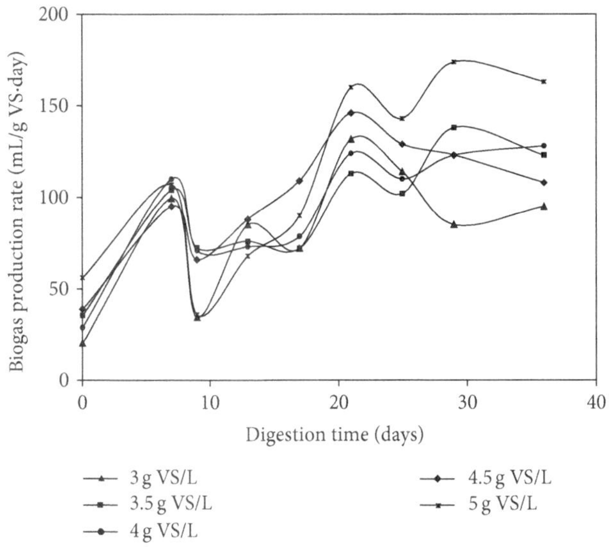

Figure 6: Biogas production rate mL/g VS for grass.

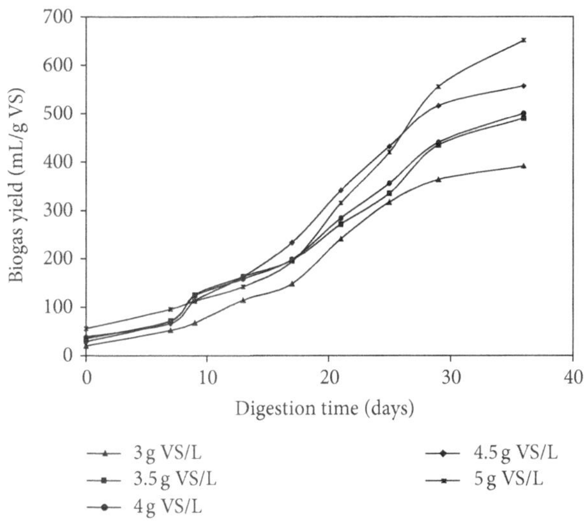

Figure 7: Biogas yield mL/g VS for grass waste.

corresponds to 199, 250, 256, 284, and 332 mL CH_4/g VS. The results gave a biogas yield higher than that found in batch experiments with grass carried out by Liu et al. [26], who obtained a yield of 372 mL/g VS. Grass and leaves contain waxes and lipids that were not quantified. However, these compounds would not be as degradable as carbohydrates [22]. The percentage of methane in the biogas produced from grass was found to be 51%.

3.6. Fed Batch Studies for Coconut Oil Cake

3.6.1. Biogas Production. Based on the previous results, the anaerobic biodegradability of coconut oil cake was evaluated in fed batch mode in a 5 L anaerobic reactor. The substrate was added at 4 g VS/L/batch for all the batches. The volume of biogas produced was determined using water displacement

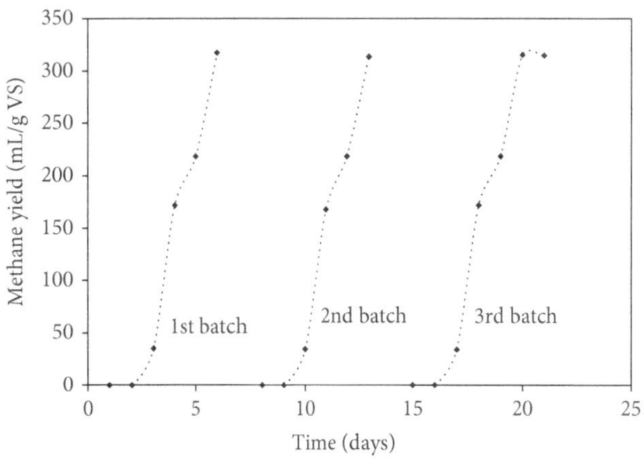

Figure 8: Methane yield for three sequential batches for Coconut oil cake.

method, and results obtained are presented in Figure 8 which represents the percentage of methane in the biogas produced with time during a typical batch. A significant lag phase was seen for all the batches during the start of each batch. The maximum biogas was produced between 5 and 13 days of digestion. The first cycle of fed batch was operated for 15 days after which there was no production of biogas. Following this, the second batch of substrate addition was made. The reactor was operated for three batches with feeding after indication of endogenous respiration. On all three accounts of substrate addition, the maximum methane yield was 320 mL/g VS. The results indicate that there is no accumulation of inhibiting chemicals [27] when operated in fed batch mode. This aspect can be taken into account for designing pilot scale anaerobic digester for coconut oil cake.

3.6.2. Methane Production Potential. The maximum biodegradability (Bo) is given as m^3 CH_4 produced per kg VS. The theoretical maximum methane production is 0.35 m^3 CH_4/kg COD, and one gram of VS for "biological" sludge is normally considered equal to 1.4 g COD [28]. Thus, the theoretical maximum methane production will be 0.49 m^3 CH_4/kg VS. Thus, the Bo for the substrates tested that is, coconut oil cake, and grass cuttings except for cashew apple were 0.36, and 0.33 m^3 CH_4/kg VS, respectively. As the methane production potential and hence maximum biodegradability of coconut oil cake were higher compared to grass cuttings it was taken for fed batch studies.

The local oil mill which processes coconuts handles about 500 kg of dried coconut copra per day, and after extraction of oil, about 200 kgs of coconut oil cake is available as waste residue whose market as cattle feed is on the decline. Based on the fed batch studies we will be able to produce 320 mL/g VS of methane or in other words 64000dL/200 kg VS or 6400 L/200 kg of raw coconut oil cake (@ 10% TS). Therefore this mill could produce 6.43 of biogas per day which could replace firewood used for cooking for 6 families considering each family has 5-6 people and does cooking thrice a day. The amount of biogas thus produced will also replace 355 kg of CO_2/m^3 of biogas.

4. Conclusions

The potentialities of three different residues as substrates for anaerobic digestion have been investigated in this study. For coconut oil cake biogas was produced only at 4 and 4.5 g VS/L. Coconut oil cake has a high potential for biogas production. However, they contain slowly biodegradable organic matter, and the loading rates should remain low to avoid any accumulation of slowly biodegradable solids in the digesters. The potential of cashew apple for anaerobic digestion is very low. For cashew apple as a substrate phase separation resulted in high process stability and significant biogas productivity compared with a single-stage reactor. The data obtained from this study would be used for designing large scale anaerobic digesters for treatment of agro waste. Our future work is focused on pilot scale anaerobic digestion of coconut oil cake in a 0.5-tonne horizontal plug flow reactor (already constructed).

Abbreviations

AD:	Anaerobic digestion
BMP:	Biochemical methane potential
TS:	Total solids
VS:	Volatile solids
COD:	Chemical oxygen demand
TCOD:	Total chemical oxygen demand
HRT:	Hydraulic retention time
OLR:	Organic loading rate.

Acknowledgment

The authors thank University Grants Commission, Government of India, for funding this Project (Ref. no. SR/33-139/2007).

References

[1] R. Zhang, H. M. El-Mashad, K. Hartman et al., "Characterization of food waste as feedstock for anaerobic digestion," *Bioresource Technology*, vol. 98, no. 4, pp. 929–935, 2007.

[2] G. Lastella, C. Testa, G. Cornacchia, M. Notornicola, F. Voltasio, and V. K. Sharma, "Anaerobic digestion of semi-solid organic waste: biogas production and its purification," *Energy Conversion and Management*, vol. 43, no. 1, pp. 63–75, 2002.

[3] D. P. Chynoweth, J. M. Owens, and R. Legrand, "Renewable methane from anaerobic digestion of biomass," *Renewable Energy*, vol. 22, no. 3, pp. 1–8, 2001.

[4] J. M. Owens and D. P. Chynoweth, "Biochemical methane potential of municipal solid waste (MSW) components," *Water Science and Technology*, vol. 27, no. 2, pp. 1–14, 1993.

[5] T. L. Hansen, J. E. Schmidt, I. Angelidaki et al., "Method for determination of methane potentials of solid organic waste," *Waste Management*, vol. 24, no. 4, pp. 393–400, 2004.

[6] I. Angelidaki, M. Alves, D. Bolzonella et al., "Defining the biomethane potential (BMP) of solid organic wastes and energy crops: a proposed protocol for batch assays," *Water Science and Technology*, vol. 59, no. 5, pp. 927–934, 2009.

[7] D. P. Chynoweth, C. E. Turick, J. M. Owens, D. E. Jerger, and M. W. Peck, "Biochemical methane potential of biomass and waste feedstocks," *Biomass and Bioenergy*, vol. 5, no. 1, pp. 95–111, 1993.

[8] V. N. Gunaseelan, "Biochemical methane potential of fruits and vegetable solid waste feedstocks," *Biomass and Bioenergy*, vol. 26, no. 4, pp. 389–399, 2004.

[9] C. Cavinato, F. Fatone, D. Bolzonella, and P. Pavan, "Thermophilic anaerobic co-digestion of cattle manure with agro-wastes and energy crops: comparison of pilot and full scale experiences," *Bioresource Technology*, vol. 101, no. 2, pp. 545–550, 2010.

[10] B. Demirel and P. Scherer, "Bio-methanization of energy crops through mono-digestion for continuous production of renewable biogas," *Renewable Energy*, vol. 34, no. 12, pp. 2940–2945, 2009.

[11] APHA, *Standard Methods for the Examination of Water and Waste Water*, American PublicHealth Association, Washington, DC, USA, 18th edition, 1998.

[12] F. Raposo, M. A. de la Rubia, R. Borja, and M. Alaiz, "Assessment of a modified and optimised method for determining chemical oxygen demand of solid substrates and solutions with high suspended solid content," *Talanta*, vol. 76, no. 2, pp. 448–453, 2008.

[13] S. Singh, S. Kumar, M. C. Jain, and D. Kumar, "Increased biogas production using microbial stimulants," *Bioresource Technology*, vol. 78, no. 3, pp. 313–316, 2001.

[14] M. Moorty and K. Vishwanathan, "Nutritive value of extracted coconut (*Cocos nucifera*) meal," *Research Journal of Agriculture and Biological Sciences*, vol. 5, no. 4, pp. 515–517, 2009.

[15] K. V. Nagaraja and K. R. M. Bhuvaneshwari, "Biochemical characterization of cashew (*Anacardium occidentale* L.) apple juice and pomace in India," *FAO Newletter*, vol. 149, pp. 9–13, 2007.

[16] T. L. Honorato and S. Rodrigues, "Dextransucrase stability in cashew apple juice," *Food and Bioprocess Technology*, vol. 3, no. 1, pp. 105–110, 2010.

[17] C. P. M. L. Fontes, T. L. Honorato, M. C. Rabelo, and S. Rodrigues, "Kinetic study of mannitol production using cashew apple juice as substrate," *Bioprocess and Biosystems Engineering*, vol. 32, no. 4, pp. 493–499, 2009.

[18] C. M. A. Chagas, T. L. Honorato, G. A. S. Pinto, G. A. Maia, and S. Rodrigues, "Dextransucrase production using cashew apple juice as substrate: effect of phosphate and yeast extract addition," *Bioprocess and Biosystems Engineering*, vol. 30, no. 3, pp. 207–215, 2007.

[19] T. H. S. Rodrigues, M. A. A. Dantas, G. A. S. Pinto, and L. R. B. Gonçalves, "Tannase production by solid state fermentation of cashew apple bagasse," *Applied Biochemistry and Biotechnology*, vol. 137, no. 1–12, pp. 675–688, 2007.

[20] T. L. Honorato, M. C. Rabelo, L. R. B. Gonçalves, G. A. S. Pinto, and S. Rodrigues, "Fermentation of cashew apple juice to produce high added value products," *World Journal of Microbiology and Biotechnology*, vol. 23, no. 10, pp. 1409–1415, 2007.

[21] M. S. Silveira, C. P. M. L. Fontes, A. A. Guilherme, F. A. N. Fernandes, and S. Rodrigues, "Cashew apple juice as substrate for lactic acid production," *Food and Bioprocess Technology*, vol. 5, no. 3, pp. 947–953, 2012.

[22] S. Yang, I. Kataeva, S. D. Hamilton-Brehm et al., "Efficient degradation of lignocellulosic plant biomass, without pre-treatment, by the thermophilic anaerobe "Anaerocellum thermophilum" DSM 6725," *Applied and Environmental Microbiology*, vol. 75, no. 14, pp. 4762–4769, 2009.

[23] I. Angelidaki and W. Sanders, "Assessment of the anaerobic biodegradability of macropollutants," *Reviews in Environmental Science and Biotechnology*, vol. 3, no. 2, pp. 117–129, 2004.

[24] W. E. Eleazer, W. S. Odle, Y. Wang, and M. A. Barlaz, "Biodegradability of municipal solid waste components in laboratory-scale landfills," *Environmental Science and Technology*, vol. 31, no. 3, pp. 911–917, 1997.

[25] A. J. Ward, P. J. Hobbs, P. J. Holliman, and D. L. Jones, "Optimisation of the anaerobic digestion of agricultural resources," *Bioresource Technology*, vol. 99, no. 17, pp. 7928–7940, 2008.

[26] G. Liu, R. Zhang, H. M. El-Mashad, and R. Dong, "Effect of feed to inoculum ratios on biogas yields of food and green wastes," *Bioresource Technology*, vol. 100, no. 21, pp. 5103–5108, 2009.

[27] A. Sharma, B. G. Unni, and H. D. Singh, "A novel fed-batch digestion system for biomethanation of plant biomasses," *Journal of Bioscience and Bioengineering*, vol. 87, no. 5, pp. 678–682, 1999.

[28] G. Zeeman and S. Gerbens, CH_4 Emissions from Animal Manure. Good Practice Guidance and Uncertainty Management in National Greenhouse Gas Inventories.

A Low-Cost Polytetrafluoroethylene-Framed TiO_2 Electrode Decorated with Oleic Acid-Capped CdSe Quantum Dots for Solar Cell

Delele Worku Ayele,[1] Wein-Nien Su,[2] and Bing-Joe Hwang[1,3]

[1] *Department of Chemical Engineering, National Taiwan University of Science and Technology, Taipei 106, Taiwan*
[2] *Graduate Institute of Engineering, National Taiwan University of Science and Technology, Taipei 106, Taiwan*
[3] *National Synchrotron Radiation Research Center, Hsinchu 300, Taiwan*

Correspondence should be addressed to Wein-Nien Su; wsu@mail.ntust.edu.tw and Bing-Joe Hwang; bjh@mail.ntust.edu.tw

Academic Editor: Ching Yuan Chang

Colloidal CdSe QDs have been assembled, as quantum dot-sensitized solar cells (QDSSCs), on a novel architecture comprising a polytetrafluoroethylene- (PTFE-) framed TiO_2 electrode for the first time. CdSe QDs are anchored on the surface of the film using a linker molecule (3-mercaptopropionic acid, MPA). The resulting photoelectrode comprises a TiO_2 compact layer and a PTFE-framed structural layer with average respective thicknesses of 2 μm for the compact layer and either 23 μm or 28 μm for the PTFE-framed structural layer. UV-vis absorption spectra show that more CdSe quantum dots are anchored on the surface of the modified with MPA TiO_2 film compared to direct absorption onto an unmodified film. Energy conversion efficiencies of up to 0.18% can be achieved with cells prepared from a TiO_2 (25 μm)/MPA/CdSe QD electrode. Electrochemical impedance measurements show that the recombination resistance is relatively higher for a cell assembled with TiO_2 (25 μm)/MPA/CdSe QD photoanode than with TiO_2 (25 μm)/CdSe QD resulting in an increase of cell efficiency. The PTFE-framed structure along with the compact layer is a new approach to QDSSC application that provides a tunable film thickness and a cost-effective preparation technique for the large-scale production of the photoanode.

1. Introduction

Dye-sensitized solar cells (DSSCs) have attracted much attention as a promising alternative energy source to conventional solid-state junction solar cells due to their low cost, low impact on the environment, and high efficiency [1]. In DSSCs, the dye adsorbed on the surface of mesoporous TiO_2 films [2] is responsible for the light absorption and subsequent electron transfer reactions and is thus regarded as one of the key factors determining the conversion efficiency from light to electricity [3, 4]. Over 11% efficiency has been reported with the use of ruthenium-based sensitizer complexes [1]. In recent years, in addition to the organic sensitizers commonly used for DSSCs, inorganic semiconductors which absorb visible light have also been used as light-harvesting materials for solar cells [3–5]. Inorganic semiconductor quantum dots (QDs), such as CdS and CdSe, have been employed as sensitizers, mainly due to their specific advantages over conventional dyes [6–9]; for instance, the size quantization effect allows tuning of the band energy and visible response by simply varying the QD size [10, 11]. Another advantage is that these QDs open new ways to utilize hot electrons or generate multiple charge carriers with one single photon through the impact ionization effect [12, 13]. Thus, it is possible to obtain a much higher conversion efficiency by utilizing the multiple excitons generation possibilities offered by QDs [14]. Despite the advantage known for a QD sensitizer, up-to-date studies on quantum dot-sensitized solar cells (QDSSCs) are relatively few compared with conventional Ru-based DSSCs, and their efficiencies are also so far unsatisfactory. Recently, QDSSCs with power conversation efficiency more than 4% have been reported [15–17]. The energy conversion efficiencies reported for an oleic acid-capped QD-sensitized DSSCs are still low (less than 1%) [18]. The problems encountered in fabricating a QDSSC include the assembly of the QDs into a mesoporous TiO_2 matrix [3, 19] and the selection of an efficient electrolyte in which the QDs can run stably without serious degradation [20]. Compared to iodide/triiodide (I^-/I^{3-}) electrolyte, using polysulfide electrolyte and CdS as a sensitizer shows better stability with a much higher short circuit current and IPCE (~80%) value [20]. Two methods are commonly employed to assemble QDs into a mesoporous thin film. The first is to assemble preprepared QDs using bifunctional linkers [3], and the other is by *in situ* deposition from precursor solutions using successive chemical bath deposition [8].

Chemical bath deposition [8, 21, 22] involving nucleation and growth leads to a high coverage of QDs on the TiO_2 surface but has the disadvantage of not giving good control of the size distribution of the deposited QDs. On the other hand, the attachment of colloidal QDs using a molecular linker leads to precise morphological characteristics (i.e., shape and size) of the semiconductor nanocrystals—leading to enhanced properties when compared to CBD [3, 23–25].

Similar to DSSC, under the illumination of solar light; QDs are excited and electrons are produced. To generate meaningful electrical power from QDSSC, the electrons need to pass four important interfaces of QDSSCs that are TiO_2/FTO, QDs/TiO_2, QDs/electrolyte, and the electrolyte/counter electrode. The nanoporous nature of the TiO_2 layer provides a high surface area that is of great importance to the efficient photon-to-electricity conversion because it first enhances QD loading and then solar light absorption [26]. However, it also provides abundant TiO_2 surface sites (direct route) and bare FTO conducting sites (indirect route), where the photoinjected electrons may recombine with species of electrolyte [27]. Such recombination will cause the loss of the photocurrent so that the photovoltaic performance of the device is seriously decreased [28].

In previous reports, it was suggested that putting an additional compact layer of TiO_2 between the fluorine-doped tin oxide (FTO) glass and the mesoporous TiO_2 layer to prevent the recombination of the photogenerated charges at the boundary seen in DSSC would also increase the contact between FTO and TiO_2 films [27–29]. Therefore, we applied a hybrid film comprising a structural layer (a mixture of polytetrafluoroethylene (PTFE) and TiO_2) on the compact layer of TiO_2 on the FTO glass.

The PTFE-framed structure helped to increase dye absorption to achieve an energy conversion efficiency of 9.04% for the dye-sensitized solar cell [30, 31]. Moreover, the use of PTFE in the structural layer allows the cost-effective preparation of large surface area TiO_2 electrodes with a tunable thickness [30, 31]. To the best of our knowledge, a photoanode with such an architecture represents a new approach for QDSSC applications.

We prepared different QDSSC using two strategies to attach the colloidal CdSe QDs onto the PTFE-framed TiO_2 surface—with a linker molecule (3-mercaptopropionic acid (MPA)) and by direct adsorption without linker. We further discussed the electrochemical properties of the cells prepared by the two methods of anchoring the CdSe QDs onto the TiO_2 surface and the resulting effects on the performance of QDSSCs. In this report, polysulfide electrolyte was used as the redox couple to regenerate the photoexcited holes in the QDs with Pt being used as a counter electrode.

2. Experimental

2.1. Chemicals. Selenium powder (99.70%), sodium sulfite (>95%), and n-hexane (>99%) were all from Acros. Sulfur (99.99%), potassium chloride (99%), sodium sulfide (90%), anhydrous acetonitrile (99.9%), 3-mercaptopropionic acid (MPA, 99%), N-dimethylformamide (>99%), titanium butoxide (97%), isopropanol (99.7%), ethanol (>99%), and oleic acid (65–88%) were all from Aldrich. Cadmium acetate dihydrate (>99%) was from J. T. Baker, and sodium hydroxide (>95%) was from Shimakyus Pure Chemicals. All chemicals were used as received without further purification. Fluorine-doped tin oxide (FTO) glass (≤14 Ω/square) was from Pilkington. Polytetrafluoroethylene (PTFE, DuPont) and TiO_2 P25 (Degussa) were also used.

2.2. Synthesis of CdSe QDs. CdSe QDs were prepared using a microwave-based protocol developed by our group [32]. In a typical synthesis, NaOH (0.72 g) and oleic acid (8 mL) were dissolved in aqueous solvents. $Cd(Ac)_2{\cdot}2H_2O$ (0.27 g) was separately dissolved in 10 mL deionized water and added to this solution under stirring, followed by addition of Na_2SeSO_3 (10 mL) solution. The resulting solution was heated in a microwave system at 100 W and 80°C, for 5 mins to grow the CdSe QDs. The oleic acid-capped CdSe QDs were washed and collected via centrifugation and decantation of the supernatant.

2.3. Preparation of TiO_2 Nanoparticles. TiO_2 nanoparticles were fabricated by a typical hydrothermal method [30, 33]. A 9.25 mL sample of tetrabutyl titanate ($Ti(OC_4H_9)_4$) was mixed with 2-propanol (2.5 mL), and the mixture was slowly added to a solution comprising deionized water (62 mL) and acetic acid (20 mL) in an ice bath with stirring. The solution was then heated to 80°C and stirred vigorously for 8 h. TiO_2 nanoparticles, to be used as a TiO_2 paste, were grown under hydrothermal conditions in an autoclave for 12 h with the autoclave temperatures set at 190°C.

2.4. Fabrication of CdSe QDSSCs. For the preparation of photoelectrodes, we used the technique developed by our group for the preparation of FTO/compact TiO_2 layer/PTFE-farmed TiO_2 films [18, 34]. To fabricate the PTFE-framed TiO_2 films, the glass plate was coated with the TiO_2 paste using a glass rod, dried in oven at 80°C for 15 min and then sintered at 500°C for 30 min to form a compact layer. On top of the compact layer, a second structural layer was built by coating a paste of PTFE and TiO_2 P25 mixed with some ethanol. The coated plate was dried in oven at 80°C for 15 min and then sintered at 500°C for 60 min to form the PTFE-framed TiO_2 film. A photoanode (i.e., a TiO_2 film comprising the compact and structural layers) with different average

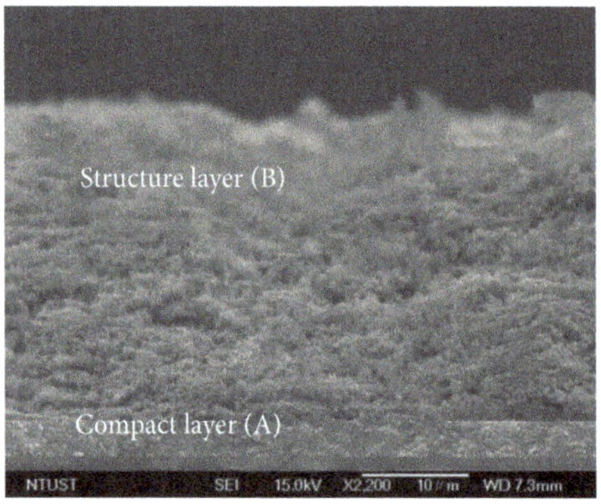

(a)

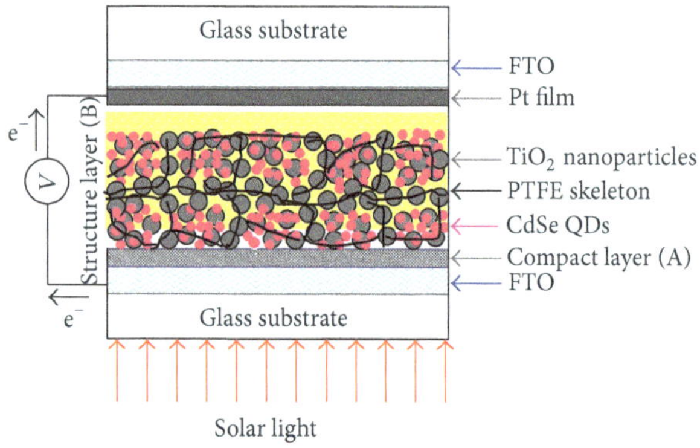

(b)

FIGURE 1: (a) SEM cross-section of compact layer (A) and structure layer (B) of the film. Scale bare 10 μm. (b) A schematic representation of QDSSC device (FTO/compact TiO_2 layer/PTFE-framed TiO_2 film/CdSe QDs).

thicknesses of the structural layer (23 μm and 28 μm) and the same average thickness of compact layer (2 μm) giving total thicknesses of 25 and 30 μm was prepared. Two modes of attachment for CdSe QD on the films were employed. The first used linker molecules to attach CdSe QDs to the surface. The film's surface was modified by immersion in a 0.5 M MPA acetonitrile solution for 12 hrs prior to rinsing with acetonitrile followed by further immersion in 0.26 mM CdSe QD toluene solution for 48 hrs and a final rinse in toluene: this electrode was designated as TiO_2/MPA/CdSe QD. The second electrode was prepared without linker molecules. The unmodified TiO_2 film was directly immersed in a CdSe QD (0.26 mM) toluene solution for 48 hrs and then rinsed in toluene: this electrode was designated as TiO_2/CdSe QD. Polysulfide electrolyte solutions were prepared following the literature procedure [20]. The typical composition of the polysulfide electrolyte was Na_2S (0.5 M), S (2 M), and KCl (0.2 M) in the mixture of solvents (3 mL of deionized water and 7 mL of methanol). Finally, a sandwich-type cell was fabricated by clamping the polysulfide electrolyte between the QD-sensitized photoanode and the Pt counter electrode (a platinum sputtered FTO plate) with two clips. The active area of all cells was 0.5 cm^2.

2.5. Characterization. UV-visible absorption spectra were obtained using Jasco-560 UV/vis Spectrophotometer. The cross-section of the electrode was measured using scanning electron microscopy (FE-SEM, JEOL JSM-6500F, Tokyo, Japan). The photocurrent-voltage (I-V) curves were measured under an illumination from a solar simulator at one sun (AM 1.5, 100 mW cm^{-2}). An Oriel 500 W xenon arc lamp and Keithley 2400 electrometer were used during the measurements of IPCE. Electrochemical impedance spectra (EIS) were measured with an Autolab, PGSTAT302N Potentiostat/Galvanostat (frequency range: 10 micro-Hz to 1 MHz, maximum output voltage: 30 V, and current range: 10 nA to 1 A) under illumination (AM 1.5 100 mW cm^{-2}) by applying different potentials. Electrochemical impedance spectra were fitted with use of Z-View software (Solartron Analytical). X-ray diffraction (XRD) measurements (D2 Phaser, Bruker, Germany) were recorded using a Cu Kα radiation source.

3. Results and Discussion

In Figure 1(a), the SEM cross-section shows that the compact (A) and structural layers (B) have average thicknesses of 2 μm and 23 μm, respectively. Figure 1(b) shows a schematic representation of the QDSSC device, depicting the FTO, the compact TiO_2 layer, the PTFE skeleton in PTFE-framed TiO_2 structure layer, the absorbed CdSe QDs on a TiO_2 film, and the counter electrode. Basically, the compact layer was constructed to create more contact between FTO and the PTFE-framed TiO_2 structure and also to avoid recombination and short circuiting at the FTO. The PTFE-framed TiO_2 structural layer (B) provides a large surface area [31] on which the CdSe quantum dots are absorbed and where light harvesting and charge separation take place. The separated electrons and holes are, respectively, conducted through TiO_2 to the FTO and through polysulfide electrolytes to the Pt cathode.

The sintering effect on the PTFE-framed TiO_2 structural layer was also analyzed by comparing with the original TiO_2 paste. The X-ray diffraction patterns of the TiO_2 paste used for compact layer (A) formation and the structural layer (B) are shown in Figure 2. Anatase and rutile phase peaks are marked as "A" and "R," respectively. Both samples consist of anatase and rutile phase. However, the rutile phase of TiO_2 films is more pronounced in the structure layer (B) than in the TiO_2 paste as a result of sintering. In general, the XRD patterns of both samples are almost identical.

In the past years, PTFE-framed TiO_2 structure for DSSC has been developed in our group [30, 31]. The study shows that the pure PTFE powder exhibits particle size in the range of 30–100 nm before sintering, and the morphology becomes column-like after sintering at 500°C for 60 minutes. PTFE formed as a highly porous skeleton which enables to hold more TiO_2 particles, a requirement for the decoration of QDs. The optimal mixing proportion of TiO_2 and PTFE was determined as 50/50 wt.% [31]. The PTFE skeleton is also

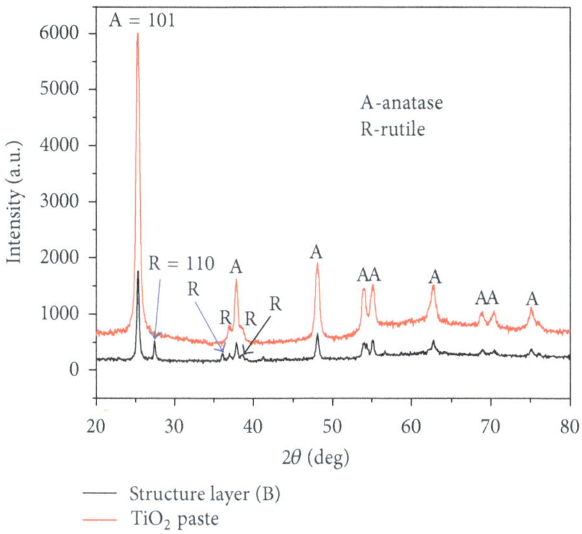

FIGURE 2: The XRD spectra of TiO_2 paste used for the preparation of compact layer and sintered structure layer.

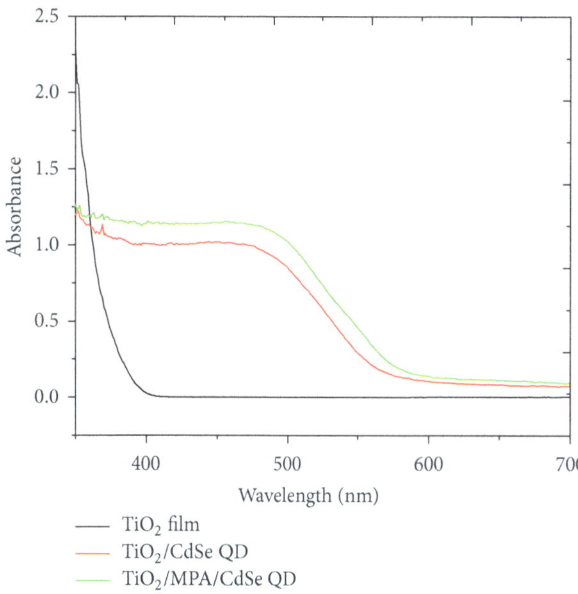

FIGURE 3: UV-visible absorption spectra of pristine TiO_2 film, TiO_2/CdSe QD electrode, and TiO_2/MPA/CdSe QD electrode. The total TiO_2 film thickness is 25 μm.

supposed to contribute the necking of TiO_2 in the PTFE-framed TiO_2 electrode. In this report, the structure of PTFE remained mainly intact for sintering temperatures up to 500°C [31].

The assembly of presynthesized colloidal QD is a popular method used to fabricate QD-sensitized electrodes. For maximum CdSe QD loading, the TiO_2 film was first modified with mercaptopropionic acid MPA linker, and then it was immersed in CdSe QD solution. The mechanism is based on well-known hard and soft acids and bases theories [35], and the ligands exchange between OA and MPA linker takes place. The hard-hard or soft-soft molecules or ions form a strong binding, owing to the polarizability of electron cloud. The Cd^{2+} and Se^{2-} ions are classified as soft ions and have stronger interaction with soft "–SH (from MPA)" but not hard "–COOH (from OA or MPA) [35]." Therefore, during the period of dynamic dissociation and association of ligand, the MPA molecule can form rather strong bonding with CdSe (soft –SH from MPA and soft Cd^{2+} or Se^{2-} ions) than –COOH. The different strengths of bonding enhance the ligand exchange between OA and MPA. After OA and MPA ligand exchange, MPA links QD and TiO_2 by a thiol (–SH) and carboxylate group (–COOH), respectively. Hence, the modification of CdSe QDs capping plays a key role to improve CdSe QDs loading on the TiO_2 films and then the performance of QDSSC.

Thus, for comparison, colloidal CdSe QDs were deposited on 3-mercaptopropionic acid- (MPA-) modified and pristine TiO_2 films separately, denoted as TiO_2/MPA/CdSe QD and TiO_2/CdSe QD, respectively. The absorption of CdSe QD on a TiO_2 film was evaluated using UV-vis absorbance spectroscopy. Figure 3 shows representative UV-vis absorbance spectra of CdSe-sensitized TiO_2 electrodes, prepared with and without MPA modification. It is found that the absorbance of the TiO_2/MPA/CdSe QD electrode was higher in comparison with that of TiO_2/CdSe QD under the same deposition conditions, indicating that a bifunctional molecular linker such as 3-mercaptopropionic acid (MPA, HOOC–CH_2–CH_2–SH) can improve the coverage of TiO_2 surface with quantum dots [3, 19, 36]. 3-mercaptopropionic acids (MPA) have both carboxylate and thiol functional groups at the opposite terminals of the structure, which can facilitate the binding between CdSe quantum dots and TiO_2 surfaces [3, 19, 36]. The presence of thiol terminal groups helps incorporating and stabilizing CdSe QDs, decreasing the desorption events of CdSe QDs during rinsing, and thereby increasing the deposition amount of CdSe QDs [3, 19, 36]. Use of such linker molecules ensures monolayer coverage of the CdSe QDs film within the TiO_2 network. For the direct deposition of OA-capped CdSe QDs on the surface of PTFE-framed TiO_2 nanoparticles, since the hydrophilic (–COOH) part of OA binds colloidal CdSe QDs with the hydrophobic part of oleic acid (OA) towards the surface of TiO_2 nanoparticles, it results in a weak binding between CdSe QDs and TiO_2 surface. As a result, most of the CdSe QDs were removed during rinsing, leading to a lower UV-visible absorption compared with MPA-modified TiO_2 surface (in Figure 3). Therefore, it is reasonable to suggest that surface modification of TiO_2 films by MPA enhances the coverage of TiO_2 with quantum dots that have shown an increase in UV-visible absorption of CdSe QDs (Figure 3). The incorporated amount of CdSe QDs in the TiO_2 film can be evaluated by the absorbance at the first absorption peak using Beer-Lambert's law [18].

Photocurrent-voltage (I-V) curves for QDSSCs were measured under the illumination of one sun (AM 1.5, 100 mW cm^{-2}) as presented in Figure 4(a). These cells were made with different total thicknesses (as indicated in Figure 4(a)) of the TiO_2 film (compact and structure layers)

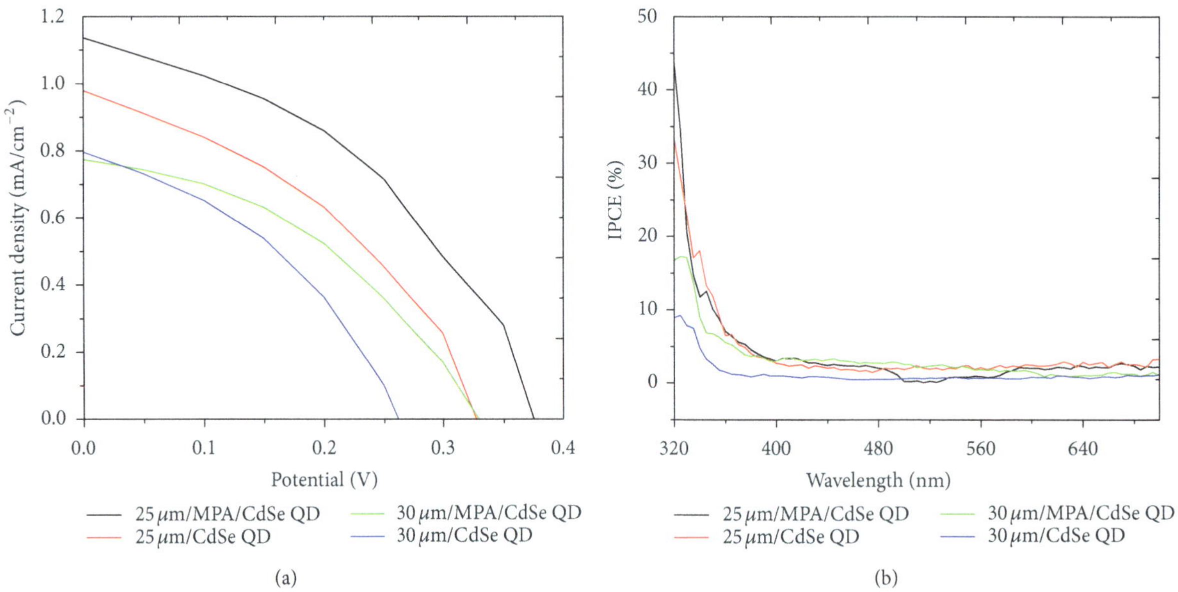

FIGURE 4: *I-V* characteristics (a) and the IPCE spectra (b) of the TiO_2 (25 and 30 μm)/MPA/CdSe QD and TiO_2 (25 and 30 μm)/CdSe QD cells in polysulfide electrolyte.

TABLE 1: Parameters obtained from photocurrent-voltage (*I-V*) measurement of the TiO_2/CdSe QD and TiO_2/MPA/CdSe QD QDSSC cells with different TiO_2 film thicknesses indicated in brackets.

Electrode	I_{max} (mA)	V_{max} (mV)	I_{sc} (mA)	P_{max} (mW)	J_{sc} (mA/cm^2)	V_{oc} (mV)	FF (%)	η (%)
TiO_2 (25 μm)/MPA/CdSe QD	0.36	249.94	0.57	0.09	1.14	399.93	39.25	0.18
TiO_2 (25 μm)/CdSe QD	0.32	199.94	0.49	0.06	0.98	349.91	36.90	0.13
TiO_2 (30 μm)/MPA/CdSe QD	0.26	199.93	0.38	0.05	0.77	349.93	38.67	0.11
TiO_2 (30 μm)/CdSe QD	0.27	149.91	0.39	0.04	0.80	249.96	40.60	0.08

and with different methods of attaching CdSe QDs to their surfaces. The thickness of the compact layer (2 μm) was constant for all photoanodes, while the thickness of the structural layer was either 23 or 28 μm. The solar energy-to-electricity conversion efficiency (η) of QDSSC was calculated by the following equation:

$$\eta = \frac{FF J_{sc} V_{oc}}{P_{in}} \times 100\%, \quad (1)$$

where J_{sc} is the short-circuit photocurrent density, V_{oc} is the open circuit voltage, P_{in} is the light power per unit area, and FF is the fill factor which is calculated by the following equation:

$$FF = \frac{P_{max}}{J_{sc} V_{oc}} = \frac{I_{max} V_{max}}{J_{sc} V_{oc}}, \quad (2)$$

where I_{max} and V_{max} are the current and potential at the maximum power point, respectively, in *I-V* curves of the solar cells. The *I-V* characterization was conducted under P_{in} of 100 mW cm^{-2} for the η calculation. The open circuit potential (V_{oc}), short circuit current density (J_{sc}), fill factor (FF), and the total energy conversion efficiency (η) of these cells are summarized in Table 1. For the cell in which the electrode was prepared by direct absorption, TiO_2 (25 μm)/CdSe QD cell, the efficiency (η) was 0.13%, but a higher value, η = 0.18%, was obtained for the cell in which its electrode is made via the MPA modification, TiO_2 (25 μm)/MPA/CdSe QD cell. This result is attributed to the higher incorporation and better coverage of CdSe QDs on MPA-modified TiO_2 surface as seen in UV-vis absorption spectra, which leads to higher values of I_{sc}, V_{oc}, and FF of the MPA-modified cell as compared with the corresponding unmodified cell. For example, the FFs measured for the TiO_2 (25 μm)/MPA/CdSe cells are 39.25% higher than those of the TiO_2 (25 μm)/CdSe cells (FF = 36.90%) and are responsible for the higher efficiency of the MPA-modified cells. It is inferred that the presence of the MPA layer helped to achieve a better coverage of CdSe QD on the MPA-TiO_2 surface and reduced the direct contact of TiO_2 to the electrolyte, preventing recombination of excited electrons from the conduction band of TiO_2 with the electrolyte, and thereby leading to the higher FF. Furthermore, a better light to electricity conversion efficiency was obtained for a solar cell assembled with TiO_2 film thicknesses of 25 μm than with 30 μm. This is may be due to the difficulty of electron transfer from CdSe QDs to FTO as a result of the higher thickness of the films. The electron collection efficiency and the electron recombination ratio can be determined from the IPCE value [29]. The IPCE is defined as the ratio of the number of electrons in the external circuit produced by an incident photon at a given wavelength. The IPCE was measured as a function of wavelength. The number of incident photons on the device was calculated for each wavelength by using a calibrated Si-diode as a reference. The incident photon-to-current conversion efficiency (IPCE) of QDSSC, a measure of the external quantum efficiency, is defined as [19, 29]

$$\text{IPCE} = \frac{\text{number of collected electrons}}{\text{number of incident photons}} = \frac{hcJ_{sc}}{e\lambda P_{in}},$$
$$\text{IPCE}(\%) = \frac{1240 \times J_{sc}\left(\text{A/cm}^2\right)}{\lambda\ (\text{nm}) \times P_{in}\ (\text{W/cm}^2)} \times 100, \tag{3}$$

where λ, e, h, and c are the incident wavelength, elementary charge, Planck constant, and speed of light, respectively [29]. Using (3), the IPCE values of the QDSSCs with four different photoanodes as a function of the illumination wavelength are shown in Figure 4(b). The QDSSC with the TiO_2 (25 μm)/MPA/CdSe electrode showed the highest IPCE value among all other electrodes. The reasons for this difference are varied and may be linked to the factors that determine IPCE, namely, the light-harvesting efficiency (LHE) of the CdSe quantum dots, the charge injection yield or efficiency (φ_{inj}) from the excited CdSe quantum dots to the TiO_2 conduction band, and the charge collection efficiency (η_{coll}) of the system. Therefore, the IPCE in terms of these parameters can be rationalized using the following equation [29, 37]:

$$\text{IPCE}(\%) = \text{LHE}(\%)\,\varphi_{inj}\,\eta_{coll}. \tag{4}$$

Based on (4), three major parameters such as light harvesting, charge injection, and charge collection determine the IPCE value of QDSSC [38]. The parameters LHE and φ_{inj} are directly related to the CdSe QDs loading on the TiO_2 surface of the photoanode. As aforementioned, a photoanode with MPA-modified TiO_2 films showed a higher CdSe QDs absorption (Figure 3). Although the bifunctional linker (MPA) helps the coverage of TiO_2 surface by CdSe quantum dots, the longer molecular chain in MPA leads to the decrease of electron injection [25, 39, 40]. This effect has been shown clearly in the value of J_{sc} for the cell with TiO_2 (30 μm)/CdSe electrode, in which a cell with TiO_2 (30 μm)/CdSe electrode has higher J_{sc} than TiO_2 (30 μm)/MPA/CdSe electrode (Figure 4(a) and Table 1). Moreover, charge injection from excited state (CdSe QDs*) into the conduction band of TiO_2 occurs in competition with the radiative and nonradiative modes of deactivation of the excited state [37]. Another dominating factor that influences the IPCE value of QDSSC could be the electron collection efficiency (η_{coll}) [41]. In our case, the electron collection efficiency may be affected by the thickness of the TiO_2 films. As shown in I-V curve (Figure 4(a)) and IPCE (Figure 4(b)), QDSSC with TiO_2 (30 μm)/CdSe electrode resulted in a lower IPCE value and conversation efficiency than TiO_2 (25 μm)/MPA/CdSe electrodes. Insights of this issue and preparation of TiO_2 films with lower thickness are underway, and their results will be reported in the future. In general, the electron collection efficiency and the electron recombination ratio will determine the IPCE of the QDSSC.

In order to reveal the interfacial reactions of photoexcited electrons in our QDSSCs, electrochemical impedance spectroscopy (EIS) analysis of the as-prepared QDSSC with a total film (compact and structure layer) thickness 25 μm was measured. Figure 5 shows the typical Nyquist plots obtained at different applied voltages (0.3, 0.4, and 0.5 V) under light illumination of one sun (AM 1.5, 100 mW cm^{-2}) for TiO_2 (25 μm)/MPA/CdSe QD (a) and TiO_2(25 μm)/CdSe QD (b) cells.

Two semicircles exist in the high-frequency (10^3–10^6 Hz) region and the middle-frequency (1–10^3 Hz) region for both cells. In the high-frequency region or the first semicircle where the phase is zero, the ohmic series resistance (R_s) of the FTO layer, the Pt layer, and the electrolyte can be determined [3, 4, 30]. The middle-frequency region or the second semicircle is correlated with heterogeneous electron transfer at TiO_2/QD/electrolyte interface, which consists of the chemical capacitance and recombination resistance [31, 34].

The charge transfer resistance for the nanoparticle (TiO_2) cell has exponential dependence on bias [36]. As seen in Figure 5, the transport and charge transfer processes in the PTFE-framed TiO_2 films, prominent at low bias, show up as a straight line followed by the big arc at intermediate frequencies, respectively. Moreover, the size of a high frequency arc depends strongly on the applied bias potential [42], while an intermediate frequency range is visible at medium to higher forward bias [36, 38, 42]. As seen in Figure 5, as bias is applied, the high-frequency semicircle remains unchanged, while the lower-frequency semicircle shrinks. The values of series resistance (R_s) and recombination resistance (R_r) were evaluated by fitting the spectra, using the equivalent circuit shown in the inset of Figure 5(a) for both cells, in which the measurements were carried out at various applied voltages. The equivalent circuit used for fitting the EIS for both cells is shown in the inset Figure 5(a): R_s is the series resistance, R_1 is the recombination resistance (R_r), CPE1 is a constant phase element representing the chemical capacitance ($C\mu$), R_2 stands for the charge transfer resistance between the counter electrode and the redox couple, and CPE2 takes into account the capacitance of the electrolyte counter electrode interface [34, 43, 44].

The results in Figure 5 clearly show that the value of recombination resistance decreases with an increase in applied voltage for both cells. For further comparison and analysis between the two electrodes, the EIS spectra measured at 0.4 V applied voltage, which is nearly equal to the open circuit voltage, were taken for both cells. With both cells (TiO_2/MPA/CdSe QD and TiO_2/CdSe QD) at 0.4 V applied voltage, the values of R_s, R_1, and R_2 were found to be 89 Ω, 292 Ω, and 41.61 Ω, respectively, for the TiO_2/MPA/CdSe QD cell and 64.57 Ω, 99.2 Ω, and 34.41 Ω, respectively, for the TiO_2/CdSe QD cell. The R_s value of both cells is nearly the same. However, the value of recombination resistance at the electrode/electrolyte for TiO_2/MPA/CdSe QD cell is relatively higher than that of TiO_2/CdSe QD cell, indicating that the charge recombination in TiO_2/MPA/CdSe QD cell is suppressed as compared to TiO_2/CdSe QD cell. This implies

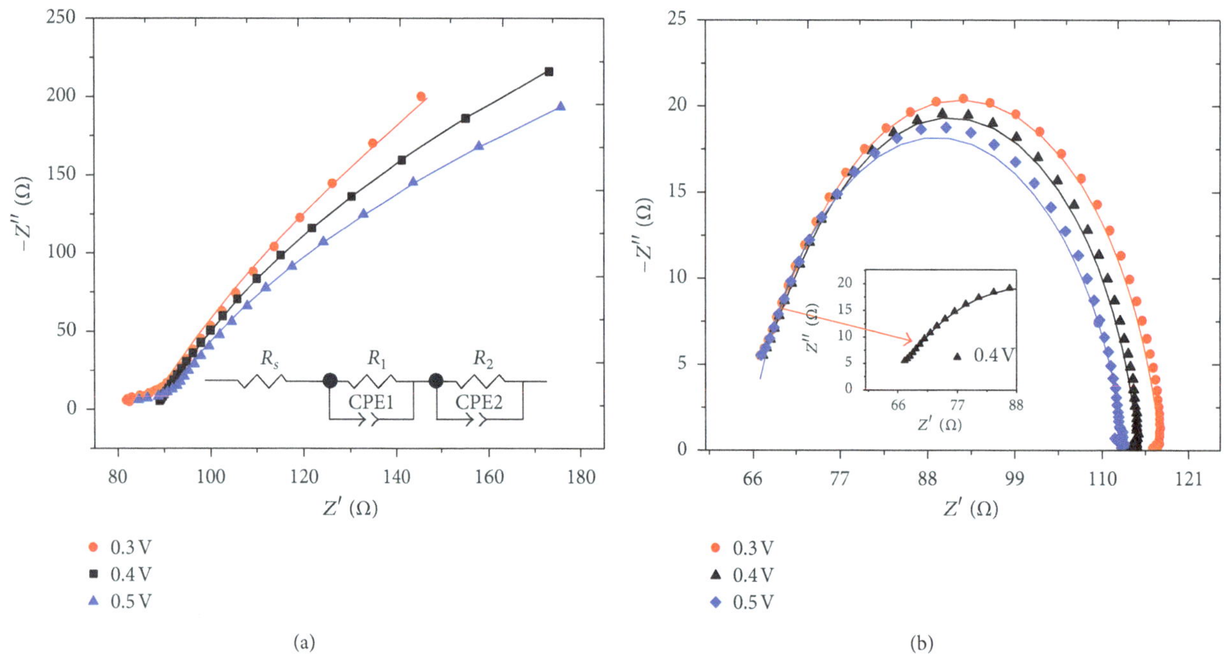

FIGURE 5: Nyquist plots of EIS spectra measured at various applied voltages under illumination of one sun (AM 1.5, 100 mW cm^{-2}) for TiO_2/MPA/CdSe QD (a) and TiO_2/CdSe QD (b) cells. The inset figure in (a) represents the equivalent circuit used for fitting EIS for both cells, while the inset figure in (b) represents the abscissa expanded in the high frequency ranges for TiO_2/CdSe QD cells. The scattered points are experimental data, and the solid lines are the fitting curves.

that the presence of a bifunctional liker (MPA) between the CdSe QDs and TiO_2 layers triggers an interfacial structure with superior ability in inhibiting charge recombination at the electrode/electrolyte interface as a result of the better coverage and higher uniformed CdSe film and the passivation effect of MPA itself, which acts as a blocking effect [3, 34, 43, 44].

The as-prepared QDSSC shows lower conversation efficiency, and one possible reason could be caused by "thick" TiO_2 films, leading the higher interfacial recombination of electrons from TiO_2 and the oxidized form of the redox couples. Another possible reason may be the conduction band alignment of TiO_2 and CdSe quantum dots and injection resistance of excited electron due to MPA. The injection of excited electrons from CdSe quantum dots into TiO_2 is influenced by the energy difference between the two conduction bands. Lower size CdSe QDs have a wider band gap which leads to the absorption of a shorter wavelength of light. However, it is suggested that use of lower size of CdSe QDs can increase the rate of excited electron injection [19]. In the future, we will try to solve the charge transport and injection resistance via preparation of different thickness of TiO_2 films, *in situ* growth of CdSe QDs, and different size of CdSe QDs with different conduction band edge position. Further research using these alternative photoanodes is planned in order to optimize QDSSCs.

4. Conclusion

We demonstrated a new type of photoanode composed of compact and structural layers for QDSSC applications. The compact layer was made of a pure TiO_2 paste, while the structural layer was prepared with PTFE to form a PTFE-framed structure. The overall thickness of the photoanode was measured by SEM: the thickness of the compact layer was ~2 μm, while the structural layer was either 23 or 28 μm. In this photoanode, colloidal CdSe QDs were attached onto the TiO_2 film, both with and without linker molecules (MPA). Different photoanodes with varying total thickness of the TiO_2 film (compact and structural layers) and with varying mode of CdSe QD attachment were prepared. The UV-vis absorption spectra of the as-prepared photoanode show that more CdSe quantum dots were anchored on the MPA surface-modified TiO_2 films suggesting that MPA helped to give a higher coverage of CdSe QDs. From its performance measurement (as shown in I-V curve), energy conversion efficiency up to 0.18% can be achieved for the cell assembled with a TiO_2 (25 μm)/MPA/CdSe QDs photoanode. The electrochemical impedance spectroscopy (EIS) measurements of the highest performance cell were carried out at different applied voltages under one sun illumination. The radius of the semicircle reduced with an increasing applied voltage. The charge recombination resistance of the solar cell prepared with TiO_2 (25 μm)/MPA/CdSe QD photoanode is higher than a cell made with TiO_2 (25 μm)/CdSe QDs photoanode indicating that MPA can help to reduce charge recombination. Furthermore, PTFE in the structural layer formation allows the flexibility to prepare a tunable thickness electrode with a large surface area, and therefore it opens a new opportunity for the simple and cost-effective preparation of TiO_2 electrodes for the large scale preparation of photoanodes.

Conflict of Interests

The authors declare that they have no conflict of interests.

Acknowledgments

The authors gratefully acknowledge the financial support from the National Science Council of Taiwan (NSC-97-2120-M-011-001 and NSC-97-2221-E-011-075-MY3) and the National Taiwan University of Science and Technology (NTUST). The authors are also grateful to the Precious Instrument Center of National Taiwan University of Science and Technology for its support with the TEM measurements.

References

[1] B. O'Regan and M. Grätzel, "A low coast, high-efficiency solar cell based on dye-sensitized colloidal TiO_2 films," *Nature*, vol. 353, pp. 737–739, 1991.

[2] H. Yu, S. Zhang, H. Zhao, B. Xue, P. Liu, and G. Will, "High-performance TiO_2 photoanode with an efficient electron transport network for dye-sensitized solar cells," *Journal of Physical Chemistry C*, vol. 113, no. 36, pp. 16277–16282, 2009.

[3] L.-W. Chong, H.-T. Chien, and Y.-L. Lee, "Assembly of CdSe onto mesoporous TiO_2 films induced by a self-assembled monolayer for quantum dot-sensitized solar cell applications," *Journal of Power Sources*, vol. 195, no. 15, pp. 5109–5113, 2010.

[4] Q. Zhang, Y. Zhang, S. Huang et al., "Application of carbon counterelectrode on CdS quantum dot-sensitized solar cells (QDSSCs)," *Electrochemistry Communications*, vol. 12, no. 2, pp. 327–330, 2010.

[5] J. H. Bang and P. V. Kamat, "Quantum dot sensitized solar cells. A tale of two semiconductor nanocrystals: CdSe and CdTe," *ACS Nano*, vol. 3, no. 6, pp. 1467–1476, 2009.

[6] H. J. Lee, J.-H. Yum, H. C. Leventis et al., "CdSe quantum dot-sensitized solar cells exceeding efficiency 1% at full-sun intensity," *Journal of Physical Chemistry C*, vol. 112, no. 30, pp. 11600–11608, 2008.

[7] Y. Xie, G. Ali, S. H. Yoo, and S. O. Cho, "Sonication-assisted synthesis of CdS quantum-dot-sensitized TiO_2 nanotube arrays with enhanced photoelectrochemical and photocatalytic activity," *ACS Applied Materials and Interfaces*, vol. 2, no. 10, pp. 2910–2914, 2010.

[8] W. Lee, S. K. Min, V. Dhas, S. B. Ogale, and S.-H. Han, "Chemical bath deposition of CdS quantum dots on vertically aligned ZnO nanorods for quantum dots-sensitized solar cells," *Electrochemistry Communications*, vol. 11, no. 1, pp. 103–106, 2009.

[9] B. Farrow and P. V. Kamat, "CdSe quantum dot sensitized solar cells. Shuttling electrons through stacked carbon nanocups," *Journal of the American Chemical Society*, vol. 131, no. 31, pp. 11124–11131, 2009.

[10] Z. A. Peng and X. Peng, "Formation of high-quality CdTe, CdSe, and CdS nanocrystals using CdO as precursor," *Journal of the American Chemical Society*, vol. 123, no. 1, pp. 183–184, 2001.

[11] W. W. Yu, L. Qu, W. Guo, and X. Peng, "Experimental determination of the extinction coefficient of CdTe, CdSe and CdS nanocrystals)," *Chemistry of Materials*, vol. 15, pp. 2854–2860, 2003.

[12] N. S. Lewis, "Toward cost-effective solar energy use," *Science*, vol. 315, no. 5813, pp. 798–801, 2007.

[13] A. J. Nozik, "Multiple exciton generation in semiconductor quantum dots," *Chemical Physics Letters*, vol. 457, no. 1–3, pp. 3–11, 2008.

[14] A. J. Nozik, "Quantum dot solar cells," *Physica E*, vol. 14, no. 1-2, pp. 115–120, 2002.

[15] J.-H. Im, C.-R. Lee, J.-W. Lee, S.-W. Park, and N.-G. Park, "6.5% efficient perovskite quantum-dot-sensitized solar cell," *Nanoscale*, vol. 3, no. 10, pp. 4088–4093, 2011.

[16] T.-L. Li, Y.-L. Lee, and H. Teng, "High-performance quantum dot-sensitized solar cells based on sensitization with $CuInS_2$ quantum dots/CdS heterostructure," *Energy and Environmental Science*, vol. 5, no. 1, pp. 5315–5324, 2012.

[17] V. González-Pedro, C. Sima, G. Marzari et al., "High performance PbS Quantum Dot Sensitized Solar Cells exceeding 4% efficiency: the role of metal precursors in the electron injection and charge separation," *Physical Chemistry Chemical Physics*, 2013.

[18] J. Chen, J. L. Song, X. W. Sun et al., "An oleic acid-capped CdSe quantum-dot sensitized solar cell," *Applied Physics Letters*, vol. 94, no. 15, Article ID 153115, 3 pages, 2009.

[19] A. Kongkanand, K. Tvrdy, K. Takechi, M. Kuno, and P. V. Kamat, "Quantum dot solar cells. Tuning photoresponse through size and shape control of CdSe-TiO_2 architecture," *Journal of the American Chemical Society*, vol. 130, no. 12, pp. 4007–4015, 2008.

[20] Y.-L. Lee and C.-H. Chang, "Efficient polysulfide electrolyte for CdS quantum dot-sensitized solar cells," *Journal of Power Sources*, vol. 185, no. 1, pp. 584–588, 2008.

[21] X. Song, X.-L. Yu, Y. Xie, J. Sun, T. Ling, and X.-W. Du, "Improving charge separation of solar cells by the co-sensitization of CdS quantum dots and dye," *Semiconductor Science and Technology*, vol. 25, no. 9, Article ID 095014, 2010.

[22] C.-H. Chang and Y.-L. Lee, "Chemical bath deposition of CdS quantum dots onto mesoscopic TiO_2 films for application in quantum-dot-sensitized solar cells," *Applied Physics Letters*, vol. 91, no. 5, Article ID 053503, 3 pages, 2007.

[23] S.-C. Lin, Y.-L. Lee, C.-H. Chang, Y.-J. Shen, and Y.-M. Yang, "Quantum-dot-sensitized solar cells: assembly of CdS-quantum-dots coupling techniques of self-assembled monolayer and chemical bath deposition," *Applied Physics Letters*, vol. 90, no. 14, Article ID 143517, 2007.

[24] Y.-L. Lee, B.-M. Huang, and H.-T. Chien, "Highly efficient CdSe-sensitized TiO_2 photoelectrode for quantum-dot-sensitized solar cell applications," *Chemistry of Materials*, vol. 20, no. 22, pp. 6903–6905, 2008.

[25] I. Mora-Seró, S. Giménez, T. Moehl et al., "Factors determining the photovoltaic performance of a CdSe quantum dot sensitized solar cell: the role of the linker molecule and of the counter electrode," *Nanotechnology*, vol. 19, no. 42, Article ID 424007, 2008.

[26] A. Zaban, O. I. Mićić, B. A. Gregg, and A. J. Nozik, "Photosensitization of nanoporous TiO_2 electrodes with InP quantum dots," *Langmuir*, vol. 14, no. 12, pp. 3153–3156, 1998.

[27] P. J. Cameron and L. M. Peter, "Characterization of titanium dioxide blocking layers in dye-sensitized nanocrystalline solar cells," *Journal of Physical Chemistry B*, vol. 107, no. 51, pp. 14394–14400, 2003.

[28] B. Peng, G. Jungmann, C. Jäger, D. Haarer, H.-W. Schmidt, and M. Thelakkat, "Systematic investigation of the role of compact TiO_2 layer in solid state dye-sensitized TiO_2 solar cells," *Coordination Chemistry Reviews*, vol. 248, no. 13-14, pp. 1479–1489, 2004.

[29] H. Yu, S. Zhang, H. Zhao, G. Will, and P. Liu, "An efficient and low-cost TiO_2 compact layer for performance improvement of dye-sensitized solar cells," *Electrochimica Acta*, vol. 54, no. 4, pp. 1319–1324, 2009.

[30] H.-F. Wang, L.-Y. Chen, W.-N. Su, J.-C. Chung, and B.-J. Hwang, "Effect of the compact TiO_2 layer on charge transfer between N3 Dyes and TiO_2 investigated by Raman spectroscopy," *Journal of Physical Chemistry C*, vol. 114, no. 7, pp. 3185–3189, 2010.

[31] H.-F. Wang, W.-N. Su, and B.-J. Hwang, "A novel architecture of poly(tetrafluoroethylene)-framed TiO_2 electrodes for dye-sensitized solar cells," *Electrochemistry Communications*, vol. 11, no. 8, pp. 1647–1649, 2009.

[32] D. W. Ayele, H.-M. Chen, W.-N. Su et al., "Controlled synthesis of CdSe quantum dots by a microwave-enhanced process: a green approach for mass production," *Chemistry A*, vol. 17, no. 20, pp. 5737–5744, 2011.

[33] C. J. Barbé, F. Arendse, P. Comte et al., "Nanocrystalline titanium oxide electrodes for photovoltaic applications," *Journal of the American Ceramic Society*, vol. 80, no. 12, pp. 3157–3171, 1997.

[34] J. Chen, C. Li, D. W. Zhao et al., "A quantum dot sensitized solar cell based on vertically aligned carbon nanotube templated ZnO arrays," *Electrochemistry Communications*, vol. 12, no. 10, pp. 1432–1435, 2010.

[35] R. G. Pearson, "Hard and soft acids and bases," *Journal of the American Chemical Society*, vol. 85, no. 22, pp. 3533–3539, 1963.

[36] K. Mukherjee, T.-H. Teng, R. Jose, and S. Ramakrishna, "Electron transport in electrospun TiO_2 nanofiber dye-sensitized solar cells," *Applied Physics Letters*, vol. 95, no. 1, Article ID 012101, 2009.

[37] I. Bedja, P. V. Kamat, X. Hua, A. G. Lappin, and S. Hotchandani, "Photosensitization of nanocrystalline ZnO films by bis(2,2′-bipyridine)(2,2′-bipyridine-4,4′-dicarboxylic acid)ruthenium(II)," *Langmuir*, vol. 13, no. 8, pp. 2398–2403, 1997.

[38] J. D. Roy-Mayhew, D. J. Bozym, C. Punckt, and I. A. Aksay, "Functionalized graphene as a catalytic counter electrode in dye-sensitized solar cells," *ACS Nano*, vol. 4, no. 10, pp. 6203–6211, 2010.

[39] E. P. A. M. Bakkers, A. L. Roest, A. W. Marsman et al., "Characterization of photoinduced electron tunneling in gold/SAM/Q-CdSe systems by time-resolved photoelectrochemistry," *Journal of Physical Chemistry B*, vol. 104, no. 31, pp. 7266–7272, 2000.

[40] I. Robel, V. Subramanian, M. Kuno, and P. V. Kamat, "Quantum dot solar cells. Harvesting light energy with CdSe nanocrystals molecularly linked to mesoscopic TiO_2 films," *Journal of the American Chemical Society*, vol. 128, no. 7, pp. 2385–2393, 2006.

[41] T. Trupke, P. Würfel, and I. Uhlendorf, "Dependence of the photocurrent conversion efficiency of dye-sensitized solar cells on the incident light intensity," *Journal of Physical Chemistry B*, vol. 104, no. 48, pp. 11484–11488, 2000.

[42] L. Etgar, T. Moehl, S. Gabriel, S. G. Hickey, A. Eychmüller, and M. Grätzel, "Light energy conversion by mesoscopic PbS quantum dots/TiO_2 heterojunction solar cells," *ACS Nano*, vol. 6, no. 4, pp. 3092–3099, 2012.

[43] J. Chen, X. Yang, J. Wu, W. Lei, and X. Zhang, "Enhanced absorption of CdS quantum dots deposited onto ZnO nanorod by using bifunctional linker," *Physica Status Solidi C*, vol. 9, no. 1, pp. 19–21, 2012.

[44] I. Mora-Seró, S. Giménez, F. Fabregat-Santiago et al., "Recombination in quantum dot sensitized solar cells," *Accounts of Chemical Research*, vol. 42, pp. 1848–1857, 2009.

A Kinetic Study of Marginal Soil Energy Plant *Helianthus annuus* Stalk Pyrolysis

Huaxiao Yan,[1,2] Hui Zhao,[1,2] Yan Zhang,[1,2] Yuanyu Tian,[1,2] and Kechang Xie[2]

[1] *College of Chemical and Environmental Engineering, Shandong University of Science and Technology, Qingdao 266590, China*
[2] *Key Laboratory of Low-carbon Energy Chemical Engineering in Universities of Shandong, Shandong University of Science and Technology, Qingdao 266590, China*

Correspondence should be addressed to Hui Zhao; zhsdust@126.com

Academic Editor: Xiumin Jiang

The pyrolytic characteristics and kinetics of new marginal soil energy plant *Helianthus annuus* stalk were investigated using thermogravimetric (TG) method from 50 to 800°C in an inert argon atmosphere at different heating rates of 5, 10, 20 and 30°C min^{-1}. The kinetic parameters of activation energy and pre-exponential factor were deduced by Popescu, Flynn-Wall-Ozawa (FWO), and Kissinger-Akahira-Sunose (KAS) methods, respectively. The results showed that three stages appeared in the thermal degradation process. The primary devolatilization stage of *H. annuus* stalk can be described by the Avrami-Erofeev function (n = 4). The average activation energy of *H. annuus* stalk was only 142.9 kJ mol^{-1}. There were minor kinetic compensation effects between the pre-exponential factor and the activation energy. The results suggest that *H. annuus* stalk is suitable for pyrolysis, and more importantly, the experimental results and kinetic parameters provided useful information for the design of pyrolytic processing system using *H. annuus* stalk as feedstock.

1. Introduction

As potential "next generation" biofuel feedstock, marginal soil plants have attracted considerable attention because of their advantages, such as high photosynthetic efficiency, maximum biomass production, fast growing, high conversion rate, ease of harvesting, and lack of arable soil requirements. *Helianthus annuus* can grow well in barren marginal lands and has good resistance to adverse situations. Moreover, it does not compete with grain crops for arable lands, which is significant for one populous country lack of arable land as China. The agriculture is producing huge amount of *H. annuus* stalks as by-products of seeds every year in China. It has been reported that the stalk yield was ranging between 5 and 14 tons per hectare. *H. annuus* has an excellent adaptability, and the growth cycle is only about 100 days. These characteristics of *H. annuus* suggest that it can be used as a good potential energy source and should be further studied as a good biofuel feedstock candidate [1–8].

Biomass pyrolysis has demonstrated itself to be a kind of biomass-energy utilization technology which transforms low-energy density biomass materials into high-energy density liquid products which can be utilized more efficiently in an environment-friendly manner. A thorough knowledge of the thermal behavior and pyrolysis kinetics of biomass are required for the proper design and operation of the pyrolysis conversion systems. Thermogravimetry (TG) analysis method was selected to investigate the thermal decomposition process. The kinetic data from TG are not only very useful for understanding the thermal degradation processes and mechanisms, but also can be used as input parameters for a thermal degradation reaction system [9–11].

In present study, *H. annuus* stalk pyrolysis was investigated using a TG/DSC instrument with nonisothermal thermogravemetric (TG) analysis method at different heating rates under an inert atmosphere. The pyrolysis characteristics and kinetics were examined by Popescu, Flynn-Wall-Ozawa (FWO) and Kissinger-Akahira-Sunose (KAS) methods and then the average activation energy, pre-exponential factors, and reaction orders were deduced. The objective of this study was to obtain the kinetic parameters of thermal decomposition and then to determine the degradation mechanism for facilitating the efficient design, operation, and modeling of pyrolysis and related thermochemical conversion systems for energy plants.

2. Materials and Methods

2.1. Materials. *H. annuus* stalk was collected from the marginal land of Shandong University of Science and Technology campus, Qingdao, China, January 2012. Sample was dried in one drying oven at 100 ± 2°C for 10 hours and pulverized in a plant disintegrator to be able to pass through a 120-mesh sieve, and then stored in a desiccator.

2.2. Proximate and Component Analysis of the Sample. Proximate analysis was carried out according to the national standard GB212-91 (China). For analysis of composition, the biomass materials were hydrolyzed using 72% H_2SO_4, 2 mol/L HCl and concentrated HCl to analyze cellulose, hemicellulose and lignin contents according to established methods, respectively [10, 12, 13]. The results were presented in Table 1. All tests were carried out in triplicate.

2.3. Pyrolysis by Thermogravimetric Analyzer. Pyrolytic characteristics were determined by using a thermal analyzer (TGA/DSC1/1600LF Mettler Toledo Co., Switzerland). In the experiment, argon gas was used as a carrier gas with a flow rate of 50 cm^3 min^{-1}. The heating rate was controlled at 5, 10, 20, 30°C min^{-1} from 50 to 800°C. The sample mass that was used for the thermogravimetric analysis in each experiment was 10 mg approximately. Mass loss and calorific changes in response to temperature were recorded and used to plot thermogravimetric (TG), derivative thermogravimetric analysis (DTG), and differential scanning calorimetric (DSC) curves [14, 15]. All experiments were replicated three times. The software Origin 8.0 was used to analyze the data and plot the curves.

2.4. Kinetic Analysis

2.4.1. Popescu Method. The Popescu method was used to evaluate the most probable mechanism of the pyrolytic reaction [12, 13]. In other words, the Arrhenius equation was selected to analyze the thermal decomposition reactions. The dynamic equation is:

$$\frac{d\alpha}{dt} = k(T) f(\alpha), \quad (1)$$

where α is the conversion rate and is defined as follows:

$$\alpha = \frac{m_0 - m}{m_0 - m_\infty} \quad (2)$$

and here, m_0, m, m_∞ are the initial, actual, and the end mass of samples, mg;
where $k(T)$ is the velocity constant, and $f(\alpha)$ is the differential coefficient mechanism function.

Consider

$$k(T) = A \exp\left(-\frac{E}{RT}\right) \quad (3)$$

TABLE 1: Proximate and component analysis of *H. annuus* stalk.

Characteristics	*H. annuus* stalk
Proximate analysis[a]/wt%	
Moisture (M_{ar})	9.93 ± 0.46
Ash (A_{ar})	4.08 ± 0.01
Volatile matter (V_{ar})	72.94 ± 0.42
Fixed carbon[b] (FC_{ar})	13.05
Component analysis/wt%	
Cellulose	21.13 ± 1.19
Hemicellulose	14.81 ± 0.37
Lignin	7.30 ± 1.52

[a] Mass percentage on as-received basis; [b] calculated by difference.

thus, with the combination of (1), (2), and (3), the rate of a solid-state reaction can generally be described by:

$$\frac{d\alpha}{dt} = A \exp\left(-\frac{E}{RT}\right) f(\alpha) \quad (4)$$

and then, if the temperature is controlled at a constant heating rate, $\beta(\text{K/min}) = dT/dt$. Therefore, (4) gives:

$$\frac{d\alpha}{dT} = \frac{1}{\beta} k(T) f(\alpha) = \frac{A}{\beta} \exp\left(-\frac{E}{RT}\right) f(\alpha). \quad (5)$$

Then the integral form of (5) is:

$$G(\alpha)_{mn} = \int_{\alpha_m}^{\alpha_n} \frac{d\alpha}{f(\alpha)} = \frac{1}{\beta}\int_{T_m}^{T_n} k(T)\,dT = \frac{1}{\beta} I(T)_{mn} = \frac{A}{\beta} H(T)_{mn}. \quad (6)$$

For this equation, where α_m, α_n are two different degrees of conversion, and T_m, T_n are their corresponding temperatures, and here $I(T)_{mn}$, $H(T)_{mn}$ and $T\delta$ can be described as follows, respectively:

$$I(T)_{mn} = \int_{T_m}^{T_n} k(T)\,dT,$$

$$H(T)_{mn} = \int_{T_m}^{T_n} \exp\left(-\frac{E}{RT}\right) = (T_n - T_m)\exp\left(-\frac{E}{RT\delta}\right), \quad (7)$$

$$T\delta = \frac{T_m + T_n}{2}.$$

Therefore, a plot of $G(\alpha)_{mn}$ versus $1/\beta_i$ gives a straight line with an intercept of zero. If the experiment data and $G(\alpha)_{mn}$ are chosen properly, a plot of $G(\alpha)_{mn}$ versus $1/\beta_i$ gives a straight line with an intercept of zero. Then, this $G(\alpha)_{mn}$ is the proper mechanism that can express the true chemical reaction process.

By logarithmic transformation of (6), a linear equation is deduced:

$$\ln\left(\frac{\beta}{T_n - T_m}\right) = \ln\left[\frac{A}{G(\alpha)}\right] - \frac{E}{RT\delta}. \quad (8)$$

A plot of $\ln(\beta/(T_n - T_m))$ versus $1/T\delta$ gives a straight line with a slope of $-E/R$ and intercept of $\ln[A/G(\alpha)]$. Therefore, E and $\ln A$ can be determined.

TABLE 2: Temperature characteristics associated with the pyrolysis process of *H. annuus* stalk.

Heating rate/°C min^{-1}	Temperature/°C				
	T_1	T_2	T_3	T_4	T_5
5	196.3	259.2	296.7	322.9	365.4
10	151.7	216.0	260.0	284.6	342.1
20	130.4	183.4	209.0	231.7	272.5
30	161.3	244.2	261.3	296.3	330.4

2.4.2. FWO and KAS Method. It is well known that flynn-wall-ozawa (FWO) method and kissinger-akahira-sunose (KAS)method were generally used to deduce the activation energy and pre-exponential factor, and the apparent activation energy is influenced by temperature T and fractional conversion α [13, 16–20]. So, activation energy can be obtained with the change of heating rate β.

Flynn-Wall-Ozawa equation:

$$\ln \beta = \ln \left[\frac{0.0048AE}{RG(\alpha)} \right] - 1.0516\frac{E}{RT}. \quad (9)$$

KAS equation:

$$\ln \left(\frac{\beta}{T^2} \right) = \ln \left[\frac{AR}{EG(\alpha)} \right] - \frac{E}{RT}. \quad (10)$$

According to (9) and (10), plots of $\ln \beta$, $\ln(\beta/T^2)$ versus $1/T$ with a certain conversion rate αgive straight lines with slopes of $-1.0516E/RT$, and $-E/R$ the activation energy can be determined. Then, the activation energy (E) can be determined. Substituting the value of E and the mechanism function $G(\alpha)$ back into (9) and (10) in conjunction with T and α, $\ln A$ can be calculated.

3. Results and Discussion

3.1. Characteristics of the Thermal Degradation Process. According to thermogravimetric (TG) and differential thermogravimetric (DTG) curves of *H. annuus* stalk (Figures 1 and 2), the results indicated that the pyrolysis process was made up of three stages. The stage I occurred as the temperature increased from room temperature to T_1, while the second started as the temperature that increased from T_1 to T_5 (II). However, the sample revealed somewhat different during stage II. Stage II included two zones for *H. annuus* stalk pyrolysis process: zone I occurred as the temperature increased from T_1 to T_3 with a maximum mass loss point at T_2, and zone II occurred as the temperature increased from T_3 to T_5 with a maximum mass loss point at T_4. The last one, stage III occurred as the temperature increased from T_5 to 800°C. The concrete contents of the temperature characteristics were shown in Table 2 and Figure 2.

The TG and DTG curves of the sample at different heating rates (Table 2) and 20°C min^{-1} (Figures 1 and 2, all profiles of different heating rates are not presented here, and 20°C min^{-1} was shown as an example for concision) showed that three stages appeared corresponding to three different pyrolytic processes. The stage I was the process of slow mass loss corresponding to the evaporation of water and other volatile components, in other words, it is that both the internal cellular water and the external water bound by surface tension were lost [12]. From the Figure 2, stage II (T_1–T_5) was the devolatilization stage, during which the main pyrolytic process occurred. In this stage, various volatile components were gradually released, and biological macromolecules were depolymerized and fractured; that resulted in a large mass loss and formation of the main pyrolytic products. The temperature range of three samples during stage II was 130.4–272.5°C (*H. annuus* stalk) at different heating rates. During the stage III, the residual was slowly decomposed [26], resulting in the formation of a loose porous end-product.

Heating rates have significant effects on the pyrolysis process. As the heating rate increasing, the mass loss rate and the average reaction rate both increased. The results were shown in Table 3.

The differential scanning calorimetry (DSC) curves of three samples at different heating rates were presented in

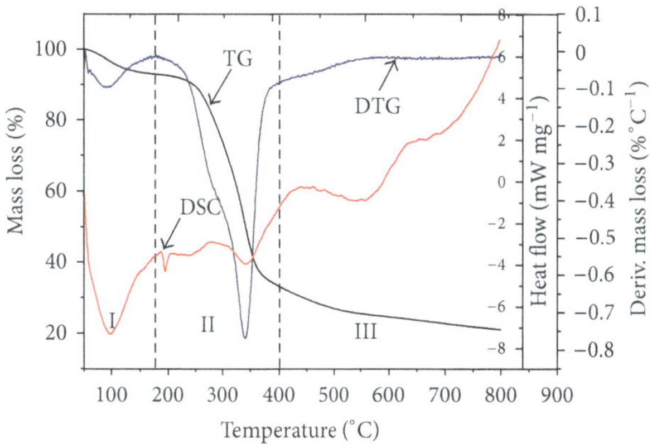

FIGURE 1: TG-DTG-DSC curves of *H. annuus* stalk at the heating rate of 20°C min^{-1} (as an example for concision).

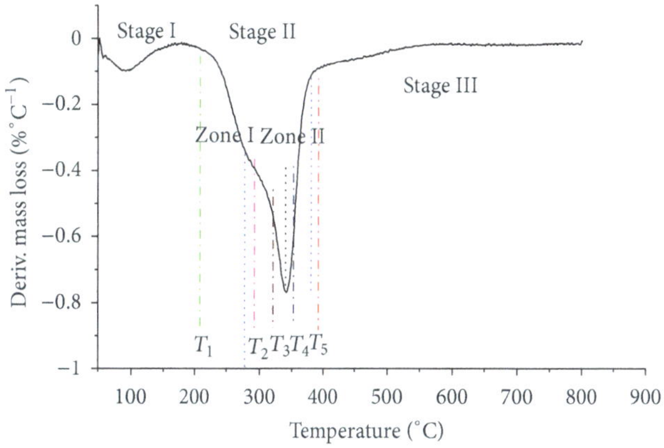

FIGURE 2: The DTG curves of *H. annuus* stalk at the different heating rate of 20°C min^{-1} with the characteristic temperature zone.

Table 3: Mass loss and average reaction rate at different stages of *H. annuus* stalk pyrolysis process.

Stage			Heating rate/°C min^{-1}			
			5	10	20	30
I	ML[a]		6.6	7.1	7.0	6.5
	AR[b]		0.2	0.6	1.1	1.5
II	Z I[c]	ML[a]	23.3	23.2	21.0	18.1
		AR[b]	1.2	1.8	3.3	4.5
	Z II[d]	ML[a]	32.9	36.8	39.3	39.2
		AR[b]	2.4	3.7	7.8	14.2
III	ML[a]		16.1	13.3	11.6	13.0
	AR[b]		0.2	0.3	0.6	0.9
Final residual at 800°C[e]			21.1	19.6	21.0	23.2

[a]Mass loss,%; [b]Average reaction rate,%/min; [c]Zone I; [d]Zone II; [e]Final residual at 800°C,%.

Table 4: The linear fitting results of kinetic mechanism functions of *H. annuus* stalk.

Function	Temperature/°C	R	SD
$[-\ln(1-\alpha)]^4$	200	0.96601	0.00145
Avrami-Erofeev	250	0.98151	0.01246
Function	300	0.99083	0.07635
$n = 4$	350	0.99305	0.13633

R: Correlation coefficient; SD: standard deviation.

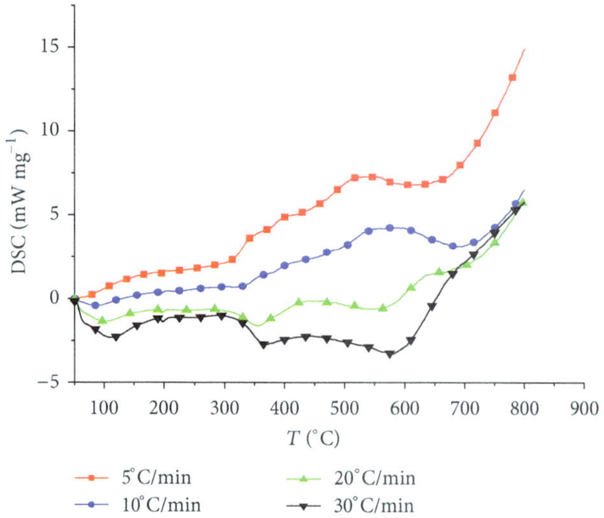

Figure 3: The DSC curves of *H. annuus* stalk at different heating rates.

Figure 3. As the heating rate increasing, an obvious exothermic effect appeared, and this finding indicated that the pyrolysis process produced heat. During the stage I, there was an obvious endothermic peak, which corresponded to moisture evaporation. In stage II (devolatiliation stage), there were some differences among these three samples. As the temperature increased, *H. annuus* stalkmaintained the exothermic effect, while as the heating rate increasing, the exothermic effect became more obvious with the peak area increased. At the temperature of 400 to 500°C, there was an exothermic peak, which showed that the charring process occurred at this moment. Generally speaking, an endothermic process can be related to depolymerization and volatilization processes, while an exothermic process is due to charring. The pyrolysis had no clear endothermic effects, which suggested that the decomposition process needs low energy and the inorganic salts promote charring and exothermic effects.

3.2. Kinetic Analysis of the Pyrolysis Process

3.2.1. Determination of $G(\alpha)$. Different conversion rates at different heating rates and temperatures were chosen to determine the pyrolysis mechanism function. These temperatures should be applied during stage II. The 41 kinds of typical mechanism functions [27] were analyzed according to the Popescu method to determine the $G(\alpha)$ function. And at last, according to these 41 kinds of typical mechanism functions, the function ($G(\alpha) = [-\ln(1-\alpha)]^4$) appeared as the most probable mechanism function for *H. annuus* stalk (Table 4).

And then, the activation energy and pre-exponential factors can be calculated by function $G(\alpha)$ [28].

3.2.2. Calculation of the Activation Energy and Pre-Exponential Factors. According to (8), a plot of $\ln(\beta/(T_n - T_m))$ versus $1/T\delta$ gives a straight line with a slope of $-E/R$ and intercept of $\ln[A/G(\alpha)]$. Therefore, E and $\ln A$ can be determined. And on basis of (9) and (10), a plot of $\ln\beta$, $\ln(\beta/T^2)$ against $1/T$ should be a linear relationship by FWO method and KAS method. Similarly, the activation energy and pre-exponential can be calculated (Figure 4). The values of activation energy and $\ln A$ was listed in Table 5. The value of R-Square was close to 0.9500, that is to say, the values of coefficient of correlation were relatively high. So the values of activation energy deduced and calculated by Popescu, FWO, and KAS methods are valid. However, the values of activation energy increased with a different fluctuation at different conversion rates. This

TABLE 5: The activation energies of *H. annuus* stalk obtained by FWO method and KAS method at different conversion rates.

Conversion rate/α	FWO method			KAS method			Popescu method			
	E/kJ Mol^{-1}	R	ln A/min^{-1}	E/kJ Mol^{-1}	R	ln A/min^{-1}	Conversion Rate/α	E/kJ Mol^{-1}	R	ln A/min^{-1}
0.1	209.9	0.9997	20.70	214.7	0.9997	19.56	0.1–0.2	88.4	0.9091	12.83
0.2	178.5	0.9976	17.54	183.2	0.9977	16.72	0.2–0.3	115.5	0.9406	20.11
0.3	148.0	0.9849	15.48	152.4	0.9855	13.03	0.3–0.4	131.1	0.9463	24.28
0.4	128.9	0.9575	13.10	133.2	0.9590	11.73	0.4–0.5	138.1	0.9457	26.43
0.5	121.3	0.9190	12.02	125.5	0.9218	12.57	0.5–0.6	128.2	0.8957	25.14
0.6	128.2	0.8968	12.56	132.5	0.9006	11.39	0.6–0.7	119.4	0.8878	24.27
0.7	139.8	0.9067	13.55	144.2	0.9105	12.22	0.7–0.8	136.4	0.8271	28.66
0.8	142.0	0.9160	14.41	146.3	0.9195	9.04	0.8–0.9	195.0	0.9644	41.34
0.9	120.6	0.9043	10.81	124.6	0.9077	10.77				
Average	146.4 ± 29.71			150.7 ± 29.94				131.5 ± 30.17		
Average	142.9 ± 10.08									

E: Activation energy; A: pre-exponential factor; R: correlation coefficient.

TABLE 6: Comparison of various kinetic parameters of pyrolysis for different biomasses.

Sample	Decomposition temperature/°C	E/kJ mol^{-1}	References
H. annuus stalk	196.3–330.4	142.9	Present study
Enteromorpha prolifera	174–551	228.1	[21]
Spirulian platensis	190–560	42.2–5.25	[22]
Wheat straw	230–400	130–175	[23]
Sunflower shell	300–600	73.81	[24]
Rice husk	225–350	79.9	[25]

TABLE 7: Kinetic compensation effects of the pre-exponential factor and the activation energy.

Method	Equation	R
FWO method	$\ln A = 0.102E - 0.3927$	0.9654
KAS method	$\ln A = 0.095E - 1.3721$	0.7603
Popescu method	$\ln A = 0.263E - 9.2219$	0.9659

may be ascribed to the complex composition of the sample and the complex reactions during pyrolysis. In the end, the average activation energy value of *H. annuus* stalkwas 142.9 kJ mol^{-1}.

In Table 6, various kinetic parameters of pyrolysis between different biomass sources were compared. Other results are not similar to the present study, which suggests that thermal behavior is influenced greatly by feedstock types [29].

In Table 7, the kinetic compensation effects of the pre-exponential factors and activation energy were shown. These results indicated that there was a partial compensation effect for A when E changed.

4. Conclusions

High priority should be given to the development and protection of biomass pyrolysis from new marginal soil energy plant with great potential, which is widely recognized as a technically feasible way for sustainability. In the present work, there were three stages in the process of *H. annuus* stalk pyrolysis, and stage II was the main pyrolysis process, and most of the organic materials were decomposed in this stage, and heating rates had a significant effect on the pyrolysis process of biomass.

The Popescu, FWO, and KAS methods were adopted to determine the kinetic parameters of the pyrolysis reaction. During the primary decomposition reactions, the most probable mechanism functions for *H. annuus* stalk was $[-\ln(1-\alpha)]^4$. The values of activation energy calculated by these three methods were similar. There was only a partial compensation effect for A when E changed. The results suggest that *H. annuus* stalk is suitable for pyrolysis, and more importantly, the experimental results and kinetic parameters provided useful information for the design of pyrolytic processing system using *H. annuus* stalk as feedstock.

Conflict of Interests

The authors declare that they have no conflict of interests.

Authors' Contribution

H. Yan and H. Zhao participated in the design and coordination of the study, interpreted the data, and drafted and

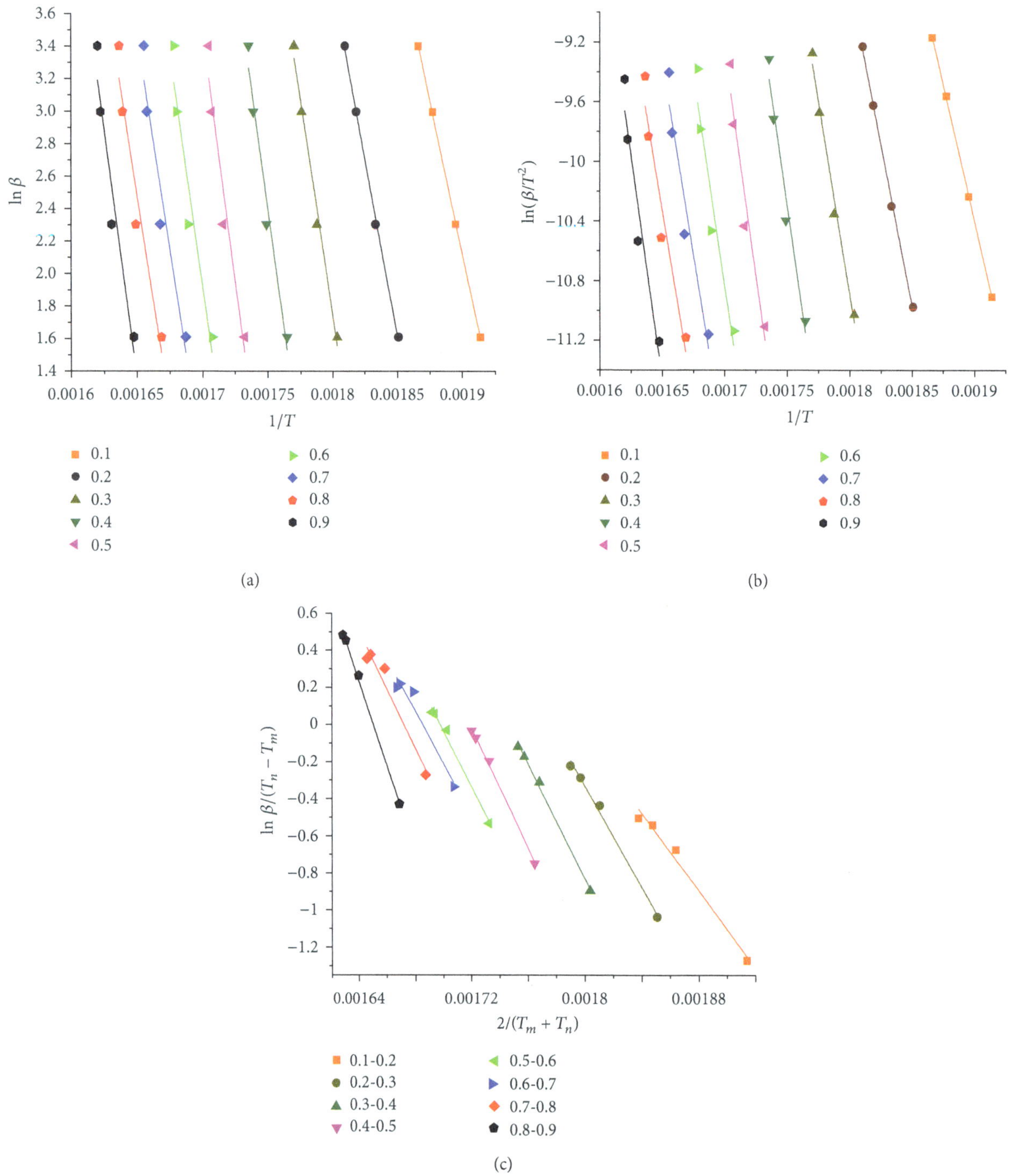

FIGURE 4: Plots for determining activation energy of Stage II of *H. annuus* stalk calculated by three methods ((a) FWO method; (b) KAS method; (c) Popescu method).

corrected the manuscript. Y. Zhang et al. participated in the interpretation of the data and corrected the draft paper. All authors read and approved the final paper.

Disclosure

H. Yan is a staff member of College of Chemical and Environmental Engineering, Shandong University of Science and Technology, Qingdao, China. Dr. H. Zhao is the leader of Biochemical Engineering and Bioresources Laboratory at College of Chemical and Environmental Engineering. His areas of specialization are biochemical engineering, energy, and environmental engineering. He has published over 10 papers in international journals. Y. Zhang is one master degree candidate. Y. Tian and K. Xie are the leaders of Key Laboratory of Low-carbon Energy Chemical Engineering in

Universities of Shandong, Shandong University of Science and Technology, Qingdao 266590, China, and contributed to supply the excellent research platform and guide research works.

Acknowledgments

This work was financially supported by National Natural Science Foundation of China (21076117, 41202165), Open Funding from the State Key Laboratory of Crop Biology in Shandong Agricultural University (2010KF06 and 2011KF14), Projects of Shandong Province Higher Educational Science and Technology Program (J10LC15), Open Project Program of Key Lab of Marine Bioactive Substance and Modern Analytical Technique (MBSMAT-2012-03), and Qingdao Government Scientific and Technological Program.

References

[1] T. Gollan, U. Schurr, and E. D. Schulze, "Stomatal response to drying soil in relation to changes in the xylem sap composition of *Helianthus annuus*. I. The concentration of cations, anions, amino acids in, and pH of, the xylem sap," *Plant, Cell & Environment*, vol. 15, no. 5, pp. 551–559, 1992.

[2] H. Chen and T. Cutright, "EDTA and HEDTA effects on Cd, Cr, and Ni uptake by *Helianthus annuus*," *Chemosphere*, vol. 45, no. 1, pp. 21–28, 2001.

[3] S. Yorgun, S. Şensöza, and O. M. Koçkar, "Flash pyrolysis of sunflower oil cake for production of liquid fuels," *Journal of Analytical and Applied Pyrolysis*, vol. 60, no. 1, pp. 1–12, 2001.

[4] J. M. Encinar, F. J. Beltrán, J. F. González, and M. J. Moreno, "Pyrolysis of maize, sunflower, grape and tobacco residues," *Journal of Chemical Technology and Biotechnology*, vol. 70, no. 4, pp. 400–410, 1997.

[5] P. Pan, C. Hu, W. Yang et al., "The direct pyrolysis and catalytic pyrolysis of *Nannochloropsis* sp. residue for renewable bio-oils," *Bioresource Technology*, vol. 101, no. 12, pp. 4593–4599, 2010.

[6] Y. Tonbul, A. Saydut, K. Yurdakoç, and C. Hamamcl, "A kinetic investigation on the pyrolysis of *Seguruk asphaltite*," *Journal of Thermal Analysis and Calorimetry*, vol. 95, no. 1, pp. 197–202, 2009.

[7] A. Szydlowska-Czerniak, K. Trokowski, and E. Szlyk, "Optimization of extraction conditions of antioxidants from sunflower shells (*Helianthus annuus* L.) before and after enzymatic treatment," *Industrial Crops and Products*, vol. 33, no. 1, pp. 123–131, 2011.

[8] A. E. Pütün, A. Özcan, H. F. Gerçel, and E. Pütün, "Production of biocrudes from biomass in a fixed-bed tubular reactor: product yields and compositions," *Fuel*, vol. 80, no. 10, pp. 1371–1378, 2001.

[9] H. R. Yuan, R. H. Liu, H. Jiang, and Y. Y. Hao, "Kinetic study of sunflower seed husk pyrolysis," *Transactions of the Chinese Society of Agricultural Engineering*, vol. 22, no. 4, pp. 220–223, 2006.

[10] M. Carrier, A. Loppinet-Serani, D. Denux et al., "Thermogravimetric analysis as a new method to determine the lignocellulosic composition of biomass," *Biomass and Bioenergy*, vol. 35, no. 1, pp. 298–307, 2011.

[11] R. French and S. Czernik, "Catalytic pyrolysis of biomass for biofuels production," *Fuel Processing Technology*, vol. 91, no. 1, pp. 25–32, 2010.

[12] H. Zhao, H. Yan, S. Dong et al., "Thermogravimetry study of the pyrolytic characteristics and kinetics of macro-algae *Macrocystis pyrifera* residue," *Journal of Thermal Analysis and Calorimetry*, 2011.

[13] H. Zhao, H. Yan, C. Zhang et al., "Pyrolytic characteristics and kinetics of phragmites australis," *Evidence-Based Complementary and Alternative Medicine*, vol. 2011, Article ID 408973, 6 pages, 2011.

[14] J. Guo and A. C. Lua, "Kinetic study on pyrolysis of extracted oil palm fiber. Isothermal and non-isothermal conditions," *Journal of Thermal Analysis and Calorimetry*, vol. 59, no. 3, pp. 763–774, 2000.

[15] S. Xiu, H. K. Rojanala, A. Shahbazi, E. H. Fini, and L. Wang, "Pyrolysis and combustion characteristics of Bio-oil from swine manure," *Journal of Thermal Analysis and Calorimetry*, vol. 107, pp. 823–829, 2012.

[16] H. Zhao, H. Yan, M. Liu, C. Zhang, and S. Qin, "Pyrolytic characteristics and kinetics of the marine green tide macroalgae, *Enteromorpha prolifera*," *Chinese Journal of Oceanology and Limnology*, vol. 29, no. 5, pp. 996–1001, 2011.

[17] A. Ji, S. Zhang, X. Lu, and Y. Liu, "A new method for evaluating the sewage sludge pyrolysis kinetics," *Waste Management*, vol. 30, no. 7, pp. 1225–1229, 2010.

[18] J. H. Flynn and L. A. Wall, "A quick direct method for determination of activation energy from thermogravimetric data," *Journal of Polymer Science B*, vol. 4, pp. 423–428, 1966.

[19] M. S. Subramanian, R. N. Singh, and H. D. Sharma, "Reaction kinetics of some actinide oxalates by differential thermal analysis," *Journal of Inorganic and Nuclear Chemistry*, vol. 31, no. 12, pp. 3789–3795, 1969.

[20] A. Aboulkas, K. El Harfi, A. El Bouadili, and M. Nadifiyine, "Study on the pyrolysis of Moroccan oil shale with poly (ethylene terephthalate)," *Journal of Thermal Analysis and Calorimetry*, vol. 100, no. 1, pp. 323–330, 2010.

[21] D. Li, L. Chen, J. Zhao et al., "Evaluation of the pyrolytic and kinetic characteristics of *Enteromorpha prolifera* as a source of renewable bio-fuel from the Yellow Sea of China," *Chemical Engineering Research and Design*, vol. 88, no. 5-6, pp. 647–652, 2010.

[22] W. Peng, Q. Wu, P. Tu, and N. Zhao, "Pyrolytic characteristics of microalgae as renewable energy source determined by thermogravimetric analysis," *Bioresource Technology*, vol. 80, no. 1, pp. 1–7, 2001.

[23] J. M. Cai and L. S. Bi, "Kinetic analysis of wheat straw pyrolysis using isoconversional methods," *Journal of Thermal Analysis and Calorimetry*, vol. 98, no. 1, pp. 325–330, 2009.

[24] A. A. Zabaniotou, E. K. Kantarelis, and D. C. Theodoropoulos, "Sunflower shells utilization for energetic purposes in an integrated approach of energy crops: laboratory study pyrolysis and kinetics," *Bioresource Technology*, vol. 99, no. 8, pp. 3174–3181, 2008.

[25] A. Sharma and T. R. Rao, "Kinetics of pyrolysis of rice husk," *Bioresource Technology*, vol. 67, no. 1, pp. 53–59, 1999.

[26] T. A. Ngo, J. Kim, and S. S. Kim, "Characteristics and kinetics of cattle litter pyrolysis in a tubing reactor," *Bioresource Technology*, vol. 101, no. 1, supplement, pp. S104–S108, 2010.

[27] R. Z. Hu, S. L. Gao, F. Q. Zhao et al., *Thermal Analysis Kinetics*, Science Press, Beijing, China, 2nd edition, 2008.

[28] Z. Shuping, W. Yulong, Y. Mingde, L. Chun, and T. Junmao, "Pyrolysis characteristics and kinetics of the marine microalgae *Dunaliella tertiolecta* using thermogravimetric analyzer," *Bioresource Technology*, vol. 101, no. 1, pp. 359–365, 2010.

[29] M. Jefferson, "Sustainable energy development: performance and prospects," *Renewable Energy*, vol. 31, no. 5, pp. 571–582, 2006.

11

Analysis and Testing of a Natural Convection Solar Dryer for the Tropics

A. O. Adelaja and B. I. Babatope

The Department of Mechanical Engineering, University of Lagos, Akoka, Yaba, Lagos 101017, Nigeria

Correspondence should be addressed to A. O. Adelaja; ao_adelaja@yahoo.com

Academic Editor: Mattheos Santamouris

Solar dryers are imperative for the tropical and sub-Saharan African countries, which are faced with the duo challenges of inadequate electrical energy supply, which has severely limited the application of conventional refrigeration as a means of preservation of agricultural produce, and the need to make produce competitive in the international market. In this study, a cost-effective natural convection solar dryer was developed; the thermal and drying analyses were done and tested to obtain some performance evaluation parameters for the system in order to examine its efficiency and effectiveness by drying some plantain fillets. The collector and system efficiencies are found to be 46.4% and 78.73%, respectively, while a percentage moisture removal of 77.5% was achieved at the 20th hour in order to give final moisture contents of 15.75% in the product, which still maintained its integrity. With a cost of about $195.00, it has been affordable for the small- and medium-scale enterprises as well as for private use in domestic applications.

1. Introduction

Sun drying is the most common, economical and environmentally friendly method of preservation of agricultural produces in the tropical and sub-Saharan African regions. Open sun drying, however, has some disadvantages, of which are, undue exposure of produce to weather elements such as rain, ultraviolet rays of the sun, and contamination by wind-borne dirt and dust. Others include infestation by insects, rodents, pests and other animals as well as degradation by mucus and bacteria. These, for a long time, had made agricultural produce consumed locally, and those meant for export to fall short of the required hygienic and international standards, and hence, they are very often uncompetitive. Obviously, lack of effective preservation technology is a major stumbling block for farmers in the tropics and sub-Saharan African regions, as electricity is either inadequate or unaffordable to the small-scale farmers, thus discouraging refrigeration of the farm produce. This has left most farmers, particularly those engaged in small- and medium-scale farming, with no other option than to depend on one of the most abundant and free primary energy resources—the sun. Faced with the challenges of inadequate supply of electricity, on one hand, and the need to make farm produce competitive in the international market, on the other hand, the closest solution option is the use of solar dryer. This drying technique is in no way a new concept. Over the past four decades, there has been a steady evolution of solar dryer technology from natural convection types (direct and passive mode) to forced convection types [1–9].

The interest in the passive mode of drying rather than the direct type is the advantage it offers over the latter as it protects the produce from the harmful effects of the ultraviolet radiation of the sun, which destroys some nutrients and vitamins in the produce. Numerous works on natural convection solar types have been done by various researchers [1–3, 5, 6, 8, 9].

In most of these studies, emphasis has been on the construction and testing of the system while, in some other cases, the drying analysis and/or simulation was also done. This study presents the thermal analysis as well as the system's evaluation during drying of plantain fillets. The next section

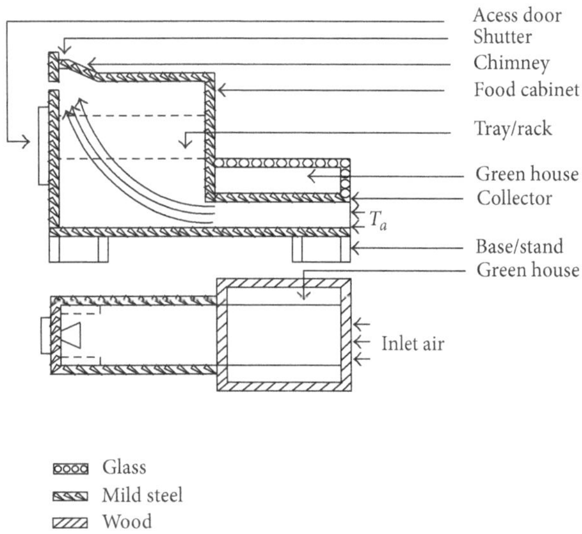

FIGURE 1: Diagram of solar dryer.

deals with the description of dryer and material selection followed by the design analysis in Section 3. Experimental procedure and testing are discussed in Section 4, while results and the discussion are in Section 5. Section 6 presents the cost analysis, and the conclusions drawn in Section 7.

2. Description of Dryer and Material Selection

The system is an L-shaped solar dryer which operates on the principle of natural convection (Figure 1). The base, which comprises the collector channel and collector case compartment, is connected to the food cabinet. The collector and cabinet are made of mild steel painted black on the inside and outside surfaces. This enables maximum absorption and emission of radiant heat energy by the system. The paint also serves to reduce the tendency of the metal to corrode. To facilitate the build-up of heat in the system, the collector is covered with a box which has a transparent glass at the top surface—greenhouse effect. The inlet air comes in through the metal collector channel heated and naturally rises through the food cabinet and exits through the shutter of the chimney, which is adjusted between 42 cm^2 and 84 cm^2 space in order to control the flow and the temperature in the cabinet. The product, plantain, is cleaned and cut into slices with a thickness of about 3 mm and weighed and spread on the first rack of the cabinet over which the hot air from the collector flow and remove moisture from the produce in the process.

This solar dryer is clearly different from most of the natural convection types in that it receives useful heat energy through indirect mode as a result of heated air flowing through the solar collector by buoyancy effect. Radiant heat energy is additionally absorbed by this cabinet without the exposure of the produce (article intended for drying) to the ultraviolet rays from the sun. However, some available indirect-mode natural convection solar dryers are made of plywood, but the collector covers are made of transparent glass [8]. This receives useful heat into the cabinet through the collector. On the other hand, the mixed-mode solar dryers [8, 9] are made of glass; both the collector and the cover of the cabinet are made of glass. The disadvantage these may have, though they seem to be more efficient than the indirect mode, is that the ultraviolet rays of the sun reaching the produce may destroy some nutrients and vitamins.

TABLE 1: Design considerations and assumptions.

S/N	Parameters	Description
1	Location	Lagos, (long 3°24′, lat 6°27′N)
2	Drying period	October
3	Solar irradiation, Lagos (October)	15.4 MJ/m^2/hr [10]
4	Annual relative humidity of Lagos taken over a ten-year period	80% [11]
5	Food products	Plantain
6	Mode of heating	Indirect
7	Number of glazing	1
8	Glazing materials	Glass
9	Loading provision	Door at the back of cabinet
10	Number of trays	2
11	Air outlet provision	Chimney at the top of cabinet
12	Air circulation mode	Natural convection
13	Drying capacity	10 kg
14	Thickness of plantain fillets	3 mm
15	Construction materials	Wood, glass, mild steel sheet
16	Insulation used	Glass fiber
17	Thickness of mild steel sheet	3 mm
18	Thickness of glass	4.5 mm
19	Thermal conductivity of glass, W/mK	0.84 [12]
20	Transmittance of glass	0.81
21	Emissivity of: glass	0.94 at 72°F [13]
22	Mild steel	0.2–0.32 [13]
23	Paint, flat black lacquer	0.96–0.98 [13]
24	Black or white lacquer	0.80–0.95 [13]
25	Thermal conductivity of mild steel, W/m°K at 38°C	50.0
26	Sky temperature	32°C
27	Dimension of collector case	1.0 m × 1.0 m × 0.2 m
28	Dimension of collector channel	1.0 m × 0.5 m × 0.1 m
29	Initial moisture content of plantain, m_i	70% [14]

The design parameters and consideration used for the construction of the equipment are listed in Table 1. The properties of air used are evaluated at 312.54 K [15], being the mean of the temperatures of air at the inlet and exit of the cabinet taken over a period of eight days during no load conditions. The average temperatures recorded in cabinet T_d, collector T_c, collector case (or greenhouse chamber) T_{ca}, and environment T_a are 41.56°C, 47.38°C, 48.81°C, and 31.69°C, respectively.

3. Design Analysis

In this work, focus will be on the thermal and drying analyses only.

3.1. Thermal Analysis. The performance of a solar system and collector is described by an energy balance that indicates the conversion of incident solar energy into useful energy gain and various losses. The thermal analysis here follows the procedures as in Duffie and Beckman [16].

Some Properties of Air. The following air properties taken from the average temperature in the cabinet are used for the analysis (extracted from Holman [15]): $\rho = 1.1324\,\text{kg/m}^3$, $c_p = 1.0066\,\text{kJ/kg·°C}$, $\mu = 1.9036 \times 10^{-5}\,\text{kg/m·s}$, $\nu = 16.9616 \times 10^{-6}\,\text{m}^2/\text{s}$, $k = 0.02719\,\text{W/m·°C}$, $\text{Pr} = 0.70524$. These properties are used to calculate the thermal characteristics and performance of the system.

Grashof Number, Gr

$$\text{Gr} = \frac{g\beta L^3 \Delta T}{\nu^2}, \tag{1}$$

where $\nu = \mu/\rho$, $\beta = 1/T$, $\Delta T = T_c - T_a$, and $T = (T_c + T_a)/2$. L depends on the section of the cabinet under analysis. The heat transfer coefficient is represented as

$$h = \frac{k\,\text{Nu}}{L}, \tag{2}$$

where the following Nusselt numbers Nu were used depending on the section or area of the cabinet under consideration.

For horizontal plate:

Above heated plate:

$$\text{Nu} = 0.54\,\text{Ra}_L^{1/4} \quad \text{for } 10^4 \le \text{Ra}_L \le 10^9, \tag{3a}$$

$$\text{Nu} = 0.15\,\text{Ra}_L^{1/3} \quad \text{for } 10^7 \le \text{Ra}_L \le 10^{11}. \tag{3b}$$

Below heated plate:

$$\text{Nu} = 0.274\,\text{Ra}_L^{1/4} \quad \text{for } 10^5 \le \text{Ra}_L \le 10^{10}. \tag{3c}$$

For vertical heated plate:

$$\text{Nu} = 0.68 + \frac{0.67\,\text{Ra}_L^{1/4}}{\left(1 + (0.492/\text{Pr})^{9/16}\right)^{4/9}} \qquad \text{Ra}_L \le 10^9, \tag{3d}$$

where $\text{Ra}_L = \text{Gr} * \text{Pr}$. However, the analysis will not be shown in detail as the values of the heat transfer are more important and are calculated using the above Nusselt numbers. In all cases, the air in the environment is assumed to be still; therefore, the heat transfer coefficients, at the outside of the dryer, are taken as zero. Also, the analysis is based on the assumption that the shutter is fully opened; that is, the cross-sectional area is 0.0084 m^2.

3.1.1. Collector Analysis. The above properties of air are used for the analysis of the collector.

Heat Absorbed by Radiation in the Collector $Q_{\text{rad,in}}$:

$$Q_{\text{rad,in}} = \alpha A_c I, \tag{4}$$

where, α is the absorptivity of the metal; A_c is the surface area of the collector exposed to sunshine, and it is equivalent to 0.7 m^2. I is the total incident solar input, and it is expressed as

$$I = \frac{\text{hourly radiation}}{\text{daily radiation}} \times \text{Solar irradiation}, \tag{5}$$

where the ratio of the hourly and daily radiation is approximately 0.17 (1/6) [10].

Heat Loss by Convection and Conduction through the Base of the Collector $Q_{\text{cb,loss}}$:

$$Q_{\text{cb,loss}} = \frac{\Delta T_{\text{cb}}}{\sum R_{\text{cb}}}, \tag{6}$$

where ΔT_{cb} is the temperature difference between the collector and the environment. $\sum R_{\text{cb}} = 1/h_{\text{cb}}A_{\text{cb}} + x/kA_{\text{cb}}$ is the thermal resistance offered by the wall and film at the inside of the collector taking the heat transfer coefficient at the outside of the wall as zero (still air). A_{cb} is the base area of the collector and its value is 0.5 m^2. h_{cb} is calculated from (2) and (3b).

Useful Heat in the Collector Q_u:

$$Q_u = \dot{m}c_p(T_c - T_a) = Q_{\text{rad,in}} - Q_{\text{cb,loss}}. \tag{7}$$

Collector Efficiency η_c:

$$\eta_c = \frac{Q_u}{Q_{\text{rad,in}}} \times 100\%. \tag{8}$$

3.1.2. Cabinet Analysis. For the analysis of the drying cabinet, the properties of the air are evaluated at the bulk temperature $T = (47.38 + 41.56)/2 + 273\,°\text{C}$, $\beta = 1/T = 3.15 \times 10^{-3}\,\text{K}^{-1}$, $\Delta T = 47.38\,°\text{C} - 41.56\,°\text{C}$, $\text{Pr} = 0.7042$, $k = 0.02756\,\text{W/m}\cdot°\text{C}$, $\rho = 1.1147\,\text{kg/m}^3$, $c_p = 1.0066\,\text{kJ/kg}\cdot°\text{C}$, $\mu = 1.9261 \times 10^{-5}\,\text{kg/m}\cdot\text{s}$, $\nu = 17.463 \times 10^{-6}\,\text{m}^2/\text{s}$.

Heat Absorbed by Radiation in the Cabinet $Q_{\text{rad,in}}$:

$$Q_{\text{rad,in}} = \alpha A_d I, \tag{9}$$

where, $A_d = 1.5416\,\text{m}^2$ is the total area of the cabinet through which radiant heat is absorbed.

Heat Transfer through the Wall of the Collector Case (Greenhouse) into the Cabinet $Q_{\text{gc,in}}$:

$$Q_{\text{gc,in}} = \frac{\Delta T_{\text{gc}}}{\sum R_{\text{gc}}}, \tag{10}$$

where, $\Delta T_{\text{gc}} = T_{\text{ca}} - T_d$ and $\sum R_{\text{gc}} = 1/h_{\text{gc}}A_{\text{gc}} + x/kA_{\text{gc}}$. h_{gc} is the heat transfer coefficient at the inside vertical wall of the cabinet and it is calculated using (2) and (3d). $A_{\text{gc}} = 0.05\,\text{m}^2$ is the area of the portion of the vertical wall separating the cabinet and the collector case.

Heat Radiated to the Environment by Cabinet $Q_{\text{rad,out}}$:

$$Q_{\text{rad,out}} = \varepsilon_m \sigma A_d \left(T_d^4 - T_a^4\right), \tag{11}$$

where, ε_m is the emissivity of the metal, $\sigma = 5.669 \times 10^{-8}\,\text{W/m}^2.\,K^4$ is the Stefan-Boltzmann constant and $A_d = 1.5416\,\text{m}^2$ is the total surface area of the cabinet through which heat is exchanged by radiation with the environment.

Heat Transfer from the Cabinet through the Side Wall Q_{vc}:

$$Q_{\text{vc}} = \frac{\Delta T_{da}}{\sum R_{\text{vc}}}, \tag{12}$$

where $\Delta T_{da} = T_d - T_a$ and $\sum R_{\text{vc}} = 1/h_{\text{vc}}A_{\text{vc}} + x/kA_{\text{vc}}$. $A_{\text{vc}} = 1.2916\,\text{m}^2$ is the area through which heat is transferred by convection and conduction from the side walls of the cabinet to the environment. h_{vc} is the heat transfer coefficient on the inside of the cabinet around the specified area and it is obtained using (2) and (3d).

Heat Transfer from the Cabinet through the Top Wall Q_{tc}:

$$Q_{\text{tc}} = \frac{\Delta T_{da}}{\sum R_{\text{tc}}}, \tag{13}$$

where, $\Delta T_{da} = T_d - T_a$ and $\sum R_{\text{tc}} = 1/h_{\text{tc}}A_{\text{tc}} + x/kA_{\text{tc}}$. $A_{\text{tc}} = 0.25\,\text{m}^2$ is the area through which heat is transferred from the lower side of the top of the cabinet to the environment. h_{tc} is the heat transfer coefficient on the inside of the cabinet around the specified area and is obtained from (2) and (3c).

Heat Transfer from the Cabinet through the Bottom Wall Q_{bc}:

$$Q_{\text{bc}} = \frac{\Delta T_{da}}{\sum R_{\text{bc}}}, \tag{14}$$

where $\Delta T_{da} = T_d - T_a$ and $\sum R_{\text{bc}} = 1/h_{\text{bc}}A_{\text{bc}} + x/kA_{\text{bc}}$. $A_{\text{bc}} = 0.25\,\text{m}^2$ is the area through which heat is transferred from the base wall of the cabinet to the environment. h_{bc} is the heat transfer coefficient at the top surface of the base wall inside

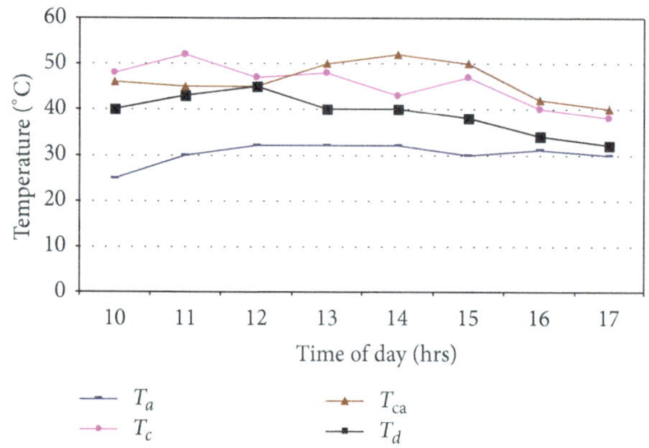

FIGURE 2: Ambient, collector, collector case, and food cabinet temperatures for Day 1.

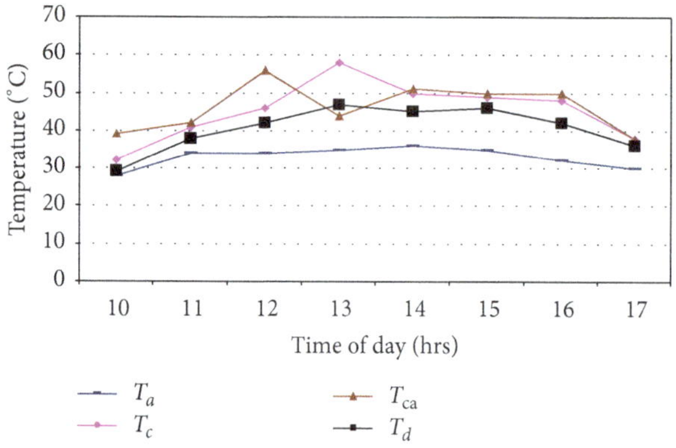

FIGURE 3: Ambient, collector, collector case and food cabinet temperatures for Day 2.

the cabinet around the specified area, and can be evaluated from (2) and (3a).

Total Heat Loss through the Walls of the Cabinet Q_{loss}:

$$Q_{\text{loss}} = Q_{\text{vc}} + Q_{\text{tc}} + Q_{\text{bc}} + Q_{\text{rad,out}}. \tag{15}$$

Useful Heat in the Cabinet:

$$Q_u = \dot{m}c_p\left(T_c - T_d\right) = Q_{\text{rad,in}} + Q_{\text{gc,in}} - Q_{\text{loss}}. \tag{16}$$

System Efficiency η_s:

$$\eta_s = \frac{Q_u}{Q_{\text{rad,in}} + Q_{\text{gc,in}}} \times 100\%. \tag{17}$$

3.2. Drying Analysis

Percentage Moisture Removed from Product, γ%:

$$\gamma\% = \frac{(m_w - m_d)}{m_w} 100\% \text{ wet basis}, \tag{18}$$

where m_w and m_d are the masses of wet and dried products, respectively.

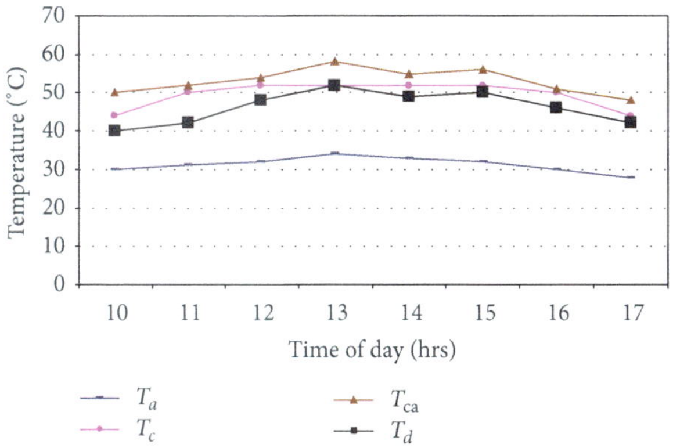

FIGURE 4: Ambient, collector, collector case, and food cabinet temperatures for Day 3.

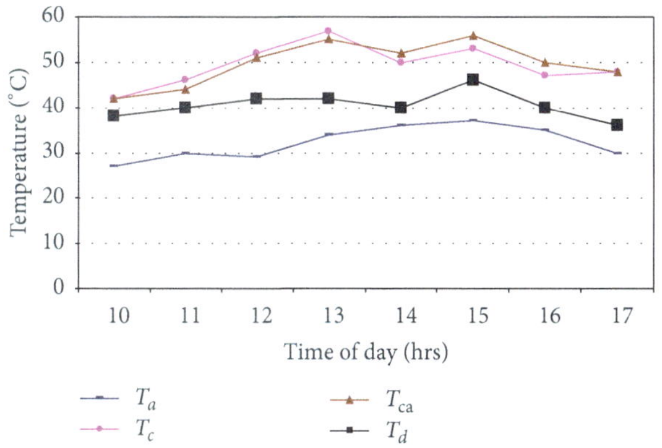

FIGURE 5: Ambient, collector, collector case, and food cabinet temperatures for Day 4.

Final Percentage Moisture Content of Product, m_f:

$$m_f = \frac{(100 - \gamma)}{100} m_i, \tag{19}$$

where m_i is the initial mass of the product.

Amount of Moisture Content Removed, m:

$$m = \frac{m_w (m_i - m_f)}{100 - m_f}. \tag{20}$$

Quantity of Heat Used in Evaporating Moisture, q_e:

$$q_e = m h_{fg}, \tag{21}$$

where h_{fg} is the latent heat of vaporisation and it is defined by Youcef-Ali et al. [19] as:

$$h_{fg} = 4.186 \times 10^3 (597 - 0.56T_d). \tag{22}$$

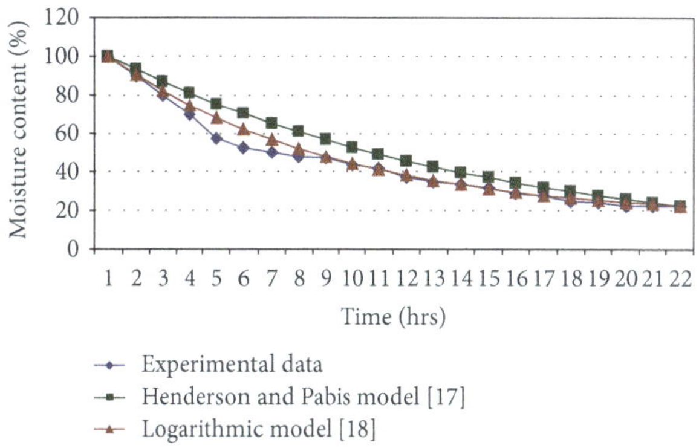

FIGURE 6: Percentage moisture content removal in the dried plantain fillet compared with two drying models.

Average Drying Rate, m_{ave}:

$$m_{ave} = \frac{m}{t}, \tag{23}$$

where t is daily sunshine hours.

Drying Time, δ:

$$\delta = \frac{m h_{fg}}{3600\, I A_T t \eta_c \eta_s}. \tag{24}$$

4. Experimental Procedure and Testing

The assembly was carefully checked to minimise heat loss. It was positioned in such a way that the food cabinet does not cast a shadow on the collector. The plantain fingers were sliced to about 3 mm thickness and were weighed to determine the initial mass. Four thermometers were placed at strategic locations on the assembly to measure the temperatures of the ambient, collector, collector case, and cabinet. The chimney shutter was adjusted to between 42 cm^2 and 84 cm^2 space in order to control the mass flow rate, hence, the temperature within the cabinet. The experiment was carried out in October. The first reading was taken at about 10:00 each day, when the sun was just coming up and subsequently on an hourly basis till around 17:00.

5. Results and Discussion

The general trend in the profiles of the temperatures obtained shows increases from 10:00 to between 13:00 and 15:00 then decline (Figures 2, 3, 4 and 5). The collector channel and collector case temperatures have higher values than the cabinet and ambient. Though the collector case temperature is always higher, the collector had higher values in Figures 2 and 5 before 14:00. Maximum temperatures of 58°C, 58°C, 52°C, and 37°C were recorded for the collector, collector case, cabinet, and ambient, respectively. Of particular interest is the temperature obtained in the

TABLE 2: Performance evaluation of the system.

S/N	Parameters	Values	Equations used
1	Heat absorbed by radiation into the collector, $Q_{\mathrm{rad,in}}$	132.35 W	(4)
2	Total solar input, I	727.22 W/m^2	(5)
3	Heat loss by conduction and convection through the base of the collector, $Q_{\mathrm{cb,loss}}$	70.99 W	(6)
4	Useful heat in *collector*, Q_u	61.36 W	(7)
5	Collector efficiency, η_c	46.4%	(8)
6	Heat absorbed by radiation in cabinet, $Q_{\mathrm{rad,in}}$	291.48 W	(9)
6	Heat transfer from collector case into cabinet, $Q_{\mathrm{gc,in}}$	0.97 W	(10)
7	Heat radiated by cabinet surface, $Q_{\mathrm{rad,out}}$	26.64 W	(11)
8	Heat transfer from cabinet side wall by cond/conv, Q_{vc}	27.94 W	(12)
9	Heat transfer from cabinet top wall by cond/conv, Q_{tc}	2.74 W	(13)
10	Heat transfer from cabinet bottom by cond/conv, Q_{bc}	4.89 W	(14)
11	Total heat loss from cabinet, Q_{loss}	62.21 W	(15)
12	Useful heat in *cabinet*, Q_u	230.24 W	(16)
13	System efficiency, η_s	78.73%	(17)
14	Percentage moisture removed from product, γ%	77.5%	(18)
15	Final moisture content of product, m_f	15.75%	(19)
16	Amount of moisture content removed, m	1.29 kg	(20)
17	Quantity of heat used in evaporation, q_e	3.1×10^6 MJ	(21)
18	Average drying rate, m_{ave}	0.184 kg/hr	(23)
19	Drying time, δ	(0.213/day)$^{-1}$	(24)

TABLE 3: Mathematical models applied to drying curves.

S/N	Model name	Model	k
1	Henderson and Pabis	$X_m = 100e^{-kt}$	0.071
2	Logarithmic	$X_m = \left(100 - m_f\right)e^{-kt} + m_f$	0.1202

cabinet, 52°C, which is in the range proposed as the optimum temperature for drying fruits and vegetables [20, 21].

From (1) to (24), the results used for the evaluation of the system are presented in Table 2. The collector efficiency obtained was 46.4% and system efficiency was 78.73%.

The collector efficiency obtained here is greater than 40% by Tarigan and Tekasakul [22]. Figure 6 shows the profile of the percentage moisture content. In general, it follows an exponentially decaying pattern before remaining constant at the 20th hour, which indicates about 78% moisture content removal. This profile pattern is exponentially decaying, and when compared with Henderson and Pabis [17] and Logarithmic models [18] constants, k was obtained as shown in Table 3. It can be seen that the Logarithmic model better predicts the moisture content profile when compared with the Henderson and Pabis model. A critical observation of the experimental data and the models shows some noticeable disparity between the second and ninth hour. A closer look further reveals that the maximum deviation takes place during the fifth and sixth hour. At the beginning of the experiment, it was easier for the produce to lose much of its water content than much later. Moreover, the rate of water moisture loss is proportional to the intensity of the heat energy, hence, the temperature of the cabinet which was always at its peak between thirteen and fifteen hours. Before and after this peak period, the rate of loss was low. It was during this initial period that the model failed to predict the experimental data. The cost of constructing a unit system from locally available materials is tabulated in (Table 4).

6. Cost Analysis

The materials were locally available, and the cost for the construction of a unit is tabulated below in the local currency, Naira, and the equivalent in US dollars. The cost per unit is affordable for small and medium manufacturing enterprises (SMMEs) [23].

7. Conclusion

The analyses are comprehensive but exclude the structural aspect, mainly because the system is not really a heavy-load carrying structure. Assessing the design based on the performance evaluation during the testing, collector and system efficiencies of 46.4% and 78.73%, percentage moisture removal of 77.5%, final moisture content of plantain chips of 15.75%, and average drying rate of 0.184 kg/hr show that the system is both efficient and effective. However, mass production of the system will further reduce the unit cost. To encourage domestic applications in the rural communities, government could subsidise or give a loan with low interest rates.

TABLE 4: Cost of constructing a unit solar dryer.

Material	Dimension	Quantity	Unit cost, N	Total, N	Total, $
Mild steel	$0.1 \times 0.2\,m^2$	2	6,600	13,200	88
Angle iron	0.2 m	3	1,500	4,500	30
Perspex	$0.12 \times 0.24\,m^2$	1/2	12,000	6,000	40
Wire gauze	$0.1 \times 0.2\,m^2$	1/2	4,000	2,000	13.3
Wood	$0.02 \times 0.1\,m^2$	4	275	1,500	10
Sealant		1	1,200	1,200	8
Paint		1	800	800	5.3
Grand total				29,200	194.6

For exchange rate of 1 US$ = N150.00.

Nomenclature

A: Surface area, m^2
c_p: Specific heat at constant pressure, kJ/kg·°C
g: Acceleration due to gravity, m/s^2
Gr: Grashof number
h: Heat transfer coefficient, $W/m^2 \cdot {}^\circ C$
h_{fg}: Latent heat of vaporisation, MJ/kg
I: Solar insolation of collector, W/m^2
k: Thermal conductivity, W/m·°C
L: Length, m
$\dot{m}$: Mass flow rate, kg/s
m: Mass, kg
Nu: Nusselt number
Pr: Prandtl number
q_e: Heat of vapourisation, MJ
Q: Heat transfer rate, W
Ra_L: Rayleigh number
x: Thickness of metal sheet, m
X_m: Percentage moisture content at any time t, %
t: Sunshine hours (Time), hrs
T: Temperature, °C
ΔT: Temperature difference, °C
ΣR: Total thermal resistance, °C/W.

Greek Letters

ν: Kinematic viscosity, $m^2 s$
μ: Dynamic viscosity, kg/m·s
β: Coefficient of thermal expansion K^{-1}
δ: Drying time, days.
τ: Transmittance
ρ: Density, kg/m^3
α: Absorptivity
ε: Emissivity
σ: Stefan-Boltzmann constant, W/m^2K^4
η: Efficiency, %
γ: Moisture removal, %

Subscript

a: Ambient
ave: Average
bc: Base of cabinet
c: Collector
ca: Collector case
d: Drying cabinet, dried product
da: Drying cabinet to environment
f: Final
gc: Collector case (greenhouse) to cabinet
i: Inlet, initial
m: Metal
loss: Heat loss to the environment
rad out: Radiation outward
rad in: Radiation into
tc: Top of cabinet
T: Total
u: Useful
vc: Vertical sides of cabinet
w: Wet product.

References

[1] B. K. Bala, M. A. Ashraf, M. A. Uddin, and S. Janjai, "Experimental and neural network prediction of the performance of a solar tunnel drier for drying jackfruit bulbs and leather," *Journal of Food Process Engineering*, vol. 28, no. 6, pp. 552–566, 2005.
[2] V. K. Sharma, A. Colangelo, and G. Spagna, "Experimental investigation of different solar dryers suitable for fruit and vegetable drying," *Renewable Energy*, vol. 6, no. 4, pp. 413–424, 1995.
[3] S. Janjai, N. Srisittipokakun, and B. K. Bala, "Experimental and modelling performances of a roof-integrated solar drying system for drying herbs and spices," *Energy*, vol. 33, no. 1, pp. 91–103, 2008.
[4] A. O. Adelaja and S. J. Ojolo, "Design, analysis and experimental evaluation of photovoltaic forced convection solar dryer for the tropics," *International Journal of Engineering Research in Africa*, vol. 3, pp. 49–61, 2010.
[5] M. M. I. Choudhury, B. K. Bala, and M. A. Hague, "Energy and exergy analysis of the solar drying of jackfruit leather," *Biosystems Engineering*, vol. 110, pp. 222–229, 2011.
[6] A. O. Adelaja, B. Y. Ogunmola, and P. O. Akolade, "Development of a photovoltaic powered forced convection solar dryer," *Advanced Materials Research*, vol. 62–64, pp. 543–548, 2009.
[7] A. O. Adelaja, O. S. Asemota, and I. K. Oshiafi, "Experimental determination of the moisture content pattern in yam during drying," *Journal of Applied Science Research*, vol. 6, no. 8, pp. 1171–1181, 2010.

[8] I. N. Simate, "Optimization of mixed-mode and indirect-mode natural convection solar dryers," *Renewable Energy*, vol. 28, no. 3, pp. 435–453, 2003.

[9] B. O. Bolaji and A. P. Olalusi, "Performance evaluation of a mixed-mode solar dryer," *AU Journal of Technology*, vol. 11, no. 4, pp. 225–231, 2008.

[10] R. O. Fagbenle, "Optimum collector tilt angles and average annual global radiation for Nigerian locations," *Nigerian Journal of Renewable Energy*, vol. 2, pp. 9–17, 1991.

[11] R. L. Fagbenle, "A comparative study of some simple models for global solar irradiation in Ibadan, Nigeria," *International Journal of Energy Research*, vol. 16, no. 7, pp. 583–595, 1992.

[12] R. S. Khurmi and J. K. Gupta, *A Textbook of Machine Design (S. I. Units), Manufacturing Consideration in Machine Design and Simple Stresses in Machine Parts*, 1979.

[13] W. H. McAdams, *Heat Transmission*, McGraw-Hill, New York, NY, USA, 3rd edition, 1954.

[14] F. B. Mukhtar, "Effect of storage temperatures on post harvest deterioration of banana and plantain (*Musa* spp.)," *International Journal of Pure and Applied Sciences*, vol. 3, no. 2, pp. 28–38, 2009.

[15] J. P. Holman, *Heat Transfer*, Tata McGraw-Hill, 9th edition, 2002.

[16] J. A. Duffie and W. A. Beckman, *Solar Engineering of Thermal Processes*, John Wiley and Sons, 2nd edition, 1980.

[17] A. Yagcioglu, A. Degirmencioglu, and F. Cagatay, "Drying characteristics of lauel leaves under different conditions," in *Proceedings of the 7th International Congress on Agricultural Mechanization and Energy (ICAME '99)*, pp. 565–569, Faculty of Agriculture, Adana, Turkey, 1999.

[18] O. Yaldiz, C. Ertekin, and H. I. Uzun, "Mathematical modeling of thin layer solar drying of sultana grapes," *Energy*, vol. 26, no. 5, pp. 457–465, 2001.

[19] S. Youcef-Ali, H. Messaoudi, J. Y. Desmons, A. Abene, and M. Le Ray, "Determinationof the average coefficient of internalmoisture transfer during the drying of a thin bed of potato slices," *Journal of Food Engineering*, vol. 48, no. 2, pp. 95–101, 2001.

[20] A. Henneman, Drying Vegetable adapted from 'So Easy to Preserve' Cooperative Extension", The University of Georgia, 1993, and Michigan Cooperative Extension Food Preservation Handbook, 1994.

[21] J. Mumba, "Development of a photovoltaic powered forced circulation grain dryer for use in the tropics," *Renewable Energy*, vol. 6, no. 7, pp. 855–862, 1995.

[22] E. Tarigan and P. Tekasakul, "A mixed-mode natural convection solar dryer with biomass burner and heat storage back-up heater," in *Proceedings of the Australia and New Zealand Solar Energy Society Annual Conference*, pp. 1–9, 2005.

[23] F. O. Okafor, "Micro-credit: an instrument for economic growth and balance development," *Journal of the Management Sciences*, vol. 3, no. 2, pp. 24–39, 2000.

12

Influences of Contact Pressure on the Performances of Polymer Electrolyte Fuel Cells

Prakash C. Ghosh

Department Energy Science and Engineering, Indian Institute of Technology Bombay, Mumbai 400076, India

Correspondence should be addressed to Prakash C. Ghosh; pcghosh@iitb.ac.in

Academic Editor: Ching Yuan Chang

Fuel cells face major challenges in sustaining the laboratory-scale performance during the scale up. The contact resistance mainly arises from the dimensional mismatch between gasket and gas diffusion layer during scale up, which may cause diminution in performance. In the present work, experiment as well as modelling is carried out for different combinations of clamping force and gasket thickness. The polarisation behaviours of PEFCs configured under different clamping torques and gasket thicknesses are analysed. The combination of 0.3 mm gasket and 0.3 mm GDL under 3 Nm and 5 Nm clamping forces offers 480 mΩ cm^2 and 148 mΩ cm^2 contact resistances, respectively. The configurations under 3 Nm and 5 Nm clamping torques with 0.2 mm thick gasket offer contact resistances as low as 23 mΩ cm^2 and 11 mΩ cm^2, respectively. The polarisation behaviour obtained from the experiment of such configurations is found to be in good agreement with the modelling results.

1. Introduction

Fuel cells show great promises in the field of portable, transport and stationary applications [1–3]. They show excellent current voltage characteristics in laboratory-scale cell. However, researchers encounter challenges during scale up due to the heterogeneous local performances inside the fuel cells [4–6]. The local performance is influenced by the operating conditions such as humidity, temperature, and reactant pressure [7–9]. Assembly parameters such as clamping force and gasket thickness influence the local and overall performances. The contact resistance that mainly arises from the dimensional mismatch between the gasket and gas diffusion layer (GDL) plays a significant role in overall performance of the polymer electrolyte fuel cell.

Among different components, mainly matching in the thicknesses of the gasket and the GDL is important to achieve best performance in the fuel cells. Mismatch may lead to the following problems as illustrated in Figure 1.

(i) Thinner gasket may lead to sealing problem causing safety issue. In addition to that, the cell will be facing mass transport-related losses (Figure 1(b)).

(ii) On the other hand, thicker gasket may result in poor contact between the bipolar plate (BPP) and the GDL, causing a rise in the contact resistance in the cell which will be reflected on the ohmic region of the current voltage characteristics (Figure 1(c)).

Hence, it is important to achieve optimum matching between the gasket and the GDL used in polymer electrolyte fuel cells. During the last decade, researchers focused on estimating the contact resistance, the polymer electrolyte fuel cells (PEFCs) [11] as well as solid oxide fuel cells (SOFCs) [12, 13]. Mishra et al. [14] presented a generalized fractal asperity-based model to predict the contact resistances as a function of contact pressure, material properties, and surface morphology at the interface. Nitta et al. [15] developed a 3D model to analyse the effect of compression pressure on fuel cell, and Liu et al. [16] studied the contact pressure distribution as a function of assembly parameter using response surface methodology (RSM). Su et al. [17] and Lee et al. [18] simulated the contact pressure distribution on the membrane electrode assembly (MEA) of the fuel cell for various assembly pressures. Chang et al. [19] studied the long-term effects of the clamping pressure on the performance of the PEFCs. Zhou et al. [20] investigated the effect of clamping force on the interfacial contact resistance and the porosity of the gas diffusion layer (GDL) in a PEFC. Barber et al. [21] studied the clamping force and the interfacial morphology on interfacial contact

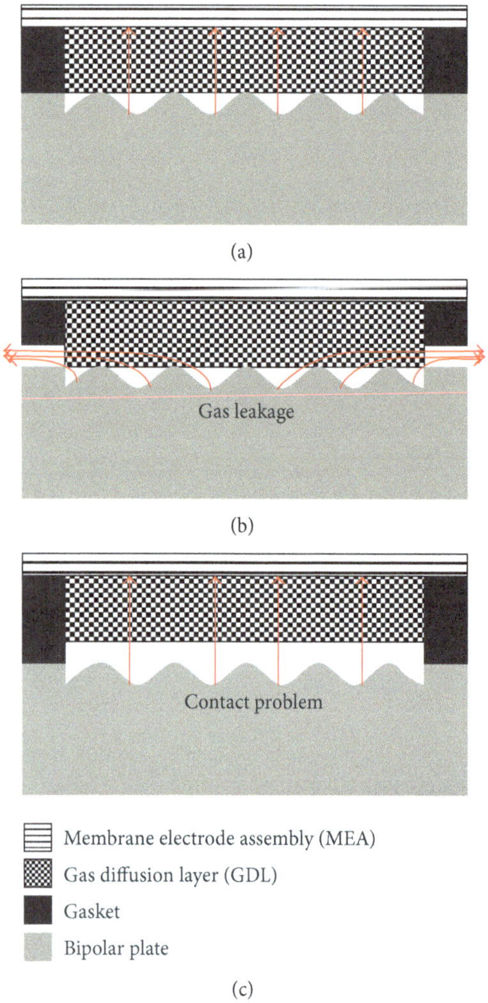

FIGURE 1: Schematic diagram of a half-cell configuration: (a) optimum, (b) with thinner gasket, and (c) with thicker gasket.

area. Xing et al. [22] estimated the optimum clamping force as a function of the operating voltages. The variation in the clamping pressure arising from the aging of the backbone of the gasket is studied [23]. Wang et al. [24] used pressure sensitive films between the MEA and the diffusion layer of a PEFC to study the effect of the end plates on the pressure distribution. Though extensive works are carried out on contact pressure inside fuel cell, author has not come across any study relating the dimensional mismatch between the GDL and gasket and performance of the fuel cells.

The present study aims at investigating the influences of the gasket system and the clamping forces on the performances of the PEFCs. In this purpose, the contact pressure distribution on the electrode surface is measured, and the area-specific contact resistances are estimated to model the performance and validated with experimental results.

2. Experimental Setup

Experimental setup consists of contact pressure measurement unit, single polymer electrolyte fuel cell, and leakage test setup unit. Commercially available pressure sensitive films are used to analyse the pressure distributions at the interfaces of different components inside the fuel cell as discussed below.

2.1. Cell Configuration. A polymer electrolyte fuel cell with an outer dimension of $13.5 \times 13.5\,\mathrm{cm}^2$ with an active area of $10 \times 10\,\mathrm{cm}^2$ is used for the present study. The flow fields on both electrodes consist of eight parallel serpentine channel of $1 \times 1\,\mathrm{mm}^2$ cross-sectional area incorporated on graphite bipolar plate. The MEA consists of cathode and anode electrode with platinum loading of $0.20\,\mathrm{mg\,cm}^{-2}$ and $0.18\,\mathrm{mg\,cm}^{-2}$, respectively. Toray carbon paper, TGP-120, of 0.3 mm thickness is used as GDL on both the electrodes, and perfluorosulfonic acid- (PFSA-) based membrane of thickness 183 μm is used as electrolyte. Silicon (0.5 mm thick), flexible graphite (0.3 mm and 0.5 mm thick), and polytetrafluoroethylene (PTFE) (0.2 mm, 0.3 mm, and 0.4 mm thick) are explored as gasket materials for interfacial contact pressure distribution analysis. The complete cell is assembled by sandwiching all the components between a pair of end plates (SS 316) using sixteen nut bolts of 6 mm diameter as shown in Figure 2(a). In the present geometry, since the gasket is placed at the periphery of the cell, they have been arranged around the cell.

2.2. Contact Pressure. Commercially available pressure sensitive films (Fujifilm, Prescale LLLW) effective in the pressure range from 0.2 MPa to 2.5 MPa are explored to map the interfacial pressure. The impressions obtained on the film are calibrated with the help of the standard software available with it, and the interfacial contact pressure distributions are estimated. During the assembly, mismatch in the thickness of the gasket and combined thickness of the electrode and GDL mainly leads to the overall resistance. Hence, the pressure sensitive films are placed between the gasket and the bipolar plate to acquire the pressure distribution between the GDL and BPP as shown in Figure 2.

2.3. Gas Sealing Characterisation. Since the actual operations of PEFCs involve hydrogen gas, proper sealing must be ensured before bringing into operation. Hydrogen-operated PEFCs pose two different types of safety concerns—escape of hydrogen from the fuel cell due to the improper sealing and mixing of hydrogen with oxygen/air inside the cell due to the defect (puncture) in the membrane. In the present investigation, an arrangement (Figure 3) is made to test the leak and puncture prior to the actual cell operation. The experimental setup is first explored to quantify the gas leakage from the cell followed by puncture test. Initially, the setup is calibrated by closing both the outlets without connecting them to fuel cell. Nitrogen/air is initially introduced in the system through the pressure regulator keeping all the ball valves in opened condition and all the needle valves in closed condition. The pressure is being controlled to achieve the actual operating condition of the fuel cell. Subsequently, the ball valves 2 and 3 are closed, and the variation in both pressure gauges is recorded with time to estimate the leakage in the setup. On achieving satisfactory sealing in the

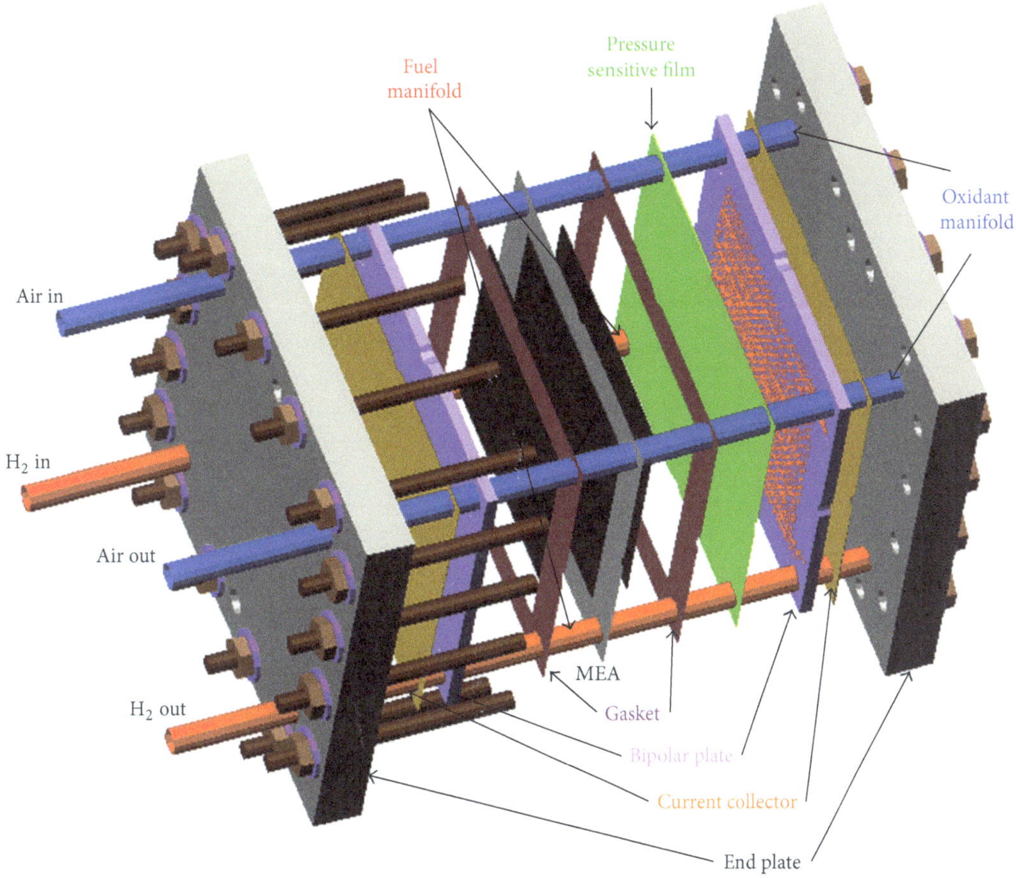

(a)

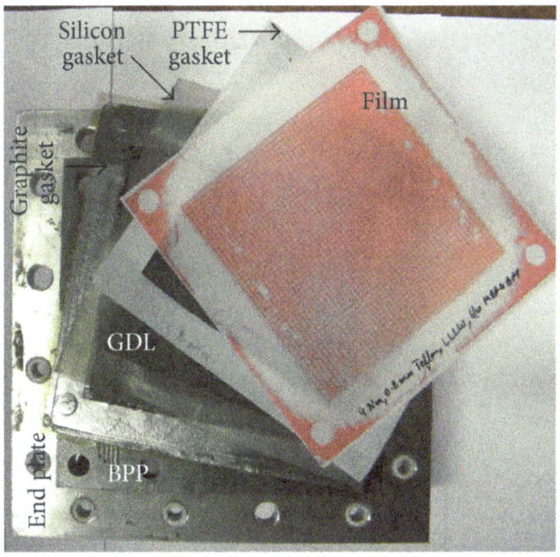

(b)

FIGURE 2: Experimental setup: (a) schematic diagram of the fuel cell with the pressure sensitive film and (b) fuel cell components.

experimental setup, it is connected to the gas inlets (anode and cathode) of the fuel cell assembly keeping the outlets at closed condition as shown in Figure 3. Afterward, the above-mentioned procedure of calibration is repeated with fuel cell assembly, and the variations in both pressure gauges are recorded with time to estimate the leakage in the setup.

On achieving satisfactory sealing in the fuel cell, the puncture test is being performed by applying a differential pressure across the membrane. For this purpose, one of the needle valve is being slightly opened for a short while to vent small amount of nitrogen from one side of the fuel cell assembly to develop the differential pressure.

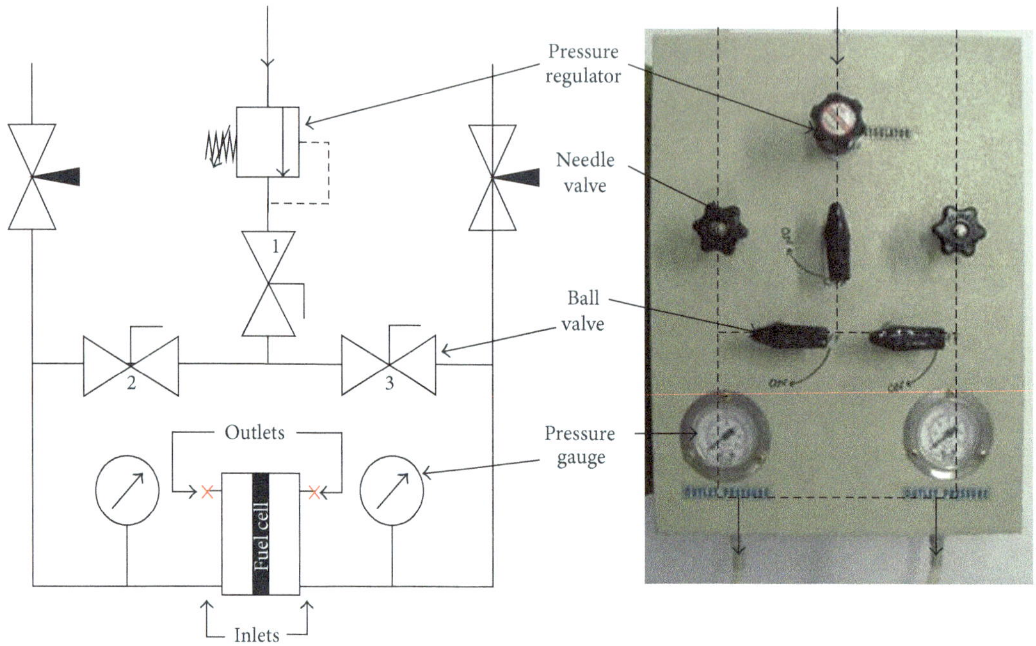

(a) (b)

FIGURE 3: Experimental arrangement for leak and puncture test: (a) schematic diagram and (b) hardware.

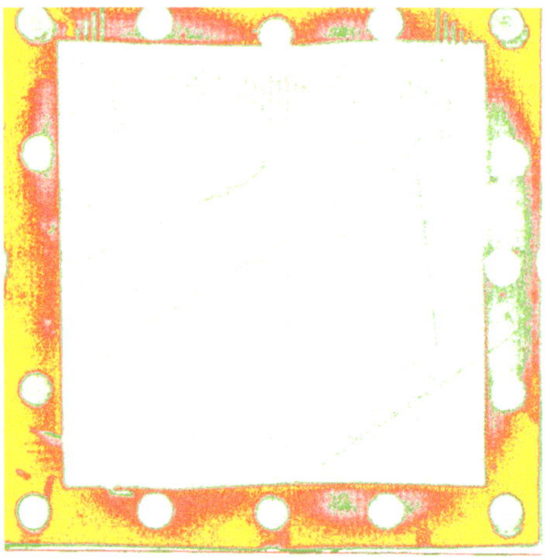

FIGURE 4: Pressure distribution between BPP and GDL under 5 Nm torque for 0.4 mm gasket [10].

3. Modelling

Performances of PEFCs are affected by activation loss, concentration loss, and ohmic loss. The activation overpotential is given by

$$\eta_{\mathrm{act}} = \frac{RT}{\alpha \cdot n \cdot F} \ln\left(\frac{i}{i_o}\right). \tag{1}$$

The overpotential arising from concentration loss is given by

$$\eta_{\mathrm{con}} = -\frac{RT}{n \cdot F} \ln\left(1 - \frac{i}{i_L}\right). \tag{2}$$

The ohmic overpotential is given by

$$\eta_{\mathrm{ohm}} = i \cdot r, \tag{3}$$

where r is the internal area-specific resistance (ASR) of the fuel cell, which is originated from bulk resistances (r_b) of the components used in fuel cell and contact resistances (r_c) between the adjacent components as expressed below:

$$r = r_c + r_b, \tag{4}$$

where

$$r_c = r_{\mathrm{GDL/BPP}} + r_{\mathrm{BPP/EP}}. \tag{5}$$

Among all the components, the ionic resistance of the electrolyte contributes with the lion's share in the bulk resistance of the of the cell. Hence, the area-specific bulk resistance can be estimated, neglecting the contribution of the electronic conductor as follows:

$$r_b = r_{\mathrm{MEM}} = \frac{L}{\sigma}. \tag{6}$$

The contact between the gas diffusion layer and the bipolar plate becomes poor due to the mismatch in the thicknesses of the gasket and the GDL leading to very high internal resistance. The contact resistance at the interface can be estimated from contact pressure ($P_{\mathrm{interface}}$) using following expression:

$$r_c = r' + r''\left(\frac{1}{P_{\mathrm{interface}}}\right)^C. \tag{7}$$

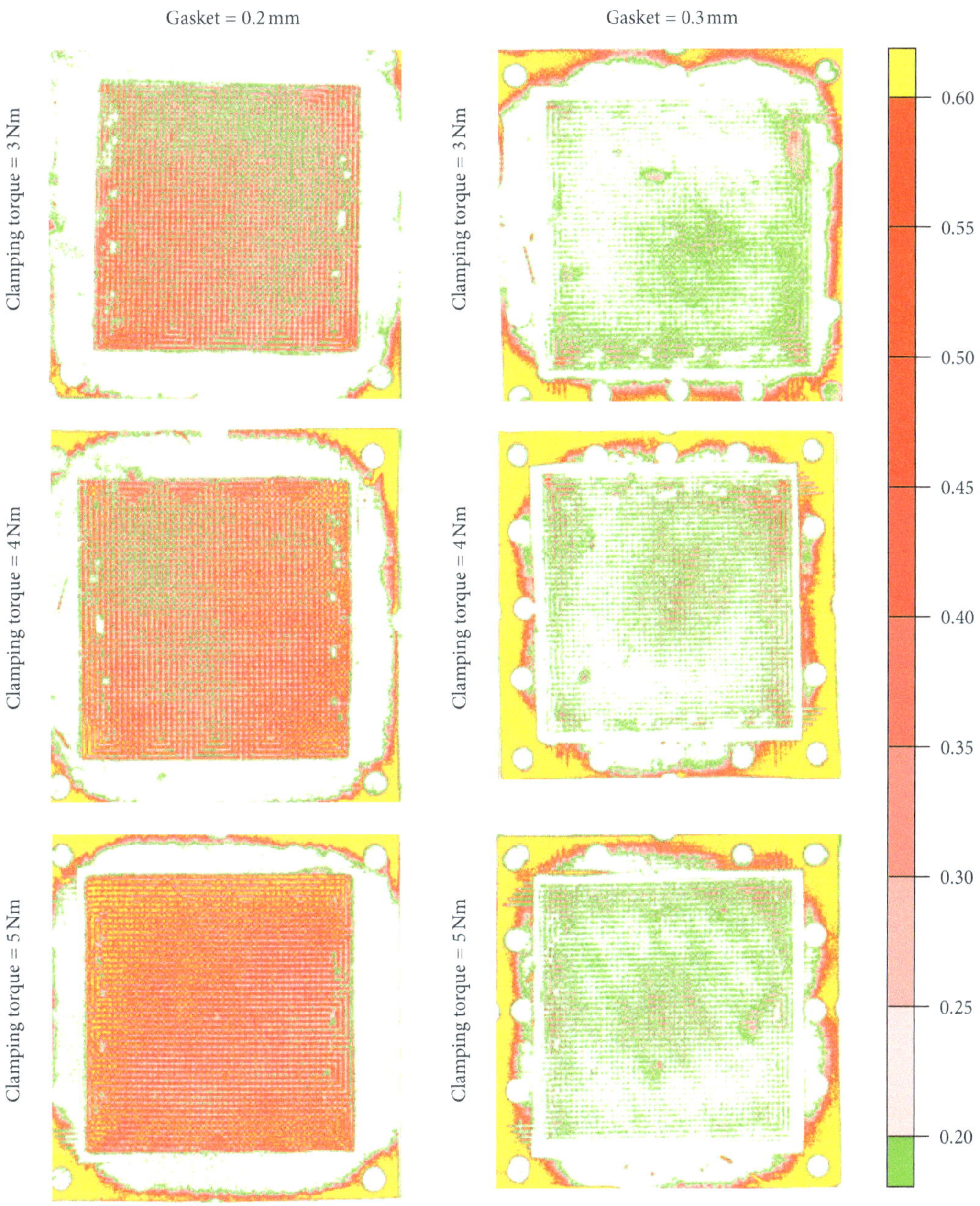

FIGURE 5: Contact pressure distribution (MPa) between GDL and BPP for different configurations.

The polarisation behaviour of a complete cell can be expressed considering all the losses given in (1)–(3) as follows:

$$E_{\text{rev}} = E^{o}_{\text{rev}} + \frac{RT}{n \cdot F} \ln \left(\frac{\prod_{i=\text{reactant}} (P_i)^{v_i}}{\prod_{j=\text{product}} (P_j)^{v_j}} \right) - \frac{RT}{\alpha \cdot n \cdot F} \ln \left(\frac{i}{i_o} \right) - i \cdot r + \frac{RT}{n \cdot F} \ln \left(1 - \frac{i}{i_L} \right). \quad (8)$$

4. Results and Discussions

The fuel cell is assembled by applying three different clamping torques of 3 Nm, 4 Nm, and 5 Nm which ensure clamping forces of 5.95 kN, 8.14 kN, and 11.18 kN, respectively, and the influences of the clamping force on the contact pressure distribution are analysed. In addition, the effect of the gasket thickness is investigated in the present study. The area-specific contact resistances (ASCRs) at the interfaces are modelled based on the measured contact pressure distributions. The sealing qualities for different gaskets are analysed and pertinently sealed fuel cell assemblies are picked out for polarisation analysis. The effect of the gasket thickness and the clamping force on the polarisation behaviour is studied and modelled.

4.1. Contact Pressure Distribution. PTFE gaskets with thickness of 0.2 mm, 0.3 mm, and 0.4 mm are preferred to assembling the cell because the thickness of the GDL is 0.3 mm.

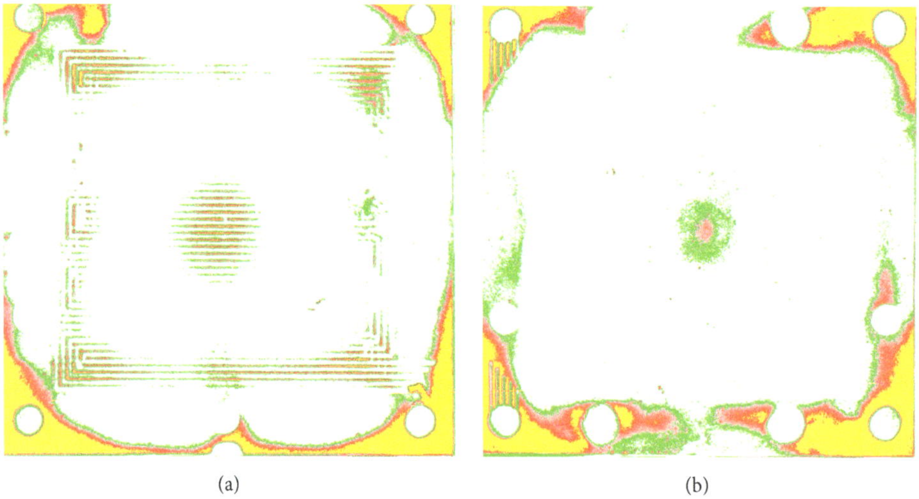

FIGURE 6: Contact pressure distribution (MPa) between (a) BPP and end plate and (b) two end plates.

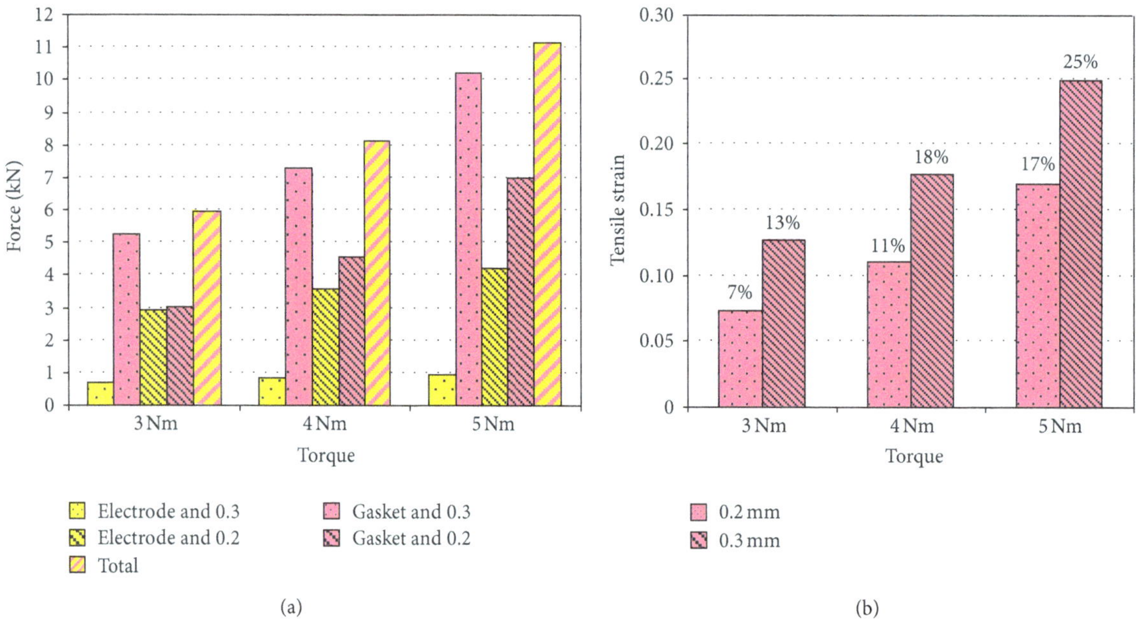

FIGURE 7: (a) Clamping force and (b) tensile strain of the PTFE gasket under different configurations.

It is observed that the assembly with 0.4 mm gasket fails to develop sufficient contact at the interface of GDL and BPP, and no impression other than the gasket is obtained on the pressure film as shown in Figure 4. Similar effects are observed in case of silicon and expanded graphite gasket. Hence, gaskets with thicknesses equal to 0.4 mm or more are discarded in further analysis. Hence, further analysis is carried out with PTFE gasket of thickness 0.3 mm and 0.2 mm and 0.3 mm expanded graphite.

In Figure 5, the contact pressure distribution between the BPP and GDL is shown for different clamping forces with 0.2 mm and 0.3 mm gaskets. The contact pressure distribution on the electrode surface is prominent for the gasket with 0.2 mm thickness. However, the impression of the gasket is not prominent, and it may result in poor reactant sealing. The assembly with 0.3 mm gasket provides very good sealing as the contact pressure distribution reaches to the maximum limit of the film. However, it fails to generate sufficient contact pressure on the electrode surface. Most of the regions experience pressures close to the lower range of the film, and some of the regions even experience pressure in the ranges below the minimum limit of the film causing no impression.

Generally, it is expected that the pressure in the middle should be minimum and increases as it approaches the edges. However, it is observed that the contact pressure is relatively higher at the centre of the electrode area in addition to the edges. It could be attributed to the marginal mechanical defect in machining the end plate resulted in

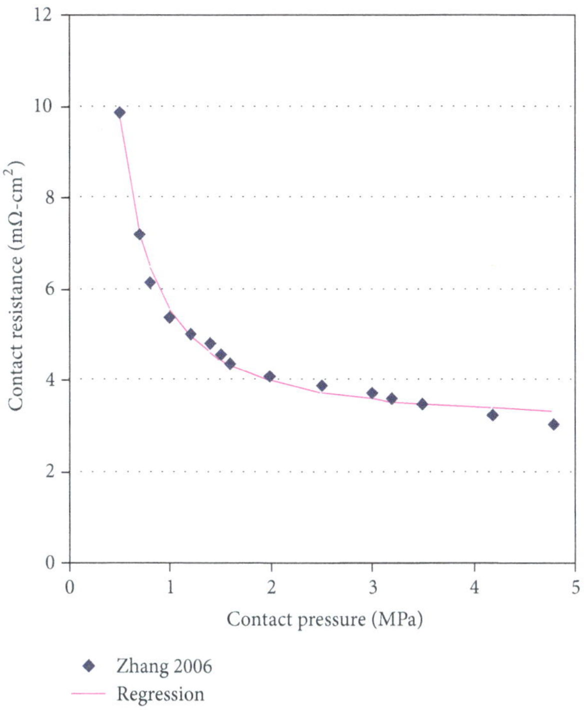

FIGURE 8: Modelling of contact resistance at the BPP and GDL interface.

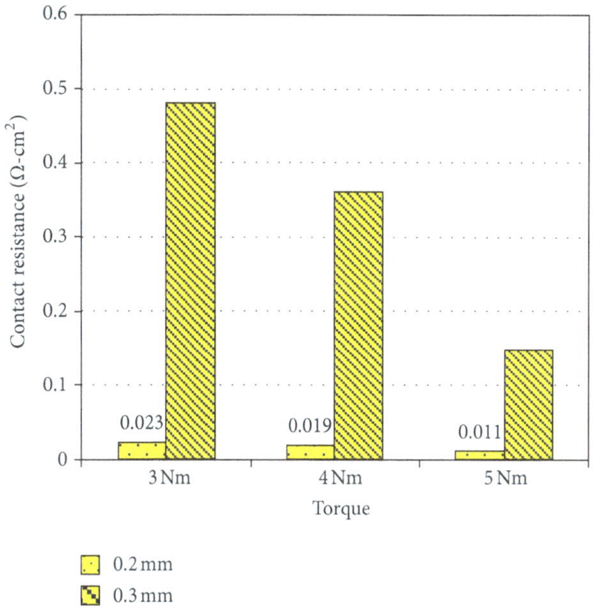

FIGURE 9: Contact resistances between the GDL and the BPP for different configurations.

slight uneven surface. The pressure distribution obtained only with end plates, and the BPPs are shown in Figure 6, which confirms the fact of machining defect. The present assembly with clamping forces around the fuel cell with a slight thicker dimension at the centre of the BPP helps to achieve relatively uniform clamping force distribution.

The share of the total clamping force, GDL, and gasket at different tightening torque is shown in Figure 7(a). For both gaskets, though the overall contact pressure increases with the increase in the clamping force, the share of the GDL decreases. In case of 0.2 mm gasket, the proportion of the force on GDL decreases from 49% to 38%, whereas, in case of 0.3 mm gasket, it decreases from 12% to 8%. Hence, it is clear that the lion's share of the force is translated to the gasket. In Figure 7(b), the tensile strains of the gaskets for different configurations are estimated considering the Young modulus of PTFE equal to 0.5 GPa.

4.2. Contact Resistance Distribution. The ASCR at the interface is modelled based on the experimental results reported by Zhang et al. [25] as shown in Figure 8. The variation in contact resistances with the contact pressure between the BPP and the carbon paper is modelled by regression analysis based on (8), which produces R^2 value of 0.992 for $r' = 3.05, r'' = 2.464$, and $c = 1.418$.

Based on the above-mentioned model, the average value of ASCR between the BPP and the GDL is estimated for different configurations as shown in Figure 9. The ASCR is found to be around 23 mΩ cm^2 and 11 mΩ cm^2 for assembly with 0.2 mm thick gasket under 3 Nm and 5 Nm, respectively. However, assembly with 0.3 mm gasket produces average ASCR around 480 mΩ cm^2 and 148 mΩ cm^2, under 3 Nm and 5 Nm, respectively. Hence, the contribution of interfacial contact resistance with 0.3 mm gasket assembly to the internal resistance becomes significant.

The ASCR distributions at the interface between the GDL and the BPP corresponding to Figure 5 are shown in Figure 10. It shows that a substantial portion of the GDL surface in the configuration with 0.3 mm gasket has very high resistance or no contact between the GDL and the BPP. On the other hand, assembly with 0.2 mm gasket generates homogeneous contact resistance on the GDL surface.

4.3. Sealing Test. The pressure film is removed from the configuration, presented in Figure 2(a), for studying the sealing and polarisation performance and assembled under different configurations as given in Figure 5. Sealing tests of the different configurations are carried out prior to the activation of the fuel cell with hydrogen and air for evaluating the polarisation characteristics. Initially, both chambers of the fuel cell are filled with air at 200 mbar (gauge), and subsequently, the variation in the gauge pressure is recorded. The leak test performances of the fuel cells at different configurations are shown in Figure 11. The configuration with 0.3 mm gasket under 5 Nm torque exhibits best sealing behaviour compared to the other configurations. Results obtained from the leak tests agree with the contact pressure distribution in Figure 5. Thus, the assembly 0.2 mm only under 5 Nm torque and all the assemblies with 0.3 mm gasket are considered for actual polarisation tests.

4.4. Polarization Behaviour. During the present investigations, the polarisation behaviour is analysed with hydrogen as fuel and air as oxidant. Humidified hydrogen and oxygen with dew point 40°C are fed at constant flow rates of 0.3 Lpm and

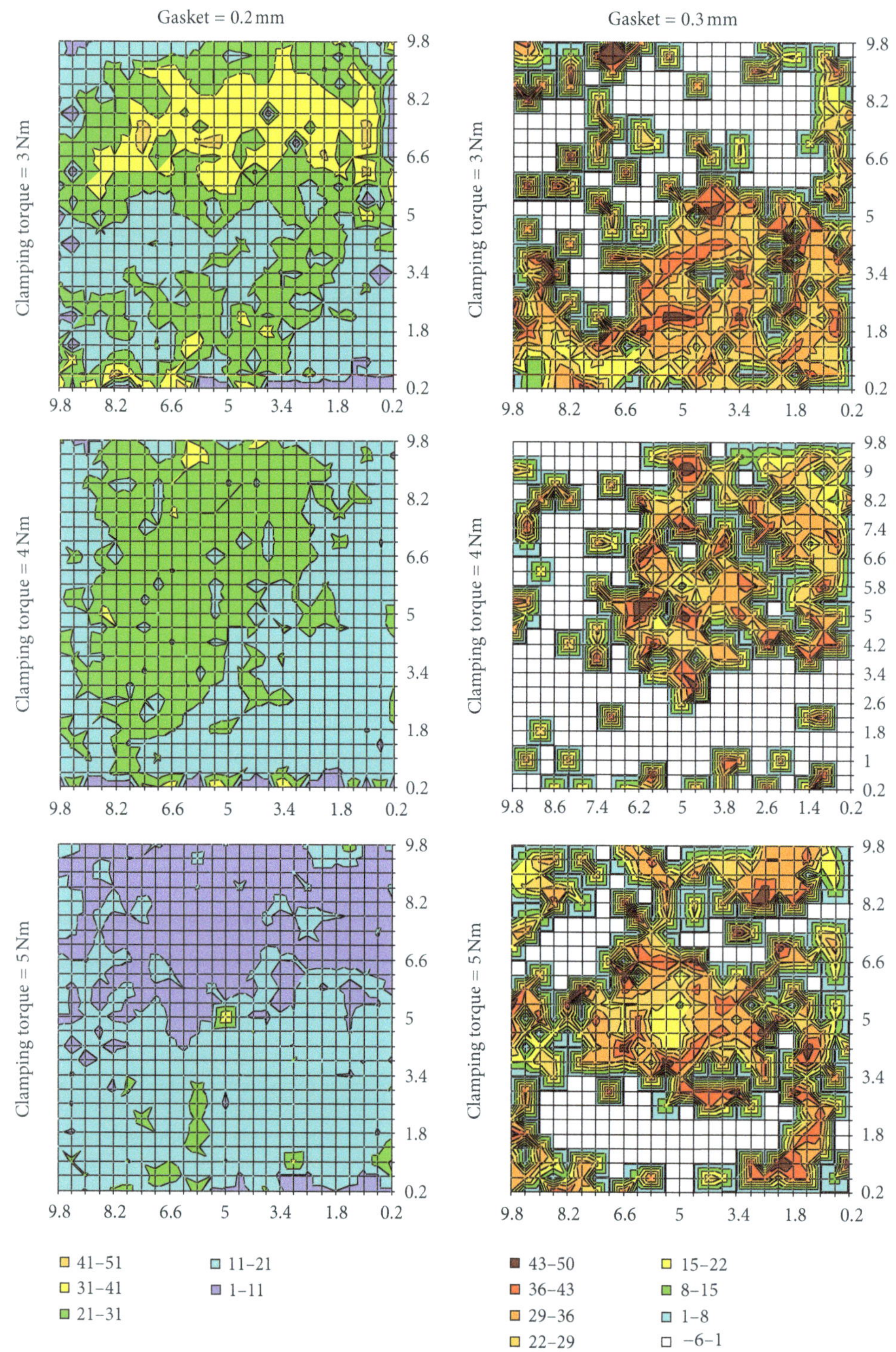

FIGURE 10: Contact resistance distribution (mΩ cm^2) between GDL and BPP for different configurations.

3 Lpm, respectively. As the configuration with 0.2 mm gasket under 3 Nm and 4 Nm torques exhibits very poor sealing, they are not considered for polarisation behaviour analysis.

The assembly with 0.2 mm gasket and 5 Nm torque exhibits best polarisation behaviour over other configurations as shown in Figure 12. Though this configuration exhibits relatively poor performance in terms of gas sealing, it offers better performance due to high contact pressure between the BPP and GDL leading to lower ASCR value compared to the 0.3 mm gasket. The performance is modelled (7) considering the estimated contact resistances based on measured contact pressures. The parameters used for modelling the polarisation behaviour are given in Table 1. The polarisation behaviours obtained experimentally with 5 Nm torque are found to be

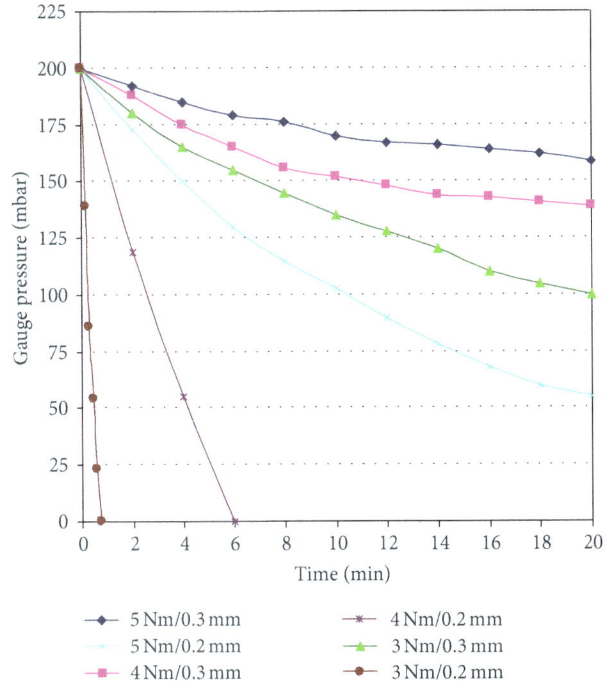

FIGURE 11: Variation in the gauge pressure of air for different fuel cell assembly with PTFE gasket.

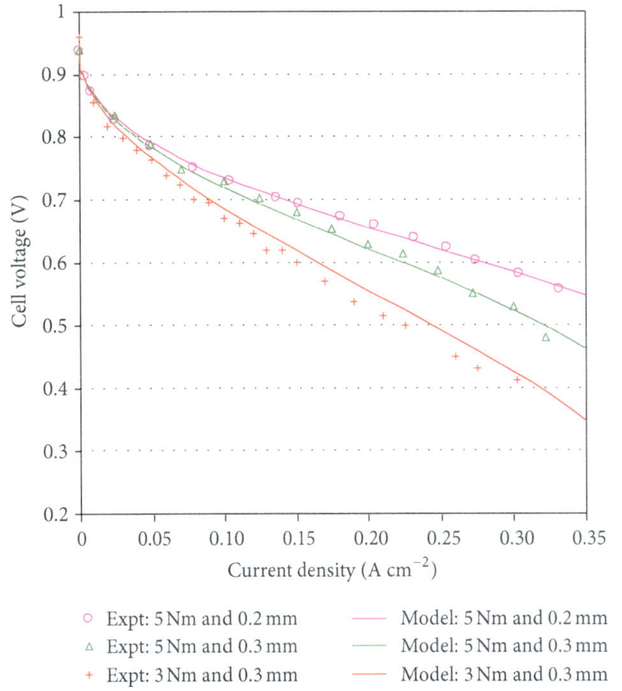

FIGURE 12: Polarisation behaviour of the fuel cell under different configurations.

in good agreement with the model. However, the assembly with 3 Nm offers relatively poor agreement. This could be due to the error in measuring the interfacial pressure in some regions where pressure lies below the minimum range of the film.

5. Conclusions

The influences of the mismatch in the thicknesses of GDL and gasket are investigated. In addition, the influences of the clamping pressure on the performances are also studied. In this purpose, the contact pressure between the GDL and the BPP is measured using pressure sensitive film, and consequently, the area-specific contact resistance is estimated. It is observed that the contact resistance plays significant role in the configuration with 0.3 mm gasket and 0.3 mm GDL. Such configuration, under 3 Nm and 5 Nm clamping forces, exhibits 480 mΩ cm^2 and 148 mΩ cm^2 contact resistance, respectively. The configurations below 3 Nm and 5 Nm clamping forces and with 0.2 mm thick gasket exhibits 23 mΩ cm^2 and 11 mΩ cm^2 contact resistances, respectively. The polarisation behaviour obtained from the experiment of such configurations agrees with the modelling results.

Nomenclature

α: Charge transfer coefficient
η_{act}: Activation overpotential (V)
i: Current density normalised based on active area (A cm^{-2})
i_o: Exchange current density (A cm^{-2})
η_{con}: Concentration overpotential (V)
i_L: Limiting current (A cm^{-2})
η_{ohm}: Ohmic overpotential (V)
r: Area-specific internal resistance (Ω cm^2)
r_c: Internal area-specific contact resistance (Ω cm^2)
r_b: Internal area-specific bulk resistance (Ω cm^2)
$r_{GDL/BPP}$: Contact resistance between BPP and GDL (Ω cm^2)
$r_{BPP/EP}$: Contact resistance between BPP and end plate (Ω cm^2)
r_{MEM}: Area-specific bulk resistance of membrane (Ω cm^2)
r_{EL}: Area-specific bulk resistance of electrode (Ω cm^2)
r_{GDL}: Area-specific bulk resistance of GDL (Ω cm^2)
r_{BPP}: Area-specific bulk resistance of BPP (Ω cm^2)
L: Thickness of the membrane (cm)
σ: Conductivity of membrane (S cm^{-1})
r': Fitting parameter for contact resistance (Ω cm^2)
r'': Fitting parameter for contact resistance (Ω cm^2)
C: Fitting parameter for contact resistance
$P_{interface}$: Contact pressure (MPa)
η_{total}: Overall total overpotential (V)
E: Operating potential (V)
E_{rev}: Reversible electrode potential (V)
E^o_{rev}: Standard electrode potential (V)
R: Universal gas constant (J mol^{-1} K^{-1})
T: Operating temperature (K)

Table 1: Parameters used for the modelling of the polarisation behaviours.

Configuration (torque and thickness)	Activation		Ohmic			Mass transfer	
	A (V)	i_o (mA cm^{-2})	r_c (Ω cm^2)	σ (S cm^{-1})	r_b (Ω cm^2)	B (V)	i_L (mA cm^{-2})
5 Nm and 0.2 mm	0.0525	0.3	0.011	0.06	0.305	0.045	500
5 Nm and 0.3 mm	0.0525	0.3	0.148	0.06	0.305	0.045	450
3 Nm and 0.3 mm	0.0525	0.3	0.480	0.06	0.305	0.045	450

n: Number moles of electrons per mole of fuel (mol)
F: Faraday's constant (C mol^{-1})
P_i: Partial pressure of ith specie
v_i: Stoichiometric coefficient for the ith reactant.

Abbreviations

ASR: Area-specific resistance
ASCR: Area-specific contact resistance
BPP: Bipolar plate
FEA: Finite element analysis
GDL: Gas diffusion layer
MEA: Membrane electrode assembly
PEFC: Polymer electrolyte fuel cell
PFSA: Perfluorosulfonic acid
PTFE: Polytetrafluoroethylene
RSM: Response surface methodology
SOFC: Solid oxide fuel cell.

Acknowledgments

The work was sponsored by the Naval Material Research Laboratory (NMRL), Defence Research and Development Organisation of India. The support from Dr. J. Rangarajan, Dr. S. Roychowdhuri, and Mrs. Suhasini Roychoudhury of NMRL is acknowledged. The contribution of Mr. Amar Nath in acquiring pressure distributions is also acknowledged.

References

[1] Y. Wang, K. S. Chen, J. Mishler, S. C. Cho, and X. C. Adroher, "A review of polymer electrolyte membrane fuel cells: technology, applications, and needs on fundamental research," *Applied Energy*, vol. 88, no. 4, pp. 981–1007, 2011.

[2] B. C. H. Steel and A. Heinze, "Materials for fuel-cell technologies," *Nature*, vol. 414, pp. 345–352, 2001.

[3] M. S. Dresselhaus and I. L. Thomas, "Alternative energy technologies," *Nature*, vol. 414, no. 6861, pp. 332–337, 2001.

[4] P. C. Ghosh, T. Wüster, H. Dohle, N. Kimiaie, J. Mergel, and D. Stolten, "Analysis of single PEM fuel cell performances based on current density distribution measurement," *Journal of Fuel Cell Science and Technology*, vol. 3, no. 3, pp. 351–357, 2006.

[5] I. Alaefour, G. Karimi, K. Jiao, and X. Li, "Measurement of current distribution in a proton exchange membrane fuel cell with various flow arrangements—a parametric study," *Applied Energy*, vol. 93, pp. 80–89, 2012.

[6] H. Dohle, J. Mergel, and P. C. Ghosh, "DMFC at low air flow operation: study of parasitic hydrogen generation," *Electrochimica Acta*, vol. 52, no. 19, pp. 6060–6067, 2007.

[7] K. Panha, M. Fowler, X.-Z. Yuan, and H. Wang, "Accelerated durability testing via reactants relative humidity cycling on PEM fuel cells," *Applied Energy*, vol. 93, pp. 90–97, 2012.

[8] A. Nishimura, K. Shibuya, A. Morimoto et al., "Dominant factor and mechanism of coupling phenomena in single cell of polymer electrolyte fuel cell," *Applied Energy*, vol. 90, no. 1, pp. 73–79, 2012.

[9] P. C. Ghosh, H. Dohle, and J. Mergel, "Modelling of heterogeneities inside polymer electrolyte fuel cells due to oxidants," *International Journal of Hydrogen Energy*, vol. 34, no. 19, pp. 8204–8212, 2009.

[10] N. Amar, *Contact pressure distribution in polymer electrolyte fuel cell [M. Tech. dissertation]*, Indian Institute of Technology Bombay, 2010.

[11] J. Ihonen, F. Jaouen, G. Lindbergh, and G. Sundholm, "A novel polymer electrolyte fuel cell for laboratory investigations and in-situ contact resistance measurements," *Electrochimica Acta*, vol. 46, no. 19, pp. 2899–2911, 2001.

[12] S. P. Jiang, J. G. Love, and L. Apateanu, "Effect of contact between electrode and current collector on the performance of solid oxide fuel cells," *Solid State Ionics*, vol. 160, no. 1-2, pp. 15–26, 2003.

[13] T. Dey, D. Singdeo, M. Bose, R. N. Basu, and P. C. Ghosh, "Study of contact resistance at the electrodeinterconnect interfaces in planar type Solid Oxide Fuel Cells," *Journal of Power Sources*, vol. 233, pp. 290–298, 2013.

[14] V. Mishra, F. Yang, and R. Pitchumani, "Measurement and prediction of electrical contact resistance between gas diffusion layers and bipolar plate for applications to PEM fuel cells," *Journal of Fuel Cell Science and Technology*, vol. 1, no. 1, pp. 2–9, 2004.

[15] I. Nitta, O. Himanen, and M. Mikkola, "Contact resistance between gas diffusion layer and catalyst layer of PEM fuel cell," *Electrochemistry Communications*, vol. 10, no. 1, pp. 47–51, 2008.

[16] D. Liu, L. Peng, and X. Lai, "Effect of assembly error of bipolar plate on the contact pressure distribution and stress failure of membrane electrode assembly in proton exchange membrane fuel cell," *Journal of Power Sources*, vol. 195, no. 13, pp. 4213–4221, 2010.

[17] Z. Y. Su, C. T. Liu, H. P. Chang, C. H. Li, K. J. Huang, and P. C. Sui, "A numerical investigation of the effects of compression force on PEM fuel cell performance," *Journal of Power Sources*, vol. 183, no. 1, pp. 182–192, 2008.

[18] S. J. Lee, C. D. Hsu, and C. Huang, "Analyses of the fuel cell stack assembly pressure," *Journal of Power Sources*, vol. 145, no. 2, pp. 353–361, 2005.

[19] W. R. Chang, J. J. Hwang, F. B. Weng, and S. H. Chan, "Effect of clamping pressure on the performance of a PEM fuel cell," *Journal of Power Sources*, vol. 166, no. 1, pp. 149–154, 2007.

[20] P. Zhou, C. W. Wu, and G. J. Ma, "Contact resistance prediction and structure optimization of bipolar plates," *Journal of Power Sources*, vol. 159, no. 2, pp. 1115–1122, 2006.

[21] M. Barber, T. S. Sun, E. Petrach, X. Wang, and Q. Zou, "Contact mechanics approach to determine contact surface area between bipolar plates and current collector in proton exchange membrane fuel cells," *Journal of Power Sources*, vol. 185, no. 2, pp. 1252–1256, 2008.

[22] X. Q. Xing, K. W. Lum, H. J. Poh, and Y. L. Wu, "Optimization of assembly clamping pressure on performance of proton-exchange membrane fuel cells," *Journal of Power Sources*, vol. 195, no. 1, pp. 62–68, 2010.

[23] J. Tan, Y. J. Chao, J. W. Van Zee, and W. K. Lee, "Degradation of elastomeric gasket materials in PEM fuel cells," *Materials Science and Engineering A*, vol. 445-446, pp. 669–675, 2007.

[24] X. Wang, Y. Song, and B. Zhang, "Experimental study on clamping pressure distribution in PEM fuel cells," *Journal of Power Sources*, vol. 179, no. 1, pp. 305–309, 2008.

[25] L. Zhang, Y. Liu, H. Song, S. Wang, Y. Zhou, and S. J. Hu, "Estimation of contact resistance in proton exchange membrane fuel cells," *Journal of Power Sources*, vol. 162, no. 2, pp. 1165–1171, 2006.

13

An Integrated Decentralized Energy Planning Model considering Demand-Side Management and Environmental Measures

Seyed Mahmood Kazemi and Masoud Rabbani

Department of Industrial Engineering, College of Engineering, University of Tehran, P. O. Box 11155-4563, Tehran, Iran

Correspondence should be addressed to Seyed Mahmood Kazemi; kazemi_m_s@yahoo.com

Academic Editor: S. Venkata Mohan

Decentralized energy planning (DEP) is looked upon as an indisputable opportunity for energy planning of villages, isolated islands, and far spots. Nonetheless, at this decentralized planning level, the value of demand-side resources is not fairly examined, despite enjoying great advantages. Therefore, the core task of this study is to integrate demand-side resources, as a competing solution against supply-side alternatives, with decentralized energy planning decisions and demonstrate the rewarding role it plays. Moreover, sustainability indicators (SIs) are incorporated into DEP attempts in order to attain sustainable development. It is emphasized that unless these indicators are considered at lower energy planning levels, they will be ignored at higher planning levels as well. Hence, to the best knowledge of the authors, this study for the first time takes into account greenhouse gas (GHG) emissions produced by utilization of renewable energies in DEP optimization models. To address the issues mentioned previously, multiobjective linear programming model along with a min-max goal programming approach is employed. Finally, using data taken from the literature, the model is solved, and the obtained results are discussed. The results show that DSM policies have remarkably contributed to significant improvements especially in terms of environmental indicators.

1. Introduction

Decentralized energy planning aims for efficient use of local resources to supply energy. DEP is an available option for satisfying energy needs of villages, isolated islands, and small-scale end users in a reliable, affordable, and environmentally sustainable way (Hiremath et al. [1]). To see some recent top researches conducted in the related area, the works of [2–7] are recommended. The DEP paramount feature is to provide an area-based DEP so that with the least economic and environmental costs develops alternate energy sources and satisfies energy requirements. References [8–11] are good example works done in the literature with focus on DEP modeling.

Energy demand management, also referred to as demand-side management (DSM), includes activities that modify end use energy demand; for example, the activities aim to reduce peak demand. It is worth mentioning that peak demand management does not necessarily lead to reduction in total energy consumption, but it defers the need for investments in establishing new power plants. The most generally accepted definition of DSM is the one presented by Gellings in 1984 [12].

As discussed by Kreith and Goswami [12], DSM alternatives must be compared to supply-side alternatives. Thus, it entails considering and evaluating both supply-side and demand-side options in energy planning efforts. In the late 1980s, Ontario Hydro of Canada estimated that satisfying its customer peak demand via supply-side measures costs four times that of demand-side measures [13]. As a result, any energy planning model without simultaneously considering both sides is incomplete.

Although developing a model which considers a DEP with evaluating both supply-side and demand-side management measures is of great importance, it is not yet inclusive. The shortcoming is related to not taking SIs into account. To overcome this, we need to have such indicators quantified, which is of course a very challenging task [14] (see also [15, 16]). Evans et al. [17] ranked four renewable energy technologies, namely, photovoltaic, wind, hydro and geothermal, against several SIs using data extracted from the literature. These indicators include electricity generation cost, GHG emissions, availability and limitations of each technology, efficiency of energy transformation, land use requirements, water consumption, and social impacts. Based on their findings, wind technology outranked the others followed by hydro, photovoltaic, and geothermal technologies, respectively. Nonetheless, it should be pointed out that the ranking was provided for the global international conditions, whereas each technology can be considerably geographically affected.

To cover the mentioned gaps in the literature, in this paper a multiobjective linear programming formulation for a DEP problem is proposed that not only takes both supply-side and demand-side alternatives into account but also considers a number of important SIs against which five energy resources, including DSM options, are evaluated. It is worth mentioning that using multi-objective formulations for energy planning is commonplace in the literature; see, for example, [10, 18–21].

In a nutshell, the contributions of this paper can be summarized as follows.

(i) It is, to the best knowledge of the authors, for the first time that in DEP models supply-side and demand-side options are treated equally.

(ii) It is, to the best knowledge of the authors, for the first time that in DEP models GHG emissions produced by renewable energy technologies during their life cycle are incorporated. It may be surprising that, for instance, for photovoltaic, the amount of produced GHG emissions is almost one quarter of natural gas emissions [22], while it had been a widespread supposition that GHG emissions of renewable energy technologies are insignificant.

(iii) A variety of important sustainable indicators, namely water consumption, land use, cost of generation, employment, and GHG emissions, are considered.

The rest of the paper is organized as follows. In the next section, materials and methods used are elaborated. In Section 3, results and discussions are provided. Finally, conclusion remarks are drawn in Section 4.

2. Materials and Methods

2.1. Materials. The data used in this paper are collected from the literature and are based on the global international conditions. To solve the model for a numerical example, Lingo 8.0 software was employed.

2.2. Mathematical Formulation. As stated before, five SIs are taken into account. These SIs are regarded as distinct objective functions. The notations used in this model along with the proposed integrated multi-objective formulation are presented below.

Notations. Sets are as follows:

I: set of renewable energy technologies;

J: set of end uses.

Parameters and variables as follows:

wc_i: water consumption of ith resource option (kg/kWh);

GHG_i: greenhouse gas emissions of the ith resource option (g $CO_{2\text{-e}}$/kWh);

lu_i: land use characteristic of ith resource option (m^2/kWh);

c_{ij}: electricity generation cost of resource i for jth end use (US$/kWh);

CDSM_j: electricity saving cost using the implemented DSM program in the jth end use (US$/kWh);

w_i: employment generation of ith resource option (jobs/kWh);

D_j: total energy demand for jth end use (kWh);

η_{ij}: conversion efficiency for the ith resource option for jth end use;

S_i: availability of the ith resource option for jth end use (kWh);

P_j: the maximum possible saving using the DSM program in the jth end use (kWh);

SOBJ_j: the required saving that is, to be achieved by the DSM program (kWh).

Decision variables:

x_{ij}: optimal amount of ith resource option for jth end use (kWh);

SDSM_j: optimal saving in jth end use using the implemented DSM program (kWh).

$$\text{Minimize total cost} = \sum_i \sum_j c_{ij} \times x_{ij} + \sum_j \text{SDSM}_j \times \text{CDSM}_j, \quad (1)$$

$$\text{Maximize total jobs created} = \sum_i \sum_j x_{ij} \times w_i, \quad (2)$$

$$\text{Minimize total water consumption} = \sum_i \sum_j x_{ij} \times wc_i, \quad (3)$$

$$\text{Minimize total greenhouse gas emissions} = \sum_i \sum_j x_{ij} \times \text{GHG}_i, \quad (4)$$

$$\text{Minimize total land used} = \sum_i \sum_j x_{ij} \times lu_i. \quad (5)$$

Which are subjected to the following:

$$\sum_i x_{ij} + \text{SDSM}_j \geq D_j, \quad \forall j \in J, \tag{6}$$

$$\sum_j \frac{x_{ij}}{\eta_{ij}} \leq S_i, \quad \forall i \in I, \tag{7}$$

$$\text{SDSM}_j \leq P_j, \quad \forall j \in J, \tag{8}$$

$$\text{SDSM}_j \geq \text{SOBJ}_j, \quad \forall j \in J, \tag{9}$$

$$x_{ij} \geq 0, \quad \text{SDSM}_j \geq 0. \tag{10}$$

The objective function given in (1) minimizes the total electricity generation costs. As can be seen, costs associated with implementing the DSM program are also considered in the first objective function, and it is treated just like other energy resources. It means that saving a unit of electricity by implementing DSM can be treated like producing a unit of electricity by a power plant, provided that the data for the associated costs are available. Equation (2) maximizes total number of jobs created due to electrifying the area. It should be noted that since the data for the number of jobs created by implementing the DSM program were not found, assumed to be negligible in this paper. The objective function given in (3), (4), and (5) minimize total water consumption, greenhouse gas emissions, and total land used, respectively. Constraints (6) and (7) are demand and resource availability constraints, respectively. Constraint (8) denotes that the maximum expected saving that could be achieved by executing the DSM program is limited to a realistic upper bound. Constraint (9) forces energy planners to implement DSM programs and meet some predetermined energy efficiency goals [23]. Constraint (10) indicates that decision variables are nonnegative real numbers.

2.3. Methods. Before we can solve the model by Lingo 8.0 software, we must find a way to cope with the multi-objective nature of the proposed formulation. To do so, a min-max goal programming approach was applied to obtain the best mix of renewable energy sources. For further study about goal programming (GP) approaches, the respected reader is referred to [24]. Now, using GP approach the model presented in the previous subsection can be easily reformulated as a single-objective linear programming formulation as presented in the following:

$$\text{Minimize max deviation} = d,$$

$$\text{Subject to} \quad \alpha_j p_j + \beta_j n_j \leq d, \quad j = 1, 2, \ldots, 5,$$

$$\sum_i \sum_j c_{ij} \times x_{ij} + \sum_j \text{SDSM}_j \times \text{CDSM}_j - p_1 + n_1 = t_1,$$

$$\sum_i \sum_j x_{ij} \times w_i - p_2 + n_2 = t_2,$$

$$\sum_i \sum_j x_{ij} \times wc_i - p_3 + n_3 = t_3,$$

Table 1: Parameter values for renewable energy technologies.

	PV	Wind	Hydro	Geothermal
w_i (10^{-4} number of jobs produced/kWh)	0.27549	0.27549	1.466	0.4[a]
wc_i	10	1	36	100
lu_i	0.046	0.072	0.411	0.05[a]
GHG_i	90	25	41	170
S_i	8000000	327000	555000	100000[a]
c_{ij} ($\forall j$)	0.398	0.02	0.039	0.03[a]
η_{ij} ($\forall j$)	0.13	0.39	0.9	0.15

[a]The value has been assumed by the authors.

$$\sum_i \sum_j x_{ij} \times \text{GHG}_i - p_4 + n_4 = t_4,$$

$$\sum_i \sum_j x_{ij} \times lu_i - p_5 + n_5 = t_5,$$

$$\sum_i x_{ij} + \text{SDSM}_j \geq D_j \quad \forall j \in J,$$

$$\sum_j \frac{x_{ij}}{\eta_{ij}} \leq S_i, \quad \forall i \in I,$$

$$\text{SDSM}_j \leq P_j, \quad \forall j \in J,$$

$$\text{SDSM}_j \geq \text{SOBJ}_j, \quad \forall j \in J,$$

$$x_{ij} \geq 0, \quad \text{SDSM}_j \geq 0, \quad n_j \geq 0, \quad p_j \geq 0. \tag{11}$$

As shown above, parameter d denotes maximum deviation from every target (t_j). The parameter values α_j and β_j are set by decision maker. In some sense, this method is similar to the weighted Tchebycheff approach except that here the optimum z^* is replaced by target value.

3. Results and Discussion

In this section, in order to demonstrate the applicability of the model to real case problems, a numerical example based on the data obtained from the literature is developed. It should be pointed out that for the unavailable data, assumed values are considered. Tables 1 and 2 show the parameter values of the model that are either taken from [10–12, 17] or assumed by the authors.

The values of parameter CDSM_j in Table 2 are taken from [12]. For example, it is stated that a successful DSM program (2002 Massachusetts Electric-Residential Lighting Program) cost was \$3.3 million while it resulted in 18037 MWh energy saving. Therefore, the unit cost of energy saving using the implemented DSM program in the domestic sector is calculated as follows:

$$\text{CDSM}_{\text{domestic}} = \frac{3.3 \times 10^6 \$}{18037 \times 10^3 \,\text{kWh}} = 0.183 \ (\text{US\$/kWh}). \tag{12}$$

TABLE 2: Parameter values for different end uses.

	Domestic	Agriculture	Community	Industry
$CDSM_j$	0.183	0.238	0.917	0.055
D_j (the demand that is to be satisfied)	258267	278915	5276	18815
Already supplied demand (ASD)	116220[a]	125512[a]	2374[a]	8467[a]
P_j (0.15 × ASD)[a]	17433	18827	356	1270
$SOBJ_j$ (0.05 × ASD)[a]	5811	6276	119	423

[a]The value has been assumed by the authors.

TABLE 3: Goal programming optimum solutions.

	Domestic (kWh)	Agriculture (kWh)	Community (kWh)	Industry (kWh)
DSM	17433	18827	119	1270
PV	100007.6	0	0	0
Wind	127530	0	0	0
Hydro	13296.4	245088	5157	17545
Geothermal	0	15000	0	0

The way of calculating the other values are the same. In this example, a rural area is assumed that is partly electrified, but the most parts are not yet provided with electricity. An energy saving target ranging from 5 to 15 percent in the already electrified parts of the area is considered.

To solve the goal programming model proposed in Section 2.3, all α_j and β_j values are set to be 0.5. The objective functions are also considered as presented below.

The first objective function:

$$\frac{\left(\sum_i \sum_j c_{ij} \times x_{ij} + \sum_j \mathrm{SDSM}_j \times \mathrm{CDSM}_j - 21528.53\right)}{21528.53}. \quad (13)$$

The second objective function:

$$\frac{\left(105.9338 - \sum_i \sum_j x_{ij} \times w_i\right)}{105.9338}. \quad (14)$$

The third objective function:

$$\frac{\left(\sum_i \sum_j x_{ij} \times wc_i - 4086100\right)}{4086100}. \quad (15)$$

The fourth objective function:

$$\frac{\left(\sum_i \sum_j x_{ij} \times \mathrm{GHG}_i - 19418390\right)}{19418390}. \quad (16)$$

The fifth objective function:

$$\frac{\left(\sum_i \sum_j x_{ij} \times lu_i - 24075.8\right)}{24075.8}. \quad (17)$$

As a decision maker, the following target values are assigned to the five objective functions mentioned above, respectively.

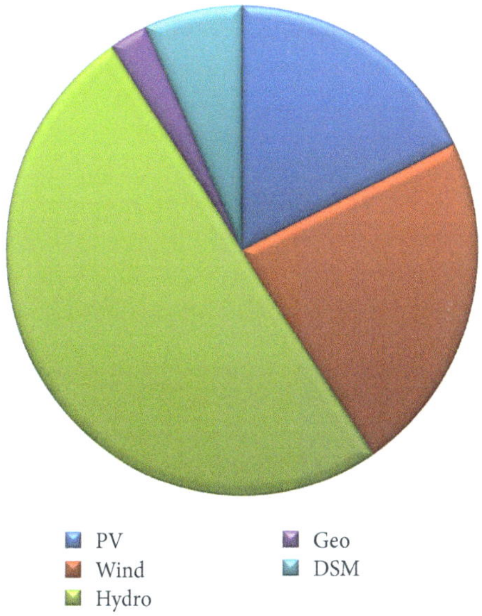

FIGURE 1: The role of each renewable energy source in electrifying the area under study.

The first objective function: 0.16.

The second objective function: 0.7.

The third objective function: 1.2.

The fourth objective function: 0.3.

The fifth objective function: 2.7.

Table 3 shows the outputs of Lingo 8.0 software for this numerical example.

Figure 1 shows the degree to which renewable energy sources have contributed to electrifying the area under study.

Using the solution given in Table 3, one can calculate the value of each individual objective function separately (Case 1). Also, by solving the original model presented in Section 2.2 once for each objective function (as a single-objective model) independently, the optimum value of each individual objective function is obtained (Case 2). Comparing the objective function values obtained in Case 1 with those obtained in Case 2 the following conclusions can be drawn.

(1) Goal programming results are more practical to implement.

(2) If the main limitations are water and land use requirements, minimizing GHG emissions and generation

costs or maximizing employment will not be an attractive goal to attain.

(3) If financial resources are scarce, minimizing water consumption and land used will not be an attractive goal to attain.

(4) If maximizing employment is of the highest importance, similar to the previous, minimizing water consumption and land used will not be an attractive goal to attain.

From the results pointed out above, it can be summarized that provided that target values are determined carefully, the model could be useful under various geographical conditions.

4. Conclusions

In this study, an integrated multi-objective linear programming formulation was developed for decentralized energy planning such that demand-side management and environmental measures were also taken into account. Moreover, a number of important sustainability indicators were utilized upon which four renewable energy technologies were compared. Using goal programming approach, the proposed model was solved by Lingo 8.0 software. The findings indicate that fine choices of sustainability indicators for doing the decision-making process are extremely affected by the limitations the policy makers encounter. For example, poor countries suffer from scarce funding resources. Thus, when planning for their energy supplies, it is not so sensible to consider minimizing the land occupied by the supply technologies as an important criterion to pursue. Also, the obtained numerical results show that in the optimum solution hydro outranked the other renewable energy technologies followed by wind, photovoltaic, DSM resources, and geothermal, respectively. It is worth pointing out that DSM alternatives have greatly contributed together with the other energy supply technologies.

Conflict of Interests

The authors declare that they have no conflict of interests.

References

[1] R. B. Hiremath, B. Kumar, P. Deepak, P. Balachandra, N. H. Ravindranath, and B. N. Raghunandan, "Decentralized energy planning through a case study of a typical village in India," *Journal of Renewable and Sustainable Energy*, vol. 1, no. 4, Article ID 043103, 24 pages, 2009.

[2] N. L. Panwar, S. C. Kaushik, and S. Kothari, "Role of renewable energy sources in environmental protection: a review," *Renewable and Sustainable Energy Reviews*, vol. 15, no. 3, pp. 1513–1524, 2011.

[3] M.-A. Hessami, H. Campbell, and C. Sanguinetti, "A feasibility study of hybrid wind power systems for remote communities," *Energy Policy*, vol. 39, no. 2, pp. 877–886, 2011.

[4] J. R. San Cristóbal, "A goal programming model for the optimal mix and location of renewable energy plants in the North of Spain," *Renewable and Sustainable Energy Reviews*, vol. 16, no. 7, pp. 4461–4464, 2012.

[5] J. Yang, W. Chen, B. Chen, and Y. Jia, "Economic feasibility analysis of a renewable energy project in the rural China," *Procedia Environmental Sciences*, vol. 13, pp. 2280–2283, 2012.

[6] K. Trapani, D. L. Millar, and H. C. M. Smith, "Novel offshore application of photovoltaics in comparison to conventional marine renewable energy technologies," *Renewable Energy*, vol. 50, pp. 879–888, 2013.

[7] C. Monteiro, I. J. Ramirez-Rosado, and L. A. Fernandez-Jimenez, "Short-term forecasting model for electric power production of small-hydro power plants," *Renewable Energy*, vol. 50, pp. 387–394, 2013.

[8] T. Senjyu, D. Hayashi, A. Yona, N. Urasaki, and T. Funabashi, "Optimal configuration of power generating systems in isolated island with renewable energy," *Renewable Energy*, vol. 32, no. 11, pp. 1917–1933, 2007.

[9] R. B. Hiremath, S. Shikha, and N. H. Ravindranath, "Decentralized energy planning; modeling and application—a review," *Renewable and Sustainable Energy Reviews*, vol. 11, no. 5, pp. 729–752, 2007.

[10] D. S. Herran and T. Nakata, "Renewable technologies for rural electrification in Colombia: a multiple objective approach," *International Journal of Energy Sector Management*, vol. 2, no. 1, pp. 139–154, 2008.

[11] A. B. Kanase-Patil, R. P. Saini, and M. P. Sharma, "Integrated renewable energy systems for off grid rural electrification of remote area," *Renewable Energy*, vol. 35, no. 6, pp. 1342–1349, 2010.

[12] F. Kreith and D. Y. Goswami, *Energy Management and Conservation Handbook*, CRC, 2007.

[13] C. Beggs, *Energy: Management, Supply and Conservation*, Spon Press, 2009.

[14] N. Lior, "About sustainability metrics for energy development," in *Proceedings of the 6th Biennal Workshop on Advances in Energy Studies*, vol. 29, pp. 390–401, Graz, Austria, 2008.

[15] H. Hashim, P. Douglas, A. Elkamel, and E. Croiset, "Optimization model for energy planning with CO_2 emission considerations," *Industrial & Engineering Chemistry Research*, vol. 44, no. 4, pp. 879–890, 2005.

[16] Varun, R. Prakash, and I. K. Bhat, "Energy, economics and environmental impacts of renewable energy systems," *Renewable and Sustainable Energy Reviews*, vol. 13, no. 9, pp. 2716–2721, 2009.

[17] A. Evans, V. Strezov, and T. J. Evans, "Assessment of sustainability indicators for renewable energy technologies," *Renewable and Sustainable Energy Reviews*, vol. 13, no. 5, pp. 1082–1088, 2009.

[18] A. Alarcon-Rodriguez, G. Ault, and S. Galloway, "Multi-objective planning of distributed energy resources: a review of the state-of-the-art," *Renewable and Sustainable Energy Reviews*, vol. 14, no. 5, pp. 1353–1366, 2010.

[19] Y. A. Katsigiannis, P. S. Georgilakis, and E. S. Karapidakis, "Multiobjective genetic algorithm solution to the optimum economic and environmental performance problem of small autonomous hybrid power systems with renewables," *IET Renewable Power Generation*, vol. 4, no. 5, pp. 404–419, 2010.

[20] J. L. Bernal-Agustín, R. Dufo-López, and D. M. Rivas-Ascaso, "Design of isolated hybrid systems minimizing costs and pollutant emissions," *Renewable Energy*, vol. 31, no. 14, pp. 2227–2244, 2006.

[21] P. S. Moura and A. T. de Almeida, "Multi-objective optimization of a mixed renewable system with demand-side management," *Renewable and Sustainable Energy Reviews*, vol. 14, no. 5, pp. 1461–1468, 2010.

[22] J. Wood, *Local Energy: Distributed Generation of Heat and Power*, IET, 2008.

[23] M. Dicorato, G. Forte, and M. Trovato, "Environmental-constrained energy planning using energy-efficiency and distributed-generation facilities," *Renewable Energy*, vol. 33, no. 6, pp. 1297–1313, 2008.

[24] K. Deb, *Multi-Objective Optimization Using Evolutionary Algorithms*, Wiley, 2009.

14

Synthesis of Petroleum-Based Fuel from Waste Plastics and Performance Analysis in a CI Engine

Christine Cleetus,[1] Shijo Thomas,[2] and Soney Varghese[2]

[1] *Department of Mechanical Engineering, NIT Calicut, Kerala 673601, India*
[2] *School of Nano Science and Technology, NIT Calicut, Kerala 673601, India*

Correspondence should be addressed to Shijo Thomas; tshijo2004@gmail.com

Academic Editor: Gang Quan

The present work involves the synthesis of a petroleum-based fuel by the catalytic pyrolysis of waste plastics. Catalytic pyrolysis involves the degradation of the polymeric materials by heating them in the absence of oxygen and in the presence of a catalyst. In the present study different oil samples are produced using different catalysts under different reaction conditions from waste plastics. The synthesized oil samples are subjected to a parametric study based on the oil yield, selectivity of the oil, fuel properties, and reaction temperature. Depending on the results from the above study, an optimization of the catalyst and reaction conditions was done. Gas chromatography-mass spectrometry of the selected optimized sample was done to find out its chemical composition. Finally, performance analysis of the selected oil sample was carried out on a compression ignition (CI) engine. Polythene bags are selected as the source of waste plastics. The catalysts used for the study include silica, alumina, Y zeolite, barium carbonate, zeolite, and their combinations. The pyrolysis reaction was carried at polymer to catalyst ratio of 10 : 1. The reaction temperature ranges between 400°C and 550°C. The inert atmosphere for the pyrolysis was provided by using nitrogen as a carrier gas.

1. Introduction

In the recent years it is quite common to find in newspapers and publications that plastics are turning out to be a menace. Days are not so far when earth will be completely covered with plastics and humans will be living over it. All the reasoning and arguments for and against plastics finally land up on the fact that plastics are nonbiodegradable in nature. The disposal and decomposition of plastics has been an issue which has caused a number of research works to be carried out in this regard. Currently the disposal methods employed are land filling, mechanical recycling, biological recycling, thermal recycling, and chemical recycling. Of these methods, chemical recycling is a research field which is gaining much interest recently, as it turns out to be that the products formed in this method are highly advantageous.

Plastic is one such commodity that has been so extensively used and is sometimes referred to as one of the greatest innovations of the millennium. There are a numerous ways in which plastic is and will continue to be used. The plastic has achieved such an extensive market due to fact that it is lightweight, cheap, flexible, reusable, do not rust or rot, and so forth. Because of this, plastics production has gone up by almost 10% every year on a global basis since 1950 [1]. Asia accounts for 36.5% of the global consumption and has been world's largest plastics consumer for several years. The major segment continues to be the packaging, which has accounted for over 35% of the global demand [2]. The global per capita consumption of plastics is shown in Figure 1.

The global production of plastics has seen an increase from around 1.3 million tonnes in 1950 to 245 MT in 2006 [1]. In recent years, significant growth in the consumption of plastic globally has been due to the introduction of plastics into newer application areas such as in automotive field, rail, transport, aerospace, medical and healthcare, electrical and electronics, telecommunication, building and infrastructure, and furniture. This significant growth in the demand for plastic and its forecast for future have certainly proved that there has been a quiet plastic revolution taking place in every sector.

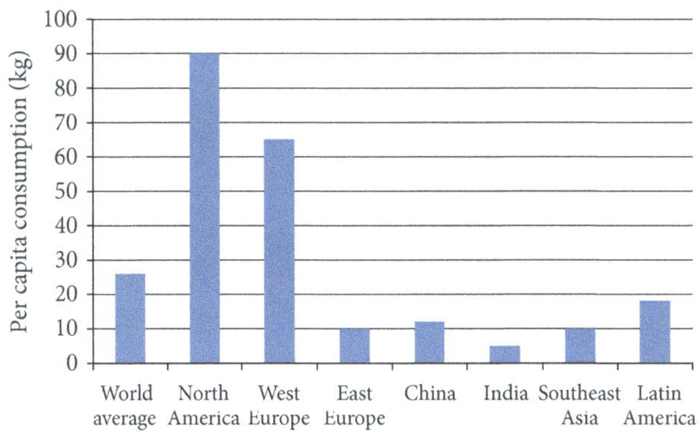

FIGURE 1: Global per capita consumption of plastics [2].

TABLE 1: Global consumption of individual plastics [2].

Type of plastic	Consumption %
Polythene (PE)	33.5
Polypropylene (PP)	19.5
Polyvinylchloride (PVC)	16.5
Polystyrene (PS)	8.5
Polyethylene terephthalate (PET) and polyurethane (PU)	5.5
Styrene copolymers (ABS, SAN, etc.)	3.5
Blends, alloys, high performance and specialty plastics, thermosetting plastics, and so forth	13

As far as the individual plastics materials are concerned, polyolefins account for 53% of the total consumption. The consumption of the individual plastic materials is shown in the Table 1. It can be seen that one-third of the global consumption of plastic is polythene. The growth in the global polythene demand is estimated to be around 4.4% annually up to 2020 [2]. This is the reason behind the selection of polythene as the source of waste plastic in this study.

The increase in the rate of plastic consumption throughout the world has led to the creation of more and more amounts of waste, and this in turn poses greater difficulties for disposal. This is because the life duration of plastic (the time period for which the plastic remains in use) is very small. About 40% of plastics consumed have duration of life smaller than one month. The service life of plastic products ranges from 1 to 35 years depending on the area of application. In India, the weighted average service life of all plastics products comes to about 8 years. This may vary among countries depending on the type of consumption. This short service life in India reflects that a major share of the plastic consumed here is short-life products. This can be accounted for as the share of plastics used in packaging which is 42% [1].

2. Different Methods of Plastic Waste Management

The suitable treatment of plastic is the most important factor in waste plastic management. This is quite important from the energetic, environmental, and economic point of view. Even though the recycling rate for postconsumer plastics has increased in the recent years, this increase has been only meager coming to only around 1.5%. This increase in the recycling is due to the strict legal regulations and growing awareness. Different techniques for the waste plastic management are being followed today [1].

The major portion of the waste plastics has been subjected to landfill. Such a disposal of the waste to landfill is now strictly regulated legally. The regulations are expected to achieve a reduction of 35% in land filling over the period from 1995 to 2020. Also the rising cost and scarcity of land, the generation of explosive greenhouse gases (such as methane), a high volume to weight ratio of plastics, and the poor biodegradability of commonly used packaging polymers also make it an unattractive option.

Reprocessing of the used plastics to form new similar products is termed as mechanical recycling. In this method, the products obtained are with almost same or less performance level than the original product. Even if the technique seems to be "a green operation," it is not cost effective as it requires high energy for cleaning, sorting, transportation, and processing to make a serviceable product. Practically it is seen that reprocessing of mixed contaminated plastics yields mechanically inferior products lacking in durability compared with the original polymers.

Biodegradable polymers are those which can be converted back to the biomass in a realistic time period. However, there are a number of difficulties over the use of degradable plastics. First, appropriate conditions are necessary for the degradation of such plastics, such as presence of light for the photodegradable plastics. Second, greenhouse gases such as methane is released when plastics degrade anaerobically. This is by enabling microorganisms in the environment to metabolize the molecular structure of plastic films to produce an inert humus-like material that is less harmful to the environment.

Incineration of plastics waste is an alternate method in which energy is recovered from waste polymers. These hydrocarbon polymers can replace fossil fuels and thereby reduce the CO_2 burden on the environment. Polyethylene is having calorific value similar to that of the fuel oil, and the energy produced by incineration of polyethylene is of the same order as that used in its manufacture making it an attractive option. However, this method produces green house gases and toxic pollutants giving it a big disadvantage.

Cracking process breaks down the long polymeric chains into useful smaller molecular weight compounds. The products of this process are highly useful and can be utilized as fuels or chemicals in various applications. The pyrolysis reaction can be carried out without or in the presence of a catalyst. If without catalyst, it is thermal cracking or thermolysis, and if in the presence of catalyst, it is catalytic pyrolysis.

Thermal cracking or pyrolysis involves the degradation or cracking of the polymeric materials by heating them to a very high temperature. The heating should be carried out in the absence of oxygen to make sure that no oxidation of the polymer takes place. The temperature ranges between 350 and 900°C. The products formed include a carbonized char (solid residues) and a volatile fraction. A portion of the volatile fraction can be condensed to give paraffins, isoparaffins, olefins, naphthenes, and aromatics, while the remaining is a noncondensable high calorific value gas. The products formed and their precise composition depend on the type of the plastic waste and the process conditions. In catalytic cracking, the same process is carried out in the presence of a catalyst. The prominent advantage of this method is that the presence of catalyst lowers the reaction temperature and time. Another added advantage is that in thermal cracking a broad variety of products are formed by the braking of the polymeric chain, while in catalytic degradation the product distribution will be a much narrower, with a peak at lighter hydrocarbons. From the economic point of view also, reducing the cost even further will make this process more attractive. Due to these reasons in the present work, this method is adopted for the synthesis of petroleum-based oil. The importance of this work lies in comparing the performance of different catalysts like barium carbonate, zeolite 1 (pore size ~4 Å), silica alumina 1 (SA1) (silica (~30 nm) 83.3%, alumina (~30 nm) 16.7%), silica alumina 2 (SA2) (silica 21.1%, alumina 78.9%), SA1 + Z1 (70% SA1, 30% Z1), and zeolite 2 (sodium Y zeolite) for the thermal cracking of waste polythene and selecting the most suitable catalyst based on the yield and thermophysical properties of the hydrocarbon oil obtained.

3. Literature Review

A lot of research has been done in noncatalytic and catalytic pyrolyses of plastics which proves that plastics waste can indeed be converted to useful chemical feedstock. The works reveal that the product distribution can be affected by a number of parameters which include the polymer source (plastic type), catalysts used, size of the catalyst, catalyst to polymer ratio, reaction temperature, reaction time, and reactor type. The effect of various process variables is described below.

3.1. Effect of Catalyst. The catalyst used in the pyrolysis of plastics definitely influences the product. The most commonly used catalysts in the literature for plastic waste pyrolysis includes silica alumina, zeolites (beta, USY, ZSM-5, REY, clinoptilolite, etc.), and MCM-41. With increasing number of acid sites, the level of the catalyst activity in polyolefin pyrolysis also increases. Thus, zeolite-based catalysts due to their high acid strength achieve higher conversion than nonzeolitic catalysts. Songip et al. [3] studied the conversion of polyethylene to transportation fuel using HY, rare earth metal-exchanged Y-type (REY), and HZSM-5 zeolites and silica-alumina (SA). It was found that REY zeolite was the most suitable catalyst producing plastic oil with the highest octane number and gasoline yield. REY had large pores and had proper acidic strength which made it the most suitable one. Y zeolite and ZSM-5 zeolite produced oils having a high research octane number comparable to that of the oil by REY, but the gasoline yield by the formers was significantly low as compared to REY. The catalytic degradation of polyethylene by ultrastable-Y zeolite was studied by Manos et al. [4]. Low pyrolysis temperature does not cause the polymer to fully degrade, and a solid residue is produced in the reaction bed. It showed that the catalyst has significantly reduced the degradation temperature as compared with pure thermal degradation in the absence of a catalyst. The products of the catalytic degradation were hydrocarbons in the C_3–C_{15} range. The catalyst was highly acidic, producing oil with high octane number. A number of research works have been done to find out the effect of silica alumina as catalyst for the pyrolysis of plastics. It can be seen that with silica alumina, high liquid yield can be obtained. The effects of silica alumina with two different SiO_2/Al_2O_3 proportions; that is, SA-1 (SiO_2/Al_2O_3 ratio of 83.3/16.7) and silica alumina SA-2 (SiO_2/Al_2O_3 = 21.1/78.9) were studied by Uddin et al. [5]. The liquid yield was found to be 68 wt% for SA-1 as compared to 77 wt% for SA-2. Therefore, the SA-1 catalyst degraded the polyethylene sample into much lighter hydrocarbon fuel oil than the SA-2 catalyst. Thus it can be concluded that the yield and composition of the liquid products can be controlled by altering the SiO_2/Al_2O_3 ratio. The liquid products are distributed in C_5–C_{20} range, that is, basically in the gasoline and diesel ranges. The effect of nonacidic catalysts for the pyrolysis of plastics was studied by Jan et al. [6]. On comparison with $MgCO_3$ when used as a catalyst under 450°C, it could be observed that the % oil yield (33.60%) is higher with $MgCO_3$ as compared to the % oil yield (29.60%) obtained with $BaCO_3$ catalyst. Similarly when $CaCO_3$ was used as a catalyst under the same reaction conditions, the obtained % oil yield was 32.20%.

3.2. Effect of Catalyst Contact Mode. There exist two methods by which catalyst can be added to the pyrolysis reactor: liquid phase contact and vapor phase contact. In the liquid phase contact, the catalyst and polymer are mixed together, and then they are placed in the reactor and heated to the reaction temperature. However, in the vapor phase contact, the polymer is first subjected to thermolysis to produce the volatile fraction. The catalyst is inserted in the path of the moving vapour, and as the vapour moves through the catalyst, the hydrocarbon vapour is degraded to get the required product distribution. However, the product yield is reported not to differ significantly with the two modes [1].

3.3. Effect of Polymer to Catalyst Ratio. Effect of polymer to catalyst ratio has been studied by Akpanudoh et al. [7]. It has been concluded that with the increase in the amount of catalyst, a direct proportionality in terms of the effectiveness is not obtained. The increase in catalyst amount increases the conversion up to particular limit, but a further increase in the catalyst percentage does not give any appreciable increase in the conversion rate. The optimum polymer to catalyst ratio as obtained from studies is 4 : 1. However, it is also found in the literature that a lesser catalyst ratio will also provide similar degradation, but only at higher reaction temperatures [8]. Some kind of optimisation has to be done with the catalyst ratio and temperature, so that the operation remains economical too.

3.4. Effect of Temperature. If the catalytic pyrolysis is taking place at higher operating temperature or at high heating rates, it causes the enhancement of bond breaking and thereby

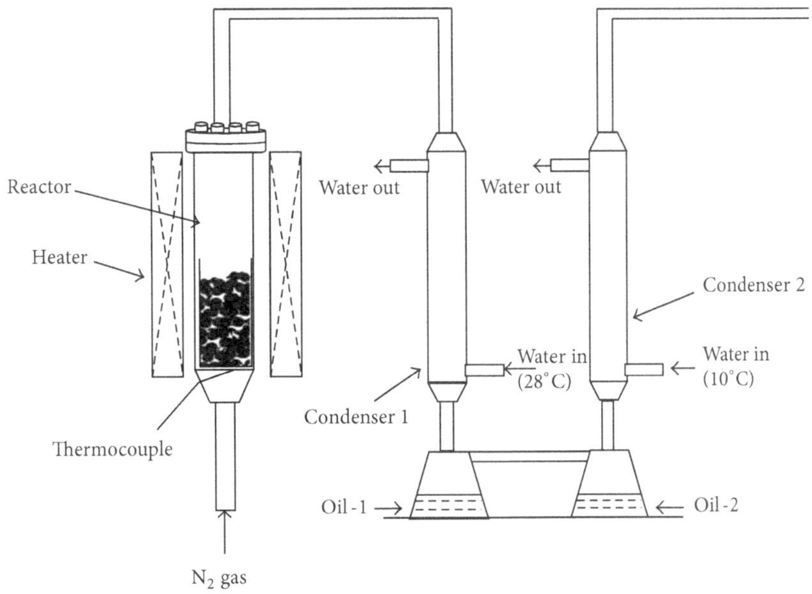

FIGURE 2: Schematic model of the experimental setup.

favouring the production of smaller molecules. The extend of conversion increases with increase of temperature, and it can be seen that with higher conversion the major products formed will be the gaseous products and the liquid yield being minimum or nil. The effect of different catalysts on the liquid yield and the product distribution becomes less significant with increasing temperature. The reaction taking place will be similar to thermal degradation [8].

3.5. Effect of Flow Rate of Nitrogen Gas. The inert gas flowing through the reaction does not affect the reaction directly, but it can produce a slight change in the liquid yield. Usually the nitrogen flow rate was chosen to be relatively high, in order to move the volatile primary products from the reactor and keep secondary reactions at a minimum. This actually favours the liquid yield. But studies of Gulab et al. [8] indicate that high carrier gas flow rate can enhance the evaporation of liquid products which are collected in the condenser. This falsifies the results of liquid yield. By course of experiments, it has been found that the optimum flow rate is 10 mL/min.

The objective of the present work is to synthesize petroleum based fuel by the catalytic pyrolysis of waste plastics using different catalysts, optimization of the yield based on catalyst and reaction conditions, and its performance analysis in an IC engine.

4. Waste Plastic Oil Production

Polythene is selected as the source of waste plastics since it comprises a prominent percentage of the waste plastic produced. The catalysts identified for the study include silica alumina, zeolites, barium carbonate, titanium chloride, and their combinations. The pyrolysis reaction is carried out in the polymer to catalyst ratio of 4 : 1. The reaction temperature ranges between 350 and 450°C. The inert atmosphere for the pyrolysis is provided by using nitrogen as a carrier gas, and the flow rate is fixed to be 10 mL/min. These selections have been made on the basis of the literature survey that has been carried out. A schematic sketch of the experimental setup is as shown in Figure 2. The setup consists of a ceramic electric heater (reactor), a steel container, two condensers, two oil collectors, and a nitrogen source (inert gas). The maximum loading capacity of the reactor was 1.5 kg of waste plastics. The reactor consists of three ceramic heaters, each having power of 2000 W, arranged in series over a cast iron pipe of 17 cm diameter and 60 cm length. The double pipe counter flow heat exchanger of length 90 cm functions as the condenser. Water at around 28°C (room temperature) is used as the coolant in the first condenser, and the temperature of water being supplied to the second heat exchanger is 10°C. The waste plastic is placed inside a steel container of 15 cm diameter and 20 cm height at a packing density of 424 kg/m^3. This container is finally placed inside the heater. The purpose of this container is to avoid the flow of melted waste plastic downwards under gravity as such a flow can block the passage of the nitrogen gas.

The waste plastic was mixed with the catalyst in a twin roll mill, before it is being supplied to the heater. During the mixing process the plastic gets heated enough to get devoid of any moisture content. The reactor was fixed vertically, and nitrogen gas was introduced into the reactor from the bottom. The flow of nitrogen replaces the air from the reactor and permits the pyrolysis reaction under anaerobic condition. Before starting the heating, nitrogen gas is allowed to flow through the heater unit to remove the oxygen that is present initially. Then heater is switched on, and the temperature controller is set to the required operational temperature. The vapor fraction formed during the pyrolysis of the plastic inside the reactor flows out along with N_2 out of the reactor. The gas mixture is first cooled in the condenser 1. Ordinary room

TABLE 2: Optimum properties of plastic oil with different catalysts.

Catalyst	None	$BaCO_3$	Z1	SA1	SA2	SA1 + Z1	Z2
Temp. (°C)	450	450	450	425	450	425	450
Liquid collected (mL)	80	50	110	145	130	130	100
After filtration (mL)	60	40	100	130	120	120	90
Calorific value (MJ/kg)	41.35	36.61	45.15	41.36	36.83	44.57	42.24
Viscosity ($\times10^{-3}$ Ns/m^2)	1.2699	1.6534	1.1891	1.2865	1.1995	1.2066	1.2245
Density (g/cc)	0.8581	0.9745	0.8635	0.9471	0.9106	0.9265	0.8785
Flash point (°C)	<32	—	<32	<32	<32	<32	<32
Fire point (°C)	≤32	—	36	34	36	35	34
Cloud point (°C)	−3	—	−2	−3	−3	−2	−2
Pour point (°C)	−13	—	−12	−12	−13	−13	−12

temperature water is supplied to the condenser for the cooling action. The low boiling fractions of the vapour fraction will be condensed and collected in a collector fitted to the condenser 1. The remaining uncondensed fraction moves to the condenser 2. The cooling water for condenser is at a temperature of 10°C. This low temperature for water is provided by an external refrigeration setup. The low boiling components moving through condenser 2 will be condensed and collected in a collector. The remaining uncondensed part escapes to the atmosphere.

The oil samples are produced without using catalysts and also in the presence of catalysts such as barium carbonate, zeolite 1 (pore size ~4 Å), silica alumina 1 (SA1) (silica (~30 nm) 83.3%, alumina (~30 nm) 16.7%), silica alumina 2 (SA2) (silica 21.1%, alumina 78.9%), SA1 + Z1 (70% SA1, 30% Z1), and zeolite 2 (sodium Y zeolite), at reaction temperatures ranging from 400°C to 550°C.

The polythene waste is first shredded to sizes of 1-2 cm and then mixed with the test catalyst (polymer to catalyst ratio of 10 : 1) uniformly in the two roll mill with friction ratio of 1 : 1.4. The mixed plastic and catalyst are then inserted into the heater assembly and are heated to the required temperature of approximately 400°C. An inert gas (nitrogen) is passed through the heater assembly so as to prevent any kind of oxidation reaction that may take place. Once the temperature is attained, it is maintained for a preset reaction time (say 3 hours). After the reaction time is over, the heater is switched off, but the reaction is allowed to take place for another one more hour, so that maximum volatile fraction formed will pass through the two heat exchangers. The condensed volatile fraction is finally collected from the collecting tanks and is filtered. Now the process is repeated for different temperatures for various catalysts.

5. Yield of Bio-Oil and Oil Properties

Out of the oil samples produced by different catalysts at different temperatures, the samples which showed optimum liquid yield for a particular catalyst are shown along with their properties in Table 2. It was noted that the liquid yield was available only at temperatures above 350°C for all catalysts. With the increase in temperature, the liquid yield decreased after a particular temperature (which was different for different catalyst). Finally, above 600°C no liquid yield was obtained for any catalyst thereby making the liquid yield gate between 400 and 550°C. Along with the oil obtained, a gel portion was also present along with the impurities which was filtered and removed. From the obtained results of liquid yield and its properties, the catalyst combination of silica alumina 1 and zeolite 1 (70% SA1, 30% Z1) is selected as the best catalyst with the optimum reaction temperature as 425°C. The oil produced using this optimum catalyst was used for GC-MS and was produced in adequate amount for the preparation of blends for testing in the engine.

6. Chemical Composition of the Oil Sample

Gas chromatography-mass spectrometry (GC-MS) of the selected oil sample was done to examine its chemical composition. Chloroform was used as the solvent for the oil sample. The mass spectrum shows a number of peaks showing that the oil is a mixture of a number of compounds. The compounds vary from C_9 to C_{19}, showing clear similarities to diesel in addition to the oil properties showing closeness to that of diesel. The mass spectrum is shown in Figure 3.

The list of compounds shown in the mass spectrum, along with the area % is listed in the Table 3. From the area %, the average chemical formula of the sample oil can be calculated and is obtained as $C_{13.18}H_{23.56}$ (the average chemical formula of diesel is $C_{12}H_{23}$). The equation for the complete combustion of the plastic oil is as follows:

$$C_{13.8}H_{23.56} + 19.07O_2 + 19.07\left(3.76N_2\right) \longrightarrow 13.18CO_2 + 11.78H_2O + 71.7N_2 \quad (1)$$

The theoretical air fuel ratio of the plastic oil is found to be 14.47. Due to the close similarities of the properties of the oil with diesel, the performance test of the oil is done in a CI engine.

7. Performance Test of the Blends in a CI Engine

The experimental setup consists of a single cylinder Kirloskar CI engine (5 hp, 1500 rpm, 4 stroke, and 500 cc), which is

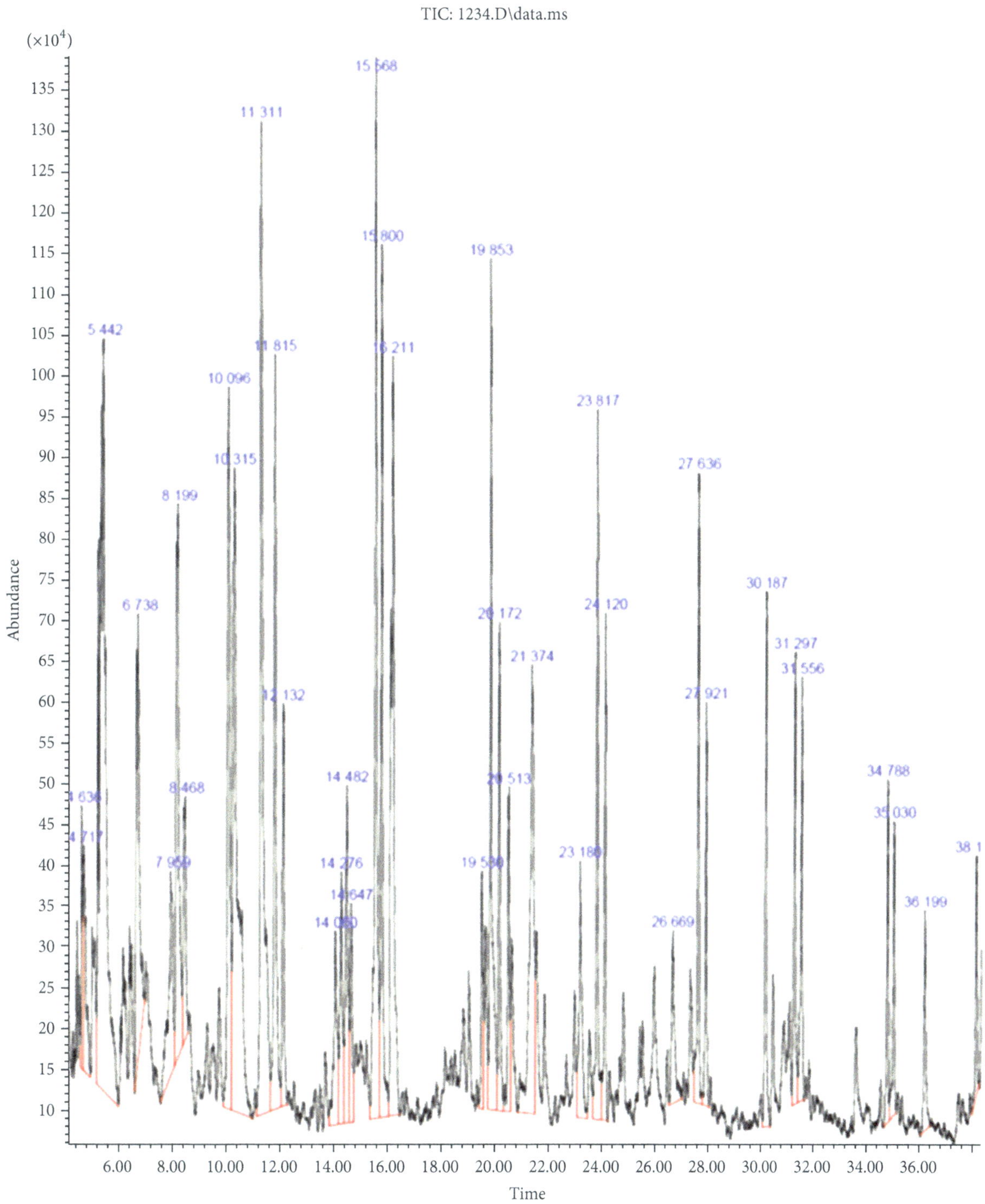

FIGURE 3: Mass spectrum of plastic oil sample (GC-MS).

mechanically loaded by means of a brake drum dynamometer. The performance of plastic oil blends in a CI engine was investigated and compared with pure diesel. Five blends of the oil were prepared. It includes B10 (10% oil and 90% diesel), B20 (20% oil and 80% diesel), B30 (30% oil and 70% diesel), B50 (50% oil and 50% diesel), and B80 (80% oil and 20% diesel). The engine is run using the above seven oils, and the results are compared.

The variation of fuel consumption with brake power is shown in Figure 4. With the increase in brake power the engine requires more energy and hence more fuel causing an increase in fuel consumption. It can be seen from the figure that with the increase in the concentration of the plastic oil in the blends, the fuel consumption goes on increasing; the reason is the plastic oil has a lesser calorific value than diesel (lower calorific value of plastic oil: 41.89 MJ/kg).

Table 3: Chemical composition of plastic oil sample.

Peak	Retention time (min.)	Composition	Chemical formula	Area %
4	6.738	Indene	C9H8	6.28
5	7.96	2-Butenyl-benzene	C10H12	3.94
6	8.201	1-Undecene	C11H22	7.39
11	11.817	1-Dodecene	C12H24	7.88
12	12.314	Dodecane	C12H26	3.39
13	14.062	1,3-Dimethyl-1H-indene	C11H14	2.91
14	14.278	1,3-Dimethyl-1H-indene	C11H12	3.39
18	15.8	1-Tridecene	C13H26	7.41
19	16.21	2-Methyl-naphthalene	C11H10	10.95
21	19.851	1-Tetradecene	C14H28	8.42
22	20.172	Tetradecane	C14H30	3.84
23	20.515	2,7-Dimethyl naphthalene	C12H12	2.79
26	23.818	1-Pentadecene	C15H30	5.99
27	24.122	Pentadecane	C15H32	3.88
29	27.636	1-Hexadecene	C16H32	4.84
30	27.919	Hexadecane	C16H34	2.89
32	31.298	3-Heptadecene	C17H34	3.57
33	31.556	Heptadecane	C17H36	3.55
34	34.787	1-Octadecene	C18H36	2.42
35	35.032	Pentadecane	C18H37	2.4
37	38.123	1-Nonadecene	C19H38	1.76

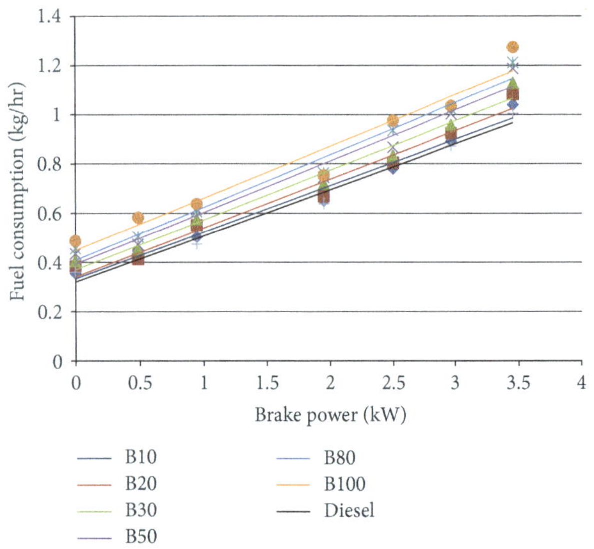

Figure 4: Fuel consumption versus brake power.

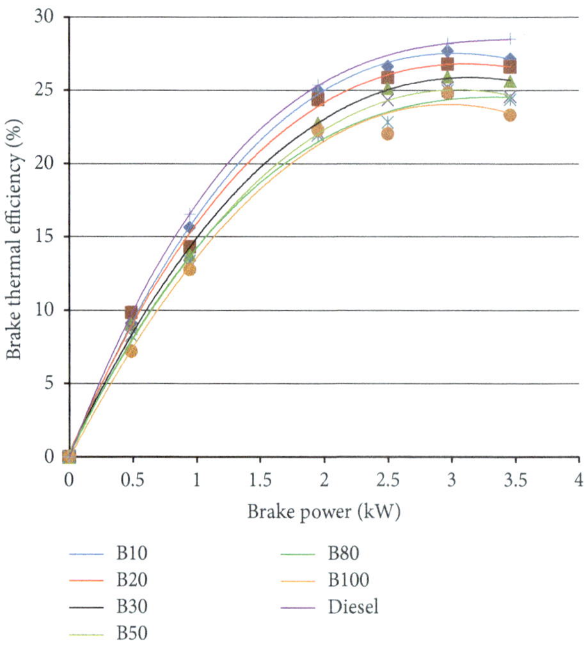

Figure 5: Brake thermal efficiency versus brake power.

The variation of brake thermal efficiency with brake power is shown in Figure 5. Brake thermal efficiency is defined as the ratio of brake power output to the net input power. Brake thermal efficiency increases with brake power only up to a limit beyond which it drops due the incomplete combustion taking place. Here, with increase in the concentration of plastic oil in the blends, the efficiency decreases which is due to the higher fuel consumption, and the pure plastic oil gives the least efficiency.

The variation of indicated thermal efficiency with brake power is shown in Figure 6. Indicated thermal efficiency is defined as the ratio of indicated power output to the input power. Even though the frictional power increases with the blend %, the blends show less efficiency because of its higher fuel consumption. Interestingly it can be noticed in the figure

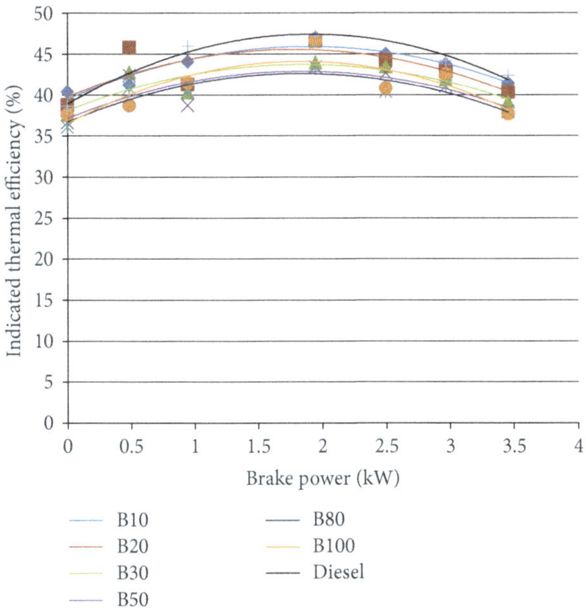

FIGURE 6: Indicated thermal efficiency versus brake power.

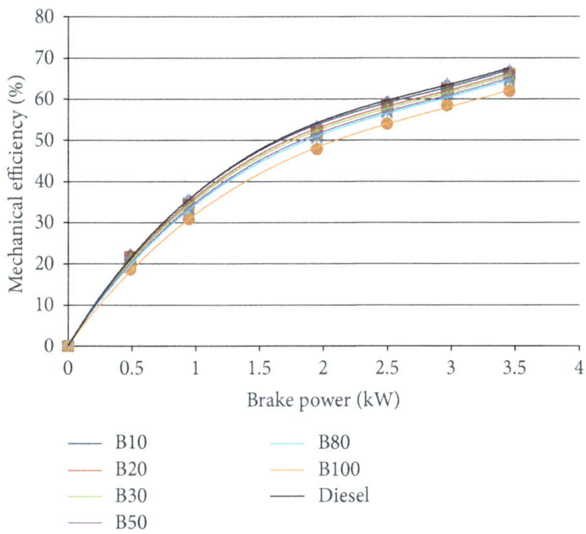

FIGURE 7: Mechanical efficiency versus brake power.

that the higher blends and pure oil show a higher efficiency than its lower counterparts. This is because the increase in frictional power is less for higher blends as compared to lesser blends.

The variation of mechanical efficiency with brake power is shown in Figure 7. Mechanical efficiency is defined as the ratio of brake power to the indicated power. It can be seen that the mechanical efficiency decreases with the increase in blend concentration. This can be attributed to the increase in frictional power with the increase in blend %.

The variation of volumetric efficiency with brake power is shown in Figure 8. Since the test engine is a constant speed engine, volumetric efficiency remains the same irrespective of the fuel used.

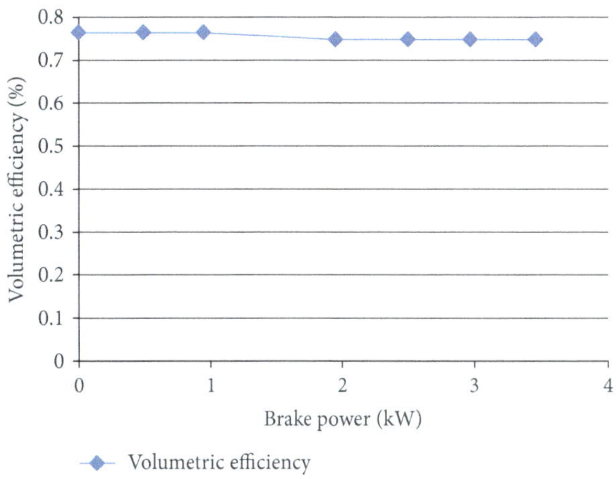

FIGURE 8: Volumetric efficiency versus brake power.

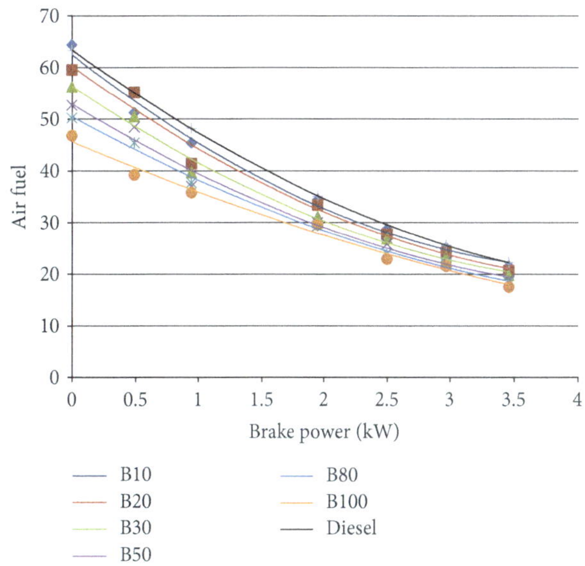

FIGURE 9: Air fuel ratio versus brake power.

The variation of air-fuel ratio with brake power is shown in Figure 9. With increase in brake power, more will be the fuel consumption, and hence the air to fuel ratio will be decreasing. From the figure it can be seen that the air fuel ratio is going on decreasing with increase in the plastic oil concentration, which is due to the increasing fuel consumption with increasing blend %.

The variation of brake specific fuel consumption with brake power is shown in Figure 10. Brake specific fuel consumption is defined as the amount of fuel required for producing unit brake power. The plastic oil and its blends have higher bsfc than diesel because of their higher fuel consumption arising from their lesser calorific value than diesel.

8. Emission Characteristics

From the performance test analysis it can be seen that the test results of B20 (20% plastic oil and 80% diesel blend) showed

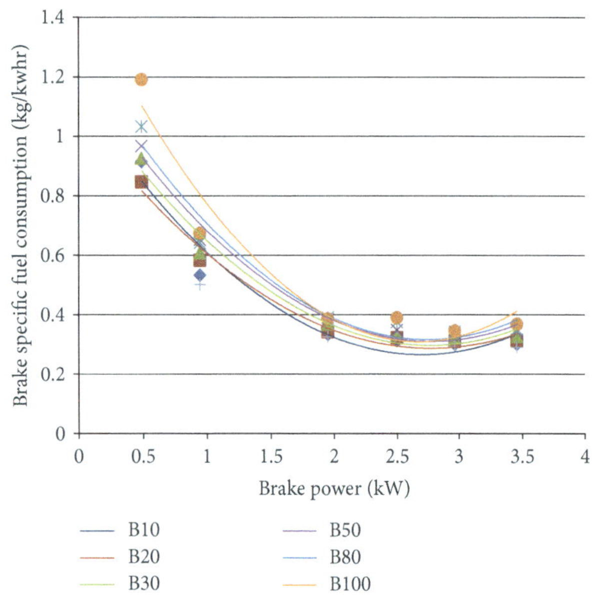

FIGURE 10: Brake specific fuel consumption versus brake power.

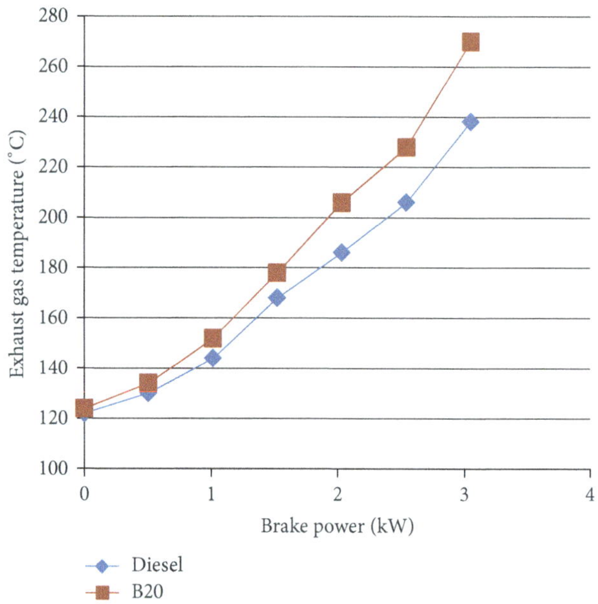

FIGURE 12: Exhaust gas temperature versus brake power.

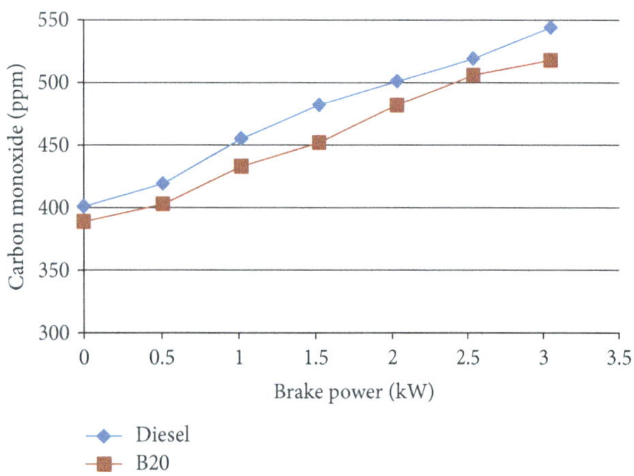

FIGURE 11: Carbon monoxide emission (ppm) versus brake power.

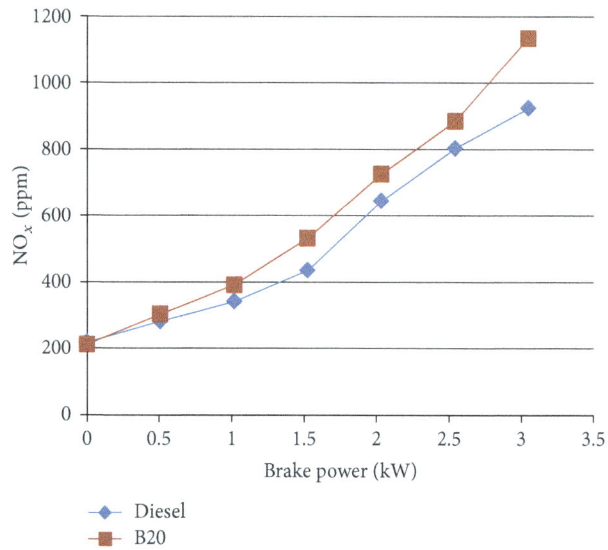

FIGURE 13: NO_x emission versus brake power.

close similarities with that of diesel. So for the analysis of the emission characteristics, the engine was run with B20 and diesel, and the emission was analysed using an exhaust gas analyser. Emissions analysed were CO emission and NO_x emission.

The variation of carbon monoxide content with brake power is compared for diesel and B20 as shown in Figure 11. It can be seen that the CO emission is lesser for B20 as compared to diesel.

From the literature, it can be seen that with plastic oil there is an ignition delay of 2.5° crank angle. This ignition delay causes a steep rise in the peak pressure causing a high exhaust temperature. This is evident from Figure 12 which shows a comparison of the exhaust gas temperatures of B20 and diesel.

The variation of NO_x emission with brake power is compared for diesel and B20 as shown in Figure 13. The exhaust gas temperature drives a direct relation with the NO_x emission. So it is clear that the B20 sample will give higher NO_x emission than with diesel.

9. Conclusion

(i) A petroleum based fuel has been produced from waste plastic (polythene).

(ii) The optimum catalyst and reactions for catalytic pyrolysis of polythene have been found. Based on the yield and thermophysical properties, the combination of silica alumina and zeolite 1 (pore size ~4 Å) was selected as the optimum catalyst.

(iii) The properties of the plastic oil and its chemical composition have been examined. The average chemical formula was found to be $C_{13.18}H_{23.56}$, and hence the performance analysis was done in a CI engine.

(iv) In the performance analysis in engine, even though the plastic oil shows inferior results as compared to diesel, the lower blends percentage oils show results close with that of diesel (B10, B20, and B30). This makes it a strong competitor in the area of alternate fuels. Also the blend B20 has low CO emissions than for diesel. However, the NO_x emissions are higher for B20.

(v) 64.15% of the production cost is accounted for the cost of catalysts. If cheaper catalysts can be employed, the production cost can be decreased considerably.

(vi) If the gaseous products and solid can be used, then the effective cost will come down even further.

(vii) Rather than considering it just as an alternate fuel, the practical importance of this method in waste plastic management adds its value as an alternate fuel.

References

[1] A. K. Panda, R. K. Singh, and D. K. Mishra, "Thermolysis of waste plastics to liquid fuel: a suitable method for plastic waste management and manufacture of value added products: a world prospective," *Renewable and Sustainable Energy Reviews*, vol. 14, no. 1, pp. 233–248, 2010.

[2] http://cipet.gov.in/plastics_statics.html.

[3] A. R. Songip, T. Masuda, H. Kuwahara, and K. Hashimoto, "Test to screen catalysts for reforming heavy oil from waste plastics," *Applied Catalysis B*, vol. 2, no. 2-3, pp. 153–164, 1993.

[4] G. Manos, A. Garforth, and J. Dwyer, "Catalytic degradation of high-density polyethylene on an ultrastable-Y zeolite. Nature of initial polymer reactions, pattern of formation of gas and liquid products, and temperature effects," *Industrial and Engineering Chemistry Research*, vol. 39, no. 5, pp. 1203–1208, 2000.

[5] M. A. Uddin, K. Koizumi, K. Murata, and Y. Sakata, "Thermal and catalytic degradation of structurally different types of polyethylene into fuel oil," *Polymer Degradation and Stability*, vol. 56, no. 1, pp. 37–44, 1997.

[6] M. R. Jan, J. Shah, and H. Gulab, "Catalytic degradation of Waste high-density polyethylene into fuel products using $BaCO_3$ as a catalyst," *Fuel Processing Technology*, vol. 91, no. 11, pp. 1428–1437, 2010.

[7] N. S. Akpanudoh, K. Gobin, and G. Manos, "Catalytic degradation of plastic waste to liquid fuel over commercial cracking catalysts: effect of polymer to catalyst ratio/acidity content," *Journal of Molecular Catalysis A*, vol. 235, no. 1-2, pp. 67–73, 2005.

[8] H. Gulab, M. R. Jan, J. Shah, and G. Manos, "Plastic catalytic pyrolysis to fuels as tertiary polymer recycling method: effect of process conditions," *Journal of Environmental Science and Health*, vol. 45, no. 7, pp. 908–915, 2010.

Energy Cost Optimization in a Water Supply System Case Study

Daniel F. Moreira[1] and Helena M. Ramos[1,2]

[1] *CEHIDRO, Instituto Superior Técnico, 1049-001 Lisbon, Portugal*
[2] *Instituto Superior Técnico, 1049-001 Lisbon, Portugal*

Correspondence should be addressed to Helena M. Ramos; hramos.ist@gmail.com

Academic Editor: S. Venkata Mohan

The majority of the life cycle costs (LCC) of a pump are related to the energy spent in pumping, with the rest being related to the purchase and maintenance of the equipment. Any optimizations in the energy efficiency of the pumps result in a considerable reduction of the total operational cost. The Fátima water supply system in Portugal was analyzed in order to minimize its operational energy costs. Different pump characteristic curves were analyzed and modeled in order to achieve the most efficient operation point. To determine the best daily pumping operational scheduling pattern, genetic algorithm (GA) optimization embedded in the modeling software was considered in contrast with a manual override (MO) approach. The main goal was to determine which pumps and what daily scheduling allowed the best economical solution. At the end of the analysis it was possible to reduce the original daily energy costs by 43.7%. This was achieved by introducing more appropriate pumps and by intelligent programming of their operation. Given the heuristic nature of GAs, different approaches were employed and the most common errors were pinpointed, whereby this investigation can be used as a reference for similar future developments.

1. Introduction

Investigating efforts in new technological innovations are aimed to improving the energy efficiency of the several production sectors, which also applies to water supply companies [1, 2]. Energy consumption due to the pumping systems represents the biggest parcel of the energy expenses in the water sector—sometimes up to 90% [3]—and there are several practical solutions which can enable a reduction in these costs. Pump and motor upgrades to more efficient solutions, either being technologically more advanced or because they are more properly adjusted to the system, often allow significant energetic savings [4]. Also, changing the pumping operational procedures is very effective since it does not need any additional investment and because the economy with the reduction of the energy cost occurs immediately. The consumption of energy in most of the water systems all over the world could be reduced by at least 25% [5], only by improving the performance in terms of energy efficiency.

Operators of the water supply networks have a complex task in taking into account the distinct goals involved in this process. Determining operational rules to improve the quality of the service and that are also energetically economical, among an extensive set of possibilities, requires the utilization of models which take into consideration these components (FILHO, 2006).

Technological advances in the computational area enabled an increase in the quality of the scientific works related to model optimization, as well as reducing the energy cost of operation. Initially, investigations related to cost optimization of pumping expenses relied on operational research techniques as, for instance, linear programming [6, 7], integral linear programming [8], nonlinear programming [9, 10], and dynamic programming [11, 12]. Wood and Reddy (1994) [13] were the first to utilize *genetic algorithms* (GA) to reduce the energetic cost of pumping systems. These algorithms can be used as heuristic optimization models for achieving the best energy policy to be applied in a water supply system (WSS), enabling the determination of the optimal scheduling of control settings of pumps throughout each day.

The goal is to analyze a WSS, model it accordingly, and simulate several scheduling scenarios with optimized pumps in order to minimize the expenditure related to energy consumption, based on the demand patterns and associated energy tariffs.

2. Background Review

2.1. Genetic Algorithms

2.1.1. Nature and Applications. Genetic algorithms are stochastic methods of search that begin with a *population* of random *chromosomes*, which represent different *solutions*. The basic principles of the GA, widespread by Goldberg [14], are inspired in Darwin's evolutionary concept of survival of the fittest [15], evaluating each generation of *individuals* and allocating reproductive opportunities in such a way that those which represent a better solution to the target problem are given more chances to reproduce than those which are poorer solutions, numerically quantified as their fitness [16]. Convergence occurs as the chromosomes evolve to best fit the *constrains* of their environment.

These algorithms are commonly used to find optimal solutions in complex systems such as the scheduling of supply systems, where the amount of variables leads to vast possible solutions. There are a certain number of initial parameters that can be tuned, prior to the simulation, that affect the quality of the convergence.

2.1.2. Initial Parameters. Variations in the parameters that allow the GA to be tuned prior to the simulations influence the convergence process. In the current study, the main parameters considered were the *population size* (PS) of each generation, the *elite population size* (EPS) in each one of them, and the *stopping criteria* (SC) of the algorithm.

PS represents the number of individuals with different chromosomes that are created each generation. Smaller PS increases optimization speed but reduces reliability as there might not be enough genetic variety and only a less optimal part of the solution space is explored. On the other hand, too many chromosomes slow down the GA convergence [17].

EPS represents the amount of the fittest individuals from each generation that will be automatically chosen for the next one without being subject to crossover. This increases the performance of the GA, as it prevents the best found solutions to be lost if they were to be recombined. Nevertheless, increasing this value too much may lead to premature convergence to local optima.

SC is heuristic stopping criteria whose goal is ending the simulation after a limit number of generations without fitness improvement.

2.1.3. Best Practices. In pump scheduling it is possible to calculate the *solution space* (S_s) given the number of possible *decisions* (n_d) and the number of pump speed *settings* (n_s), where the total solution space can be determined as being $S_s = n_s^{n_d}$. To determine the solution space for two scenarios, one with 1 pump and the other with 2 pumps, with 2 possible drive speed settings each, in a 24-hour period with 1-hour discrete steps, the total would be $2^{1\times 24}$ and $2^{2\times 24}$ possible combinations, respectively.

It is not immediately obvious that the size of the solution space in optimization problems can quickly grow to overwhelming sizes. Adding a second pump in a scenario in which they can be turned on or off in a 24-hour period with 1-hour discrete steps increases the solution space almost 17 million times. It is therefore important to consider minimizing the number of pumps, the number of speed choices, and the duration and course of the simulation [18].

Any improvement that can be made towards reducing the duration of each individual simulation directly reduces the GA's overall run time. Consideration should be taken to keep the model size small and simple.

The nature of GAs implies that it has an inherent randomness associated with it. Therefore, two optimization runs that are otherwise identical except for one minor change (e.g., larger PS) will in all likelihood produce different optimized solutions. This is more likely to be the case the larger the solution space of the problem. It is a good practice to run multiple optimizations changing nothing other than one or more genetic algorithm parameters to ensure that the best optimized solution is really the best that can be achieved.

2.2. Characteristic Curves. The *pump characteristic curve* describes the relation between *flow rate* and *total head* for a specific pump (Figure 1). Other important information is also included, such as pump curves for different *impeller diameters*, *net positive suction head curve* (NPSH), and the *efficiency* and *power* curves [19].

In the case of the pump characteristic curves used, their designation determines its specific *nominal diameter*, in couple with nominal and actual *impeller diameters* (for the top curve in Figure 1). The pump's efficiency throughout its characteristic curve varies and should not drift too much from the *best efficiency point* (BEP). The motors rotating the *shaft* that moves the impeller, whose pole number is referenced in the pump designation, also have their own efficiencies that must be considered.

The *system characteristic curve* describes graphically the variation of head (H) given different flows (Q), according to an equation which depends on the pipe diameter (d) and its specific roughness coefficient (K). In this case, the Hazen-Williams empirical formula was used (1):

$$H = \frac{10.67\ Q^{1.65}}{K^{1.85}\ d^{4.87}}. \tag{1}$$

The *operation point* in the pump curve is dependent upon the characteristics of the system in which it is operating. The *system head curve* is the graphic representation of the *energy equation*, or the relationship between flow and hydraulic losses in a system. Figure 2 shows that, by plotting both the system head curve and pump curve together, the pump's operating point (Q_1, H_1) can be determined.

When two or more pumps are arranged in parallel, their resulting performance curve is obtained by adding their flow rates at the same head as indicated in Figure 2. The operating point at the intersection of the resulting curve (Q_2, H_2) represents a higher volumetric flow rate than for a single pump and a greater system head loss. In the case of

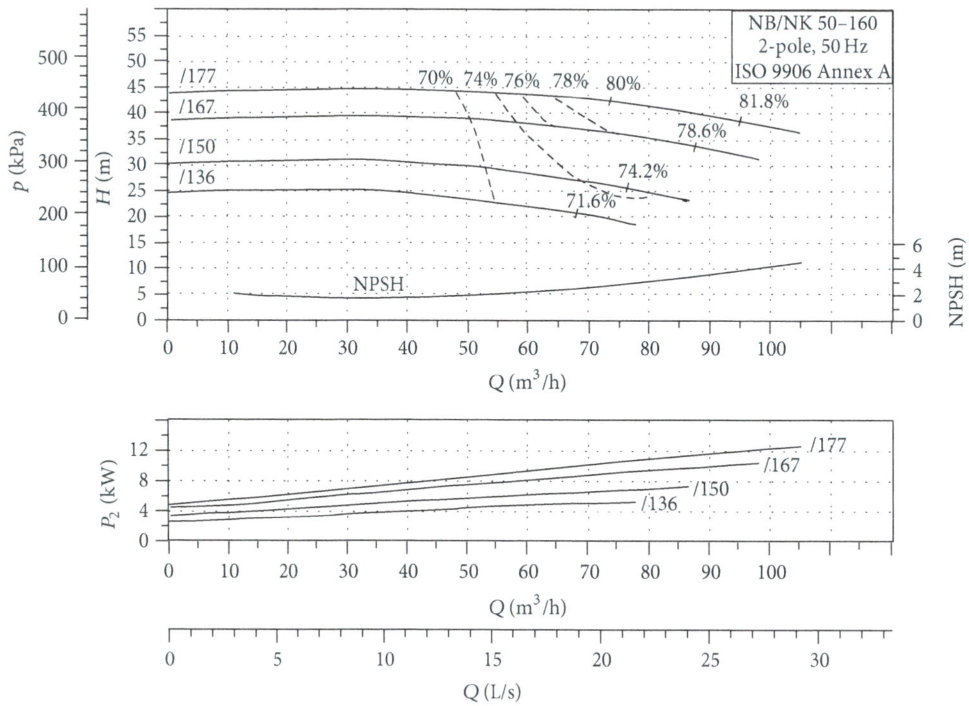

FIGURE 1: Example of a pump curve.

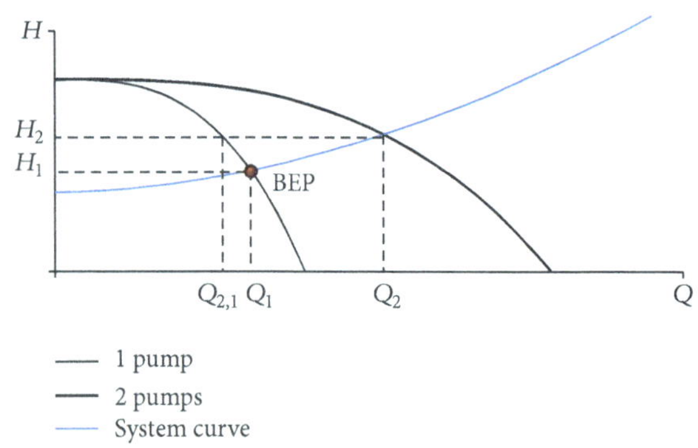

FIGURE 2: Graphic representation of one system curve intersecting two pump curves.

two similar pumps, because of the greater system head, the volumetric flow rate is actually less than twice the flow rate achieved by using single pump [19].

When using pumps with operation points next to their BEP (Q_1, H_1), it is not recommended to run them in parallel, since they will be operating at less efficient flows ($Q_{2,1}$, H_2).

The affinity laws expressed in (2) represent the mathematical relationship between the *speed* (n), *discharge* (Q), *head* (H), and *water power* (P) of a pump.

$$\frac{n_1}{n_2} = \frac{Q_1}{Q_2} = \sqrt[2]{\frac{H_1}{H_2}} = \sqrt[3]{\frac{P_1}{P_2}}. \quad (2)$$

There are different ratios between the pump's speed and the other variables. A reduction of 10% in pump speed translates into a 27% reduction in power consumption. It must be noted that this also reduces the pump head in 19%.

For *variable-speed-driven* (VSD) pumps, small speed variations of the shaft—directly proportional to the flow—translate into significant pump power variations, which can increase the efficiency of the pumping operation when compared to pumps equipped with *fixed-speed drives* (FSD). Nevertheless, this also affects the pump head, rendering it inoperable below the point where it will not cross the system curve.

According to [19], as these points come closer to each other and the system curve becomes tangent to the pump curve, the operation of the pump becomes extremely unstable if transient regimes occur. Besides, the pump curve can have a shut-off head inferior to the system curve, meaning that the pump is not able to start at that particular speed without a bypass.

3. Case Study

3.1. Water Supply System Selection. The supply system for this case study was chosen following the investigation of Costa and Ramos (2010) serving the demands of the Fátima region, in Portugal. In reality, the pumping station is comprised of two NK 65-250 type pumps with an average pumped flow of 42 l/s, with an overall efficiency of 60%. Since the goal was to test a better type of GA, the pump curve utilized had a constant discharge for all heads and the efficiency was invariable. Thus, these simplifications imply that the data obtained in terms of energy cost should be considered less than accurate, but relevant, proving that GA scheduling optimization can improve considerably the efficiency of a pumping operation.

Figure 3 represents the WSS modeling. The water is pumped through a 250 mm diameter pipe with a Hazen-Williams roughness coefficient of 145, spanning for 1,607.8 m

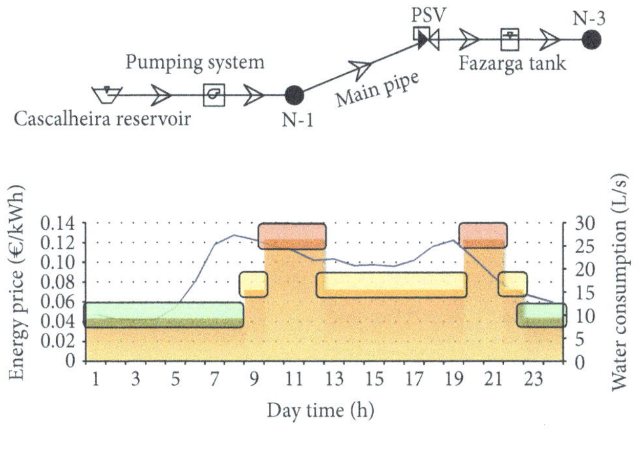

FIGURE 3: Computational model of the pump system (top), daily energy tariff and water consumption in node N-3 (bottom).

and connecting the Cascalheira reservoir (elevation: 375.3 m) to the Fazarga tank (405.0 m).

The tank has a volume of 347.0 m^3 when full, but it operates only beyond a 45.3 m^3 minimum reserved for emergencies, and therefore the effective total is of 301.7 m^3. For this study the average daily consumption is of 19 l/s, which represents an average of 4 hours and 24 minutes of available water with a full tank.

Node N-3 was considered the consumption node, simulating the daily water demand variation, based on information from sensors placed downstream the Fazarga tank. There must be also an energy tariff associated with the model, so the simulation can calculate the energy consumption.

According to lanscy and awumah (1994), increasing the number of actions per operational cycle proportionally increases the wear of the pumps. Therefore, this value was limited to a recommended daily maximum of three cycles per pump.

3.2. Simulation Scenarios. Two distinct models were considered to simulate the studied system. These have different model complexities and solution spaces, as well as particular operational modes.

3.2.1. Two Pumps with Three Starts (2 Pump/3). This scenario comprises two pumps, operating independently of each other with a maximum of three start/stop cycles per pump per day. It allows more versatile solutions as the pumps can operate in parallel with independent speeds at the same time, but it is also the option with the biggest solution space and system complexity of all. In addition, when two pumps operate in parallel, the overall operation point changes a potential less efficient one if the pumps are already correctly dimensioned to operate near their BEP.

3.2.2. One Pump with 6 Starts (1 Pump/6). This scenario comprises one pump which can start up to six times per day, emulating two pumps with three starts each but which cannot operate simultaneously. When the pumps are prevented to operate in parallel, it guarantees that the pumping occurs always at the same operating point. Therefore, the computational model has to balance the equations for only one pump, which reduces the simulation time. Furthermore, the solution space is also considerably reduced, along with versatility loss in the potential operational strategies.

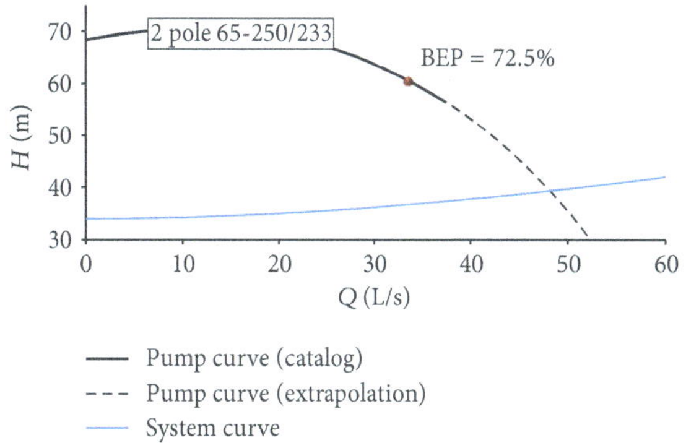

FIGURE 4: Original Veolia pump curve compared to the system curve.

TABLE 1: Solution spaces for each simulation approach for 1 h time steps in a 24 h period.

Scenario	S_s	
	Off/On	Off/I_1/I_2/I_3/I_4/On
1 pump/6	1.68E + 07	4.74E + 18
2 pump/3	2.81E + 14	2.25E + 37

Table 1 shows a comparison between solution spaces of these scenarios when combined with both FSD and VSD scenarios. These actions vary between a simple on/off operation to the possibility of intermediate speeds in a pump equipped with a VSD motor. It must be also noted that, in simulations with four intermediate speeds, the 1 pump/6 approach has almost *five quintillion times less possible solutions* than its 2 pumps/3 counterpart.

3.3. Selection of the Pump and Motor. The total supply needs of the population must be met by the operation of only one pump, whereby the discharge flow must be superior to the average daily demand (20 l/s < Q < 40 l/s), having also considered pumps with a flowup to double that value. In relation to the total head, the pumps should comprise a range which take into account the increase in roughness of the pipes over time (32 m < H < 38 m).

Based on the pump curve, Figure 4 highlights the fact that the pumps used by Veolia do not seem appropriate for this WSS and are probably operating with reduced efficiency. With pump efficiency at BEP of 72.5% and considering a motor with 92% efficiency, the total efficiency at that point is of 66.7%. Since the operating point is far from the BEP, a total efficiency of 60% was determined as reasonable. The operational cost resulting from energy consumption was of 22.2 €/day (12/07/2007). All subsequent simulations were

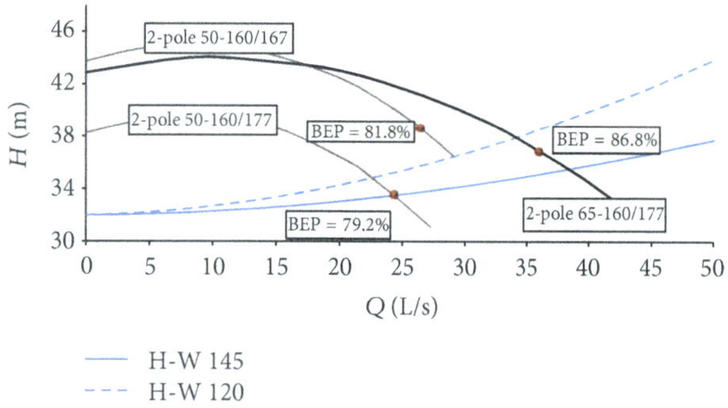

FIGURE 5: Pump curves of the considered pumps compared to system curves with two different roughness (Hazen-Williams coefficients of 145 and 120).

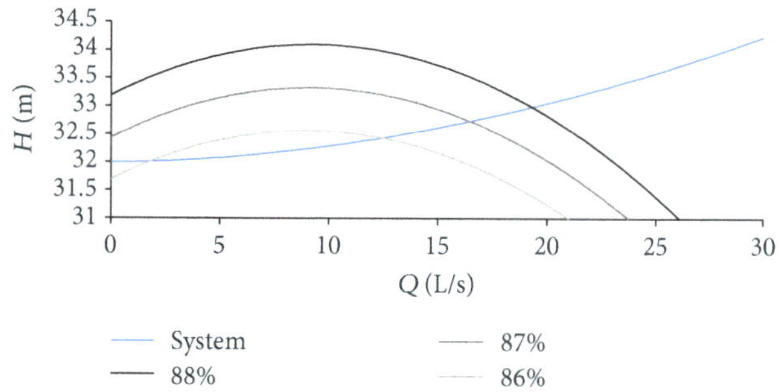

FIGURE 6: 65-160/177 pump curve operating at three different speeds when compared to the system curve.

performed using the energy tariffs and consumption curves practiced at that time for coherence in comparison.

Using the Grundfos pump catalogue, three 2-pole motor pumps which fitted the considered flow and head ranges were selected for testing. Their characteristic curves were extracted from their specification manuals and input in the modeling software. Since pipe roughness can increase over time, these curves were compared against system curves which represent both present and future scenarios (Figure 5).

(i) Pump 50-160/177 does not seem appropriate, since the pump curve extracted from the catalogue does not intersect the system curve, meaning that it would operate with flows which are not recommended by the manufacturer;

(ii) Pump 50-160/167 has a discharge flow near to the average daily demand, increasing its probability of becoming obsolete if the demand is intensified or if the flow is reduced due to a pipe roughness increase over time. The maximum efficiency is also inferior to the 65-160/177 and, if equipped with a VSD, the possible speed range is minor given its inferior total heads.

The selected pump for comparison with the current one was, therefore, the 65-160/177 equipped with a 2-pole motor (X_1).

By comparison to the system curve, it is relevant to determine the speed range in which this pump can operate so the solution space does not include unnecessarily unfeasible solutions. The affinity laws allow the determination of new curves based on the original one for different drive speeds. Figure 6 shows that in order to avoid system instability, the minimum motor speed should be of 87%.

4. Results

4.1. General Considerations. It is always preferable to operate the pumps at off-peak hours (T_1), as the pumping costs due to electricity consumption for the same volume are inferior. A sufficiently large tank would have enough storage capacity to be filled during those periods, avoiding pumping during peak hours. Two different approaches were studied in order to cope with the fact that the volume of the tank is limited. The demand curve and energy tariff have a repetitive 24 h pattern; thus a daily cycle was considered in the simulations. Thus, the tank must finish the cycle with the same water level as it began, in order to prevent volume variations throughout the days.

The commercial software used for these simulations was Bentley's WaterGEMS, which has embedded in it a genetic algorithm simulation module (Darwin Scheduler).

4.2. Manual Management. In order to be properly compared with the existing scenario, the model was tested with a simulation using the original operational scheduling using the original pump (X_0) and the 65-160/177 pump (X_1). Subsequently, a manual optimization approach was carried out with the goal of determining an operational strategy which sought to restrict operations during semipeak (T_2) and peak (T_3) hours, while guaranteeing a stable supply. The next step was to broaden those intervals during which the pumps were not operating.

The energy cost of operation obtained from the calibration (22.1€/day) was similar to the real data provided by Veolia (22.2€/day), corroborating the reliability of the hydraulic simulation model. This operational regime imposes five pump starts, which probably means that the two pumps took turns so that the daily three starts limit is not exceeded.

In Figure 7, the utilization of pumps during peak hours (12:00-13:00 and 21:00-22:00) was verified. Pump X_1 operated with a total efficiency of 78.9%, a 13.9% difference comparing to pump X_0. Pumps X_0 and X_1 operated with discharge flows of 41.8 l/s and 38.7 l/s, respectively.

Pump X_1 managed to complete the daily cycle within the maximum and minimum limits of the tank, operating with a discharge flow similar to the X_0 and finishing without tank volume variation (Figure 8). The operational cost of the pump (15.1€/day) was 32.0% less than the one practiced by Veolia even without an optimized scheduling pattern, which is directly related to the increase in pump efficiency.

After the verification of these results, the heuristic manual optimization was performed operating only one pump at each time and with FSD motors. This strategy can be inferred from Figure 9. Both pumps operated within the tank limits, completing the daily cycle without volume variation (Figure 10).

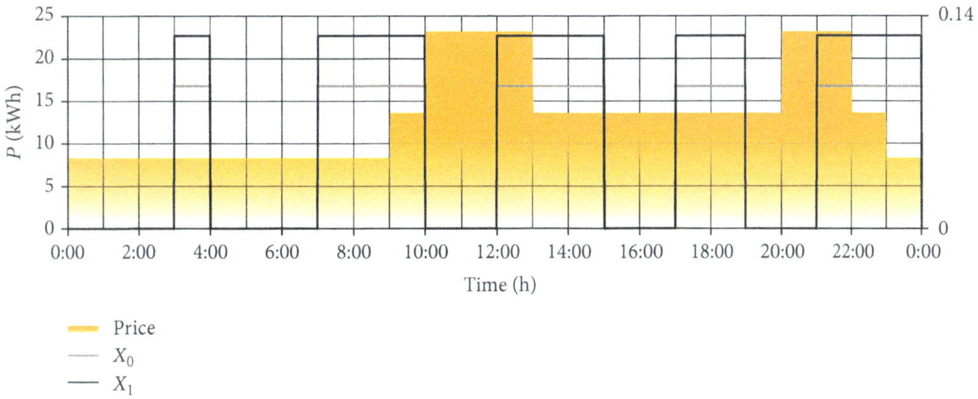

FIGURE 7: Pump power consumption for the original strategy and energy tariff.

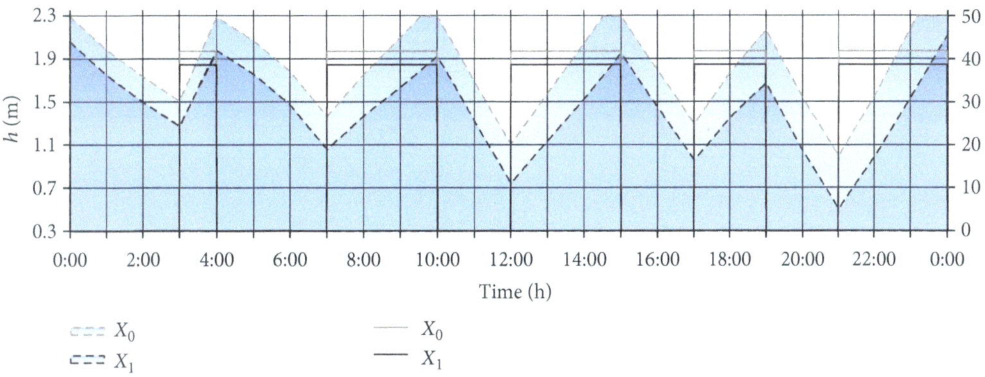

FIGURE 8: Flow and tank water level for the original strategy.

The operational costs of optimized pumps X_0 and X_1 are of 17.5 €/day and 13.1 €/day, less 21.1% and 41.0% than the initial scheduling strategy, respectively.

4.3. Genetic Algorithm Approach. This approach tests thousands of simulations in order to select the fittest of all given the system in which they are. Harnessing the applicability of these algorithms in complex scenarios, in these simulations two parallel pumps were allowed to operate, which represents an increase in versatility when compared to the manual override approach. Since the software could not simulate the original pump curve due to its maladjustment to the system at issue, no GA simulations were performed for this pump.

With VSD pumps it was allowed a total of six possible actions: pump turned off, minimum speed, three intermediate speeds, and maximum speed. The tradeoff was to allow some continuity in the speed steps while minimizing the solution space. The model was restricted by the number of pump starts and the tank minimum and maximum water levels, and the heuristic stopping criterion was a maximum of 200 generations without solution fitness improvement.

In the original Costa et al. investigation, all the daily cycles started at 0:00 h for an initial tank water level of 2.00 m. This fact implied that the algorithm should guarantee that level at that specific hour, even if it would mean to pump at least advantageous periods. Thus, four different initial water levels (h_{initial}) were tested for each pump and motor—0%, 25%, 50%, and 75% of the effective volume, that is, 0.30 m, 0.80 m, 1.30 m, and 1.80 m, respectively.

In order to overcome the inherent randomness of the convergence of the GA, four different simulations were performed for each water level and pump/motor type, varying the initial GA parameters with a PS of 100 and 200 individuals and EPS of 10 and 20 individuals. The total number of performed GA simulations resulting from these combinations was of 96.

Figures 11 and 12 show the daily cost results for each scenario and water level considered. It is immediately visible the influence of both initial tank water level and initial GA parameters from the scattered results. This is more pronounced in the FSD results, where the minimum result achieved was of 13.3 €/day by comparison to a maximum of 16.5 €/day.

Figures 13 and 14 present the results for the best solutions achieved. The daily operational cost was of 13.1 €/day and 12.5 €/day for the X_1 pump with both FSD and VSD motors, respectively. When analyzing the VSD motor results, it can be concluded that the increase in versatility of solutions allowed by this motor compensates its inferior efficiencies. The fact that the pump operates at lower speeds—it never reaches maximum speed—and only has need for 3 daily starts becomes an advantage in terms of pump life expectancy and

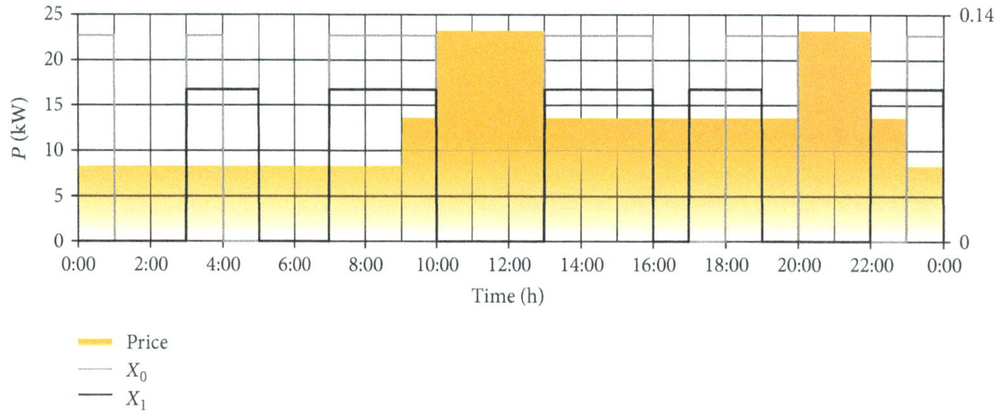

FIGURE 9: Pump power consumption for the manually optimized strategy and energy tariff.

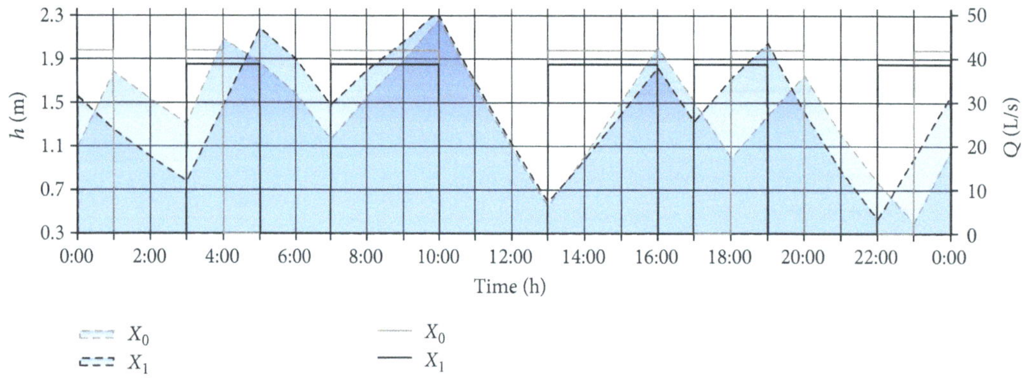

FIGURE 10: Flow and tank water level for the manually optimized strategy.

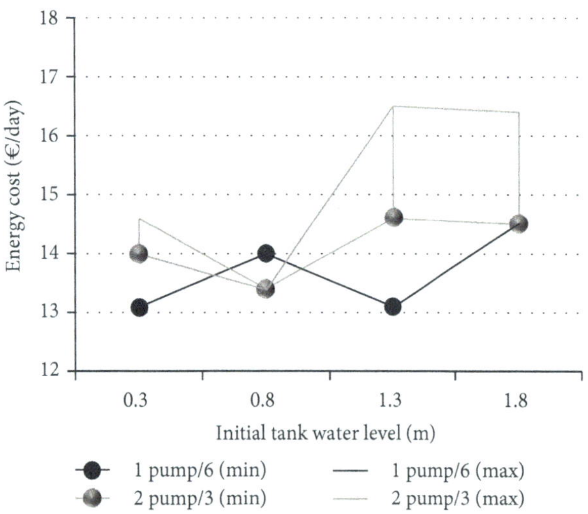

FIGURE 11: B_1 FSD.

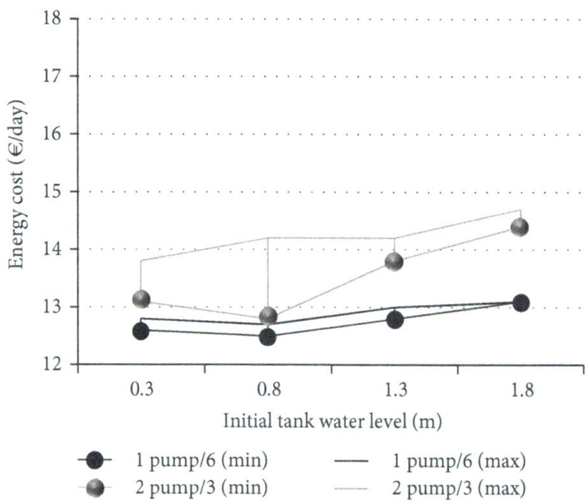

FIGURE 12: B_1 VSD.

maintenance when compared to the FSD, which operates at maximum speed and has 4 daily starts.

It is possible that the worst results obtained for the FSD are justified by the difficulty in matching a daily cycle without volume variation for 1 h control intervals. Also, the VSD can better adapt the supply to the demand by varying flow, reducing power as necessary.

Table 2 compiles data from the simulation runs considering different initial tank and water levels. It should be noted that different GA parameters produced different results.

Systematized in Table 3 is the power consumption of each pump/motor system, as well as its operating hours at different energy tariffs, the tank water level, overall pump efficiencies, and energy cost.

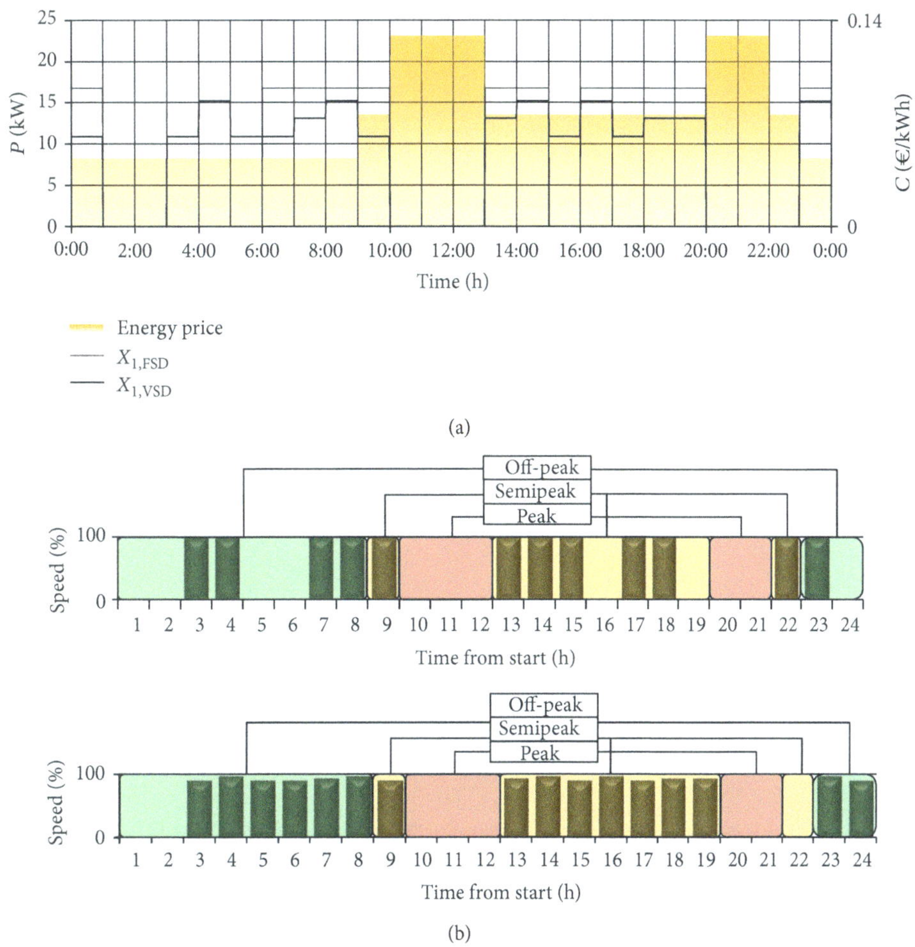

FIGURE 13: Pump power consumption for the best AG solution and energy tariff (top) and scheduling operations pattern for the X_1, FSD and the X_1, VSD pumps, with the genetic algorithm optimization.

TABLE 2: Minimum operational costs achieved for the different simulation scenarios and respective standard deviation between simulation runs with different initial parameters.

h_{initial} (m)		0.30		0.80		1.30		1.80	
C (€/day)		Min	σ	Min	σ	Min	σ	Min	σ
GA $X_{1,FDS}$	1 pump/6	13.1	0.0	14.0	0.0	13.1	0.0	14.5	0.0
	2 pump/3	14.0	0.3	13.4	0.0	14.6	1.0	14.5	0.8
GA $X_{1,VDS}$	1 pump/6	12.6	0.1	**12.5**	0.1	12.8	0.1	13.1	0.0
	2 pump/3	13.1	0.3	12.8	0.6	13.8	0.2	14.4	0.1

5. Conclusions

The solutions obtained from optimization processes produced considerable superior results when compared to the original strategy, with the best solution allowing a 43.7% savings. It is advisable to develop a cash flow analysis to the option of installing new pumps, as well as to introduce integrated dynamic monitoring systems which would allow a real-time optimization of the scheduling of the pumps, both in new projects and in existing WSS.

It is possible to conclude that these algorithms are susceptible to the initial parameters input in the beginning of the simulation and to the size of the solution space. Consequently, it is essential to perform several simulations in order to overcome this disadvantage, as well as considerations in minimizing the model complexity. A careful and critical analysis of the solutions obtained is also indispensable to achieve the most efficient strategies.

Satisfactory results were obtained with a manual override optimization process, when compared to the GA, given the fact that the criteria were primarily to avoid consumption in peak hours and to prevent volume variation in the tank. By contrast, the GA runs many simulations and it is essential to optimize more complex scenarios, such as those with interaction between pumps. Furthermore, it was demonstrated that given the stochastic nature of this algorithm, its usage requires careful considerations of all the variables involved in order to achieve the best possible results. Hence, it is vital to carefully consider good practices in GA optimization. The results obtained by the 1 pump/6 scenario are not only the best but also more consistent than the ones obtained from the 2 pump/3. This points to a systematic convergence towards an optimal solution and demonstrates the importance of model complexity and solution space in obtaining the best results.

TABLE 3: Relevant data from each scenario.

	P (kW)	T_1 (h)	T_2 (h)	T_3 (h)	Q (l/s)	h_{min} (m)	h_{max} (m)	η (%)	C (€/day)	Δ (%)
X_0	24.6	4	6	2	41.8	0.99	2.30	60.0	22.1	0.5
$X_{1,\text{ FSD}}$	16.8	4	6	2	38.7	0.50	2.12	78.9	15.1	32.0
MO X_0	24.6	5	6	0	41.8	0.40	2.26	78.9	17.5	21.2
MO $X_{1,\text{FSD}}$	16.8	5	7	0	38.7	0.43	2.30	78.9	13.1	41.0
GA $X_{1,\text{FDS}}$	16.8	5	7	0	38.7	0.38	2.13	78.9	13.1	41.0
GA $X_{1,\text{VDS}}$	12.8*	8	8	0	28.9*	0.32	2.28	74.7*	**12.5**	43.7

*Given the different possible drive speeds, this is an average value.

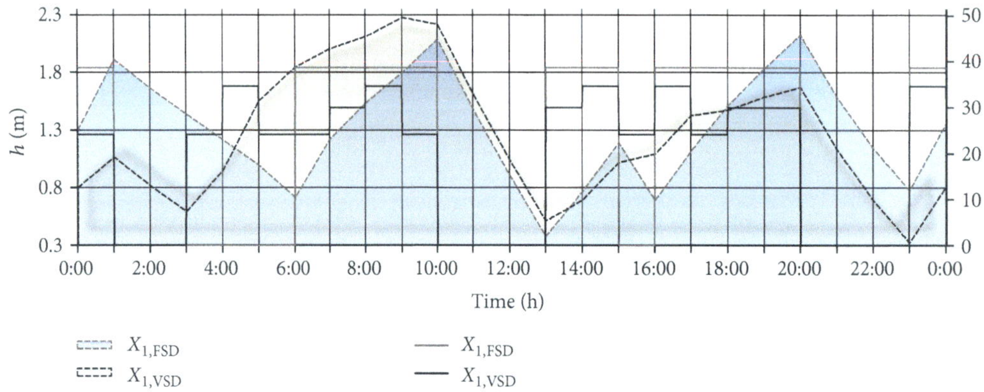

FIGURE 14: Flow and tank water level for the best AG solution.

References

[1] M. A. Braghiroli, M. B. dos Santos, and D. B. Filho, *Estação Elevatória de Água de Santana: Um Caso de Sucesso Na Redução de Perdas E Consumo de Energia No Setor de Saneamento*, Cleaner Production Initiatives and Challenges for a Sustainable World, 2011.

[2] L. H. M. Costa, H. M. Ramos, and M. A. H. Castro, "Hybrid genetic algorithm in the optimization of energy costs in water supply networks," *Engenharia Sanitaria E Ambiental*, vol. 15, no. 2, 2010.

[3] Grundfos, *Pump Handbook*, Management A/S Grundfos, 2004.

[4] M. Tsutiya, *Redução do Custo de Energia Elétrica em Sistemas de Abastecimento de Água*, ABES, 2001.

[5] M. Tsutiya, *Abastecimento De Água*, Departamento de Engenharia Hidráulica e Sanitária da USP, São Paulo, Brazil, 2nd edition, 2004.

[6] P. W. Jowitt and G. Germanopoulos, "Optimal pump scheduling in water-supply networks," *Journal of Water Resources Planning and Management*, vol. 110, no. 1, pp. 17–35, 1992.

[7] D. Burnell, J. Race, and P. Evans, "An overview of the trunk scheduling system for the London Ring Main," *Water Science and Technology*, vol. 28, no. 11-12, pp. 99–109, 1993.

[8] K. W. Little and B. J. McCrodden, "Minimization of raw water pumping costs using MILP," *Journal of Water Resources Planning and Management*, vol. 115, no. 4, pp. 511–522, 1989.

[9] D. V. Chase and L. E. Ormsbee, "Computer-generated pumping schedules for satisfying operational objectives," *Journal of the American Water Works Association*, vol. 85, no. 7, pp. 54–61, 1993.

[10] G. Yu, R. S. Powell, and M. J. H. Sterling, "Optimized pump scheduling in water distribution systems," *Journal of Optimization Theory and Applications*, vol. 83, no. 3, pp. 463–488, 1994.

[11] M. J. H. Sterling and B. Coulbeck, "A dynamic programming solution to the optimization of pumping costs, in Hybrid genetic algorithm in the optimization of energy costs in water supply networks," *ICE Proceedings*, vol. 59, no. 2, pp. 813–818, 1975.

[12] K. E. Lansey and K. Awumah, "Optimal pump operations considering pump switches," *Journal of Water Resources Planning and Management*, vol. 120, no. 1, pp. 17–35, 1994.

[13] D. J. Wood and L. S. Reddy, "Control de bombas de velocidad variable y modelos en tiempo real para minimizar fugas y costes energéticos," in *Mejora del Rendimiento y de la Fiabilidad en Sistemas de Distribución de Agua*, pp. 173–207, Universidad Politécnica de Valencia, Universitat Jaume I de Castellón, València, Spain, 1994.

[14] D. E. Goldberg, *Genetic Algorithm is Search Optimization and Machine Learning*, Addison-Wesley Longman Publishing, Boston, Mass, USA, 1989.

[15] C. Darwin, *On the Origin of Species*, Wordsworth Limit, 1998.

[16] D. Whitley, "A genetic algorithm tutorial," *Statistics and Computing*, vol. 4, no. 2, pp. 65–85, 1994.

[17] J. Koljonen and J. T. Alander, *Effects of Population Size and Relative Elitism on Optimization Speed and Reliability of Genetic Algorithms*, Department of Electrical Engineering and Automation, University of Vaasa, Vaasa, Finland, 2006.

[18] Bentley Systems, *WaterGEMS V8i User's Guide*, Bentley, 2009.

[19] A. Quintela, *Hidráulica*, Fundação Calouste Gulbenkian, Lisboa, Portugal, 2002.

16

Mitigating Climate Change by the Development and Deployment of Solar Water Heating Systems

S. T. Wara and S. E. Abe

Department of Electrical and Electronics Engineering, College of Engineering, Federal University of Agriculture Abeokuta, Ogun State, Nigeria

Correspondence should be addressed to S. T. Wara; warast@funaab.edu.ng

Academic Editor: Ching Yuan Chang

Solar energy is becoming an alternative for the limited fossil fuel resources. One of the simplest and most direct applications of this energy is the conversion of solar radiation into heat, which can be used in Water Heating Systems. Ogun State in Nigeria was used as a case study. The solar radiation for the state was explored with an annual average of 4.775 kWh/m^2 recorded. The designed system comprised storage tanks and the collector unit which comprises wooden casing, copper tube, and aluminium foil. Test results for the unlagged and lagged storage tanks for water temperature at various angles of inclination (2.500°–20.000°) were on the average 27.800°C and 28.300°C, respectively, for the inlet temperature and 60.100°C and 63.000°C for the outlet temperature, respectively. The efficiency of the Solar Water Heating System was 72.500% and the power saved 2.798 kW. The cost of the unit is put at 1121,400 ($145) as at August 2012. The unit developed can be applied for the purpose of reducing the cost of energy, dealing with environmental challenges, and improving the use of energy, hence serving as a climate mitigation process as this can be extended for water heating for domestic and other industrial purposes.

1. Introduction

Solar energy happens to be a naturally occurring renewable source of energy that is gotten directly from one of the major heavenly bodies known as the sun. It has been in existence since the creation of the world and of mankind, but its potentials have not been and could not be harnessed as a result of ignorance and lack of the right technology until now as a result of increase in the demand of power energy supply for human consumption, basically for industrial, commercial, and domestic uses.

The total deposit of fossil fuels left on earth is not up to the amount of energy that could be supplied in a couple of days by the sun which happens to be the root of all energy sources. Therefore, there is a need to changeover to renewable energy sources because of the skyrocketing cost of oil and gas in the production of electricity and the effects of pollutants on climate change (including global warming and ozone layer depletion); especially when it comes to the area of domestic and commercial water heating, solar energy can be a very good substitute to the conventional way of water heating. One of the biggest uses of electricity, gas, and oil is the heating of water in home, offices, schools, and hospitals. Solar water heating is a very simple and efficient way to tap energy from the sun and use it. Solar water heaters concentrate diffused solar radiation into thermal energy [1].

Some of the limitations are the following

(1) the sun is not usually out every time of the year; even when it is out at times, it may not be enough to cater for our energy requirements and needs depending on its intensity;

(2) the facilities required to harness this energy resource are very expensive to put in place;

(3) management of these facilities requires special attention and training on the equipment to avoid misuse.

2. Literature Review

Solar water heaters can operate in any climate. The performance of these heaters varies depending on how much solar energy is available at that locality. Jaisankar et al. [2] reported a comprehensive review on solar water heaters. They reported that the efficiency of solar thermal conversion is about 70% when compared to solar electrical direct conversion system, which has an efficiency of only 17%. Owing to its ease of operation and simple maintenance, Solar Water Heating systems play an important role in domestic as well as industrial sectors.

The first published analysis of thermosyphon solar water heater circuit was by Close [3]. There have been many other publications on the analysis of these systems but they are all based on the original formulation. For verifying the theoretical results, he tested two thermosyphon systems with different characteristics and the results conformed well to those predicted.

In communities throughout the developing world, poor households struggle to meet their hot water needs. Some households rely on biomass to heat water. In many countries demand for fuel wood is one of the principal contributors to deforestation. Others rely on electricity or liquid fuels such as propane to heat their water. These fuel options are unsustainable as they are costly to households and contribute to the build-up of greenhouse gases in the atmosphere. Many communities face limited or intermittent access to fuel and/or electricity, limiting their ability to access hot water for hygiene and domestic uses. One potential solution to this problem is the use of solar energy to heat water. Water enters at the bottom and is "baked" in the sun. As it warms, hot water travels to the top of the tank due to its lower density. When the water is ready for use, cool water is supplied to the inlet and hot water flows out the top. Access to a low cost solar water heater would provide numerous benefits to households in developing communities. Many households could reduce their fuel costs by eliminating or reducing their need for wood, gas, or electricity to heat water. Substituting traditional fuel sources with solar energy would reduce carbon emissions. Further reduction in biomass consumption would relieve stress on depleted forests. There are also health benefits associated with solar hot water due to lessened exposure to toxins and pollutants released from burning fuels. By enabling access to hot water, households could improve their health and hygiene. Nigeria happens to be one of the few countries that have not really adopted the use of solar water heating methods considering the fact that the intensity of the sun in many areas of the country is so much such as the northern parts where the sun is at the highest intensity.

3. Methodology

3.1. Solar Insolation: How Much Heat Is Available? The sun is the energy source but we need to know how much of the sun's energy reaches a particular place. Solar insolation is the amount of electromagnetic energy (solar radiation) incident on the surface of the earth [5]. Basically that means how much sunlight is shining down. By knowing the insolation levels of a particular region we can determine the size of solar collector that is required. An area with poor insolation levels will need a larger collector than an area with high insolation levels. Once you know your region's insolation level you can more accurately calculate collector size and energy output. The values are generally expressed in kWh/m^2. This is the amount of solar energy that strikes a square metre of the earth's surface in a single day. This value is averaged to account for differences in the days' length (Figure 1, Table 1). Metrological data from the National Aeronautics and Space Administration (NASA) is used. The NASA website has 22+ years of metrological data stored according to longitude and latitude (Table 5):

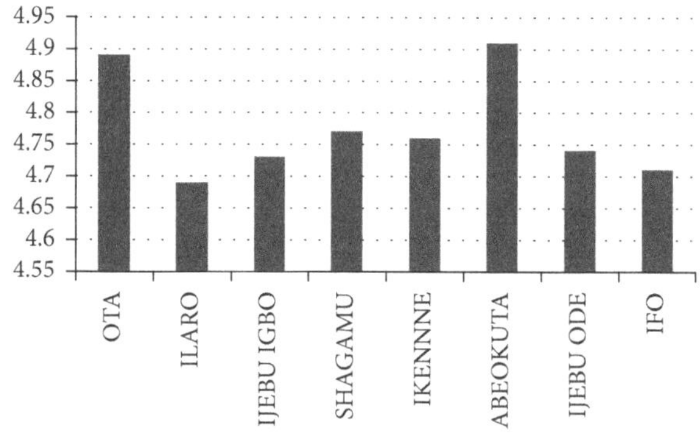

FIGURE 1: Graph showing average monthly solar insolation for Ogun State, Nigeria.

$$(4.890 + 4.690 + 4.730 + 4.770 + 4.760 + 4.910 + 4.740 + 4.710) \times 8^{-1} = \frac{38.200}{8} = 4.775 \text{ kWh/m}^2. \quad (1)$$

3.1.1. Conduction: Heat Transfer. In the construction of the Solar Heating System, thermal conductivity is a high priority consideration. The materials with the highest thermal conductivity are metals. The heat received via radiation must be conveyed to the water via conduction (Table 2).

3.1.2. Collector Heat Loss Coefficient (U_L). Heat loss is not to be taken lightly, because it determines how hot the water can become. U_L, the loss factor, is mainly determined by the number of covers and the wind speed. It also increases a little at high absorber temperatures. For one cover $U_L = 5.5\,W/m^2K$ and for two covers $U_L = 3.0\,W/m^2K$ [5]. The solar collector in the Solar Water Heating System uses one cover due to the availability and cost considerations of the cover material.

3.1.3. Transmission Coefficient of Glazing. Radiation passing through the glazing material is said to be transmitted. Transmittance not only varies for different materials but also varies with the wavelength of the radiation. It is therefore desirable to know the spectral transmittance of glazing for

Table 1: Monthly solar insolation for major towns in Ogun State [1].

	Monthly averaged insolation incident on a horizontal surface (kWh/m^2)							
	OTA	ILARO	IJEBU IGBO	SHAGAMU	IKENNNE	ABEOKUTA	IJEBU ODE	IFO
January	5.570	5.200	5.320	5.300	5.320	5.500	5.280	5.240
February	5.740	5.410	5.480	5.480	5.500	5.700	5.490	5.450
March	5.660	5.410	5.370	5.480	5.460	5.640	5.460	5.430
April	5.340	5.180	5.210	5.250	5.230	5.350	5.210	5.200
May	5.010	4.710	4.790	4.790	4.770	5.090	4.760	4.720
June	4.510	4.000	4.150	4.070	4.050	4.570	4.040	4.000
July	3.890	3.910	3.880	3.990	3.980	4.000	3.950	3.910
August	3.730	3.920	4.040	3.990	3.980	3.790	3.980	3.980
September	4.050	4.040	3.960	4.120	4.100	4.110	4.090	4.060
October	4.620	4.520	4.440	4.590	4.560	4.700	4.550	4.510
November	5.180	4.900	4.950	4.980	4.960	5.110	4.950	4.920
December	5.370	5.120	5.210	5.200	5.190	5.350	5.170	5.130
Annual average	4.890	4.690	4.730	4.770	4.760	4.910	4.740	4.710

Average insolation for all the towns is 4.775 kWh/m^2.

Table 2: Thermal conductivity of some metals at 68°F (20°C) [4].

Material	Thermal conductivity (Btu/(hr °F ft))
Pure copper	223
Pure aluminium	118
Stainless steel	7–26
Pure iron	42
Pure silver	235

Table 3: Transmission coefficients of sample materials [5].

Material	Transmission coefficient
Single 4 mm glass or acrylic sheet	0.850
Double 4 mm glass	0.720
Teflon thin film	0.950
Tedlar thin film	0.900

Transmission coefficients of sample materials;all of thesecan be reduced by 1 or 2% due to dirt, up to 8% in dry dusty areas.

Solar Water Heating Systems. Ideally, the glazing should be very transparent to incoming shortwave radiation but opaque to outgoing long wave (thermal) radiation, because radiant losses may account for over 70 per cent of collector heat loss (Table 3).

3.1.4. Absorptance and Emissivity. Radiation striking an object is absorbed if the energy is retained by the material. For example, a black hose left in the sun heats up as it absorbs energy from the sun. The percentage of incoming radiation that is absorbed by a material is referred to as its absorptance and is a measure of the ease with which a material or surface collects energy. Table 4 presents absorptance for various materials in sunlight. The best materials are those with high absorptance and very low emittance.

The high absorptance and low cost of black paint make it a good choice for the Solar Water Heating System.

Table 4: Absorptance and emissivity of sample materials [4].

Material	Absorptance	Emissivity
Black paint	0.950	0.875
White paint	0.200	0.900
Polished aluminium	0.100	0.050
Selective surfaces	0.850–0.950	0.120–0.080

3.2. Efficiency of Collector. In the design of the "Automated Solar Water Heater System" the collector area is another important design consideration. Using the efficiency equation of the flat plate collector, this is given as the heat gained by water with respect to the actual solar energy received by the flat plate collector

$$mc_p\Delta T = A_c I\eta \tag{2}$$

(see [8]) where m = mass of water in kilograms (kg), c_p = specific heat of water (J/g°C), ΔT = temperature difference (outlet temperature-inlet temperature °C), A_c = area of collector, I = intensity of solar radiation, (W/m^2), and η = collector efficiency.

The equation simply equates the heat acquired by the collector to the heat received by the water. If the collector Area (A_c) is made the subject of the formula, (2) becomes

$$A_c = \frac{mc_p\Delta T}{\eta I} \tag{3}$$

(see [8]). The equation above requires us to know the mass of water to be heated, the specific heat of water (c_p) which is a constant (4.18 J/g°C), the outlet and inlet water temperature, the solar insolation of the environment, and the efficiency of the collector. All parameters are known except the efficiency of the collector which is a value set by the materials used for construction and some constants. The efficiency of the collector therefore must be calculated.

Equation (4) shows the instantaneous thermal efficiency (η) equation for flat plate collectors:

$$\eta = F_R\tau\alpha - F_R U_L \left(\frac{T_i - T_a}{I}\right) \quad (4)$$

(see [8]), where F_R = collector heat removal factor, I = intensity of solar radiation, W/m^2,T_i = inlet fluid temperature, °C, T_a = ambient temperature, °C, U_L = collector heat loss coefficient, W/m^2 K, τ = transmission coefficient of glazing, and α = absorption coefficient of plate.

The equation shows that the efficiency of a flat plate collector is dependent on the transmission coefficient of glazing (τ), the absorption coefficient of plate, collector heat removal factor (F_R), intensity of solar radiation (I), ambient temperature (T_a), inlet fluid temperature (T_i), and the collector overall heat loss coefficient (U_L).

For the Solar Water Heating System from the efficiency equation, F_R,τ, α, and U_L are fixed by the collector design. T_i, T_a,and I are variables determined by the application. This simply implies that the solar collector efficiency can change when T_i, T_a,and I are changed. Using Ogun State as a case study the following where decided in design.

Transmission coefficient of glazing (τ) will be 0.850 because a single 4 mm sheet of glass was used in glazing. The absorption coefficient of plate was 0.950 because the collector plate was be painted black. The collector heat removal factor (F_R) was 0.900; intensity of solar radiation (I) was 4.775 kWh/m^2 from metrological data from NASA. The ambient temperature (T_a) was 25.700°C from metrological data from NASA and inlet fluid temperature (T_i) can vary at any point but it was assumed to be 22°C in the mornings and the collector overall heat loss coefficient (U_L) was 5.500 W/m^2 K because only one layer of glazing was used.

Using (4), if the whole temperature is converted to kelvin the efficiency η will be 73.100%.

3.3. Collector Tubing. Copper tubes vary from 1/2 to 1 inch. The 1/2 inch copper tube was used because of more heating per surface area. The small size of the 1/2 inch copper tube allows for more copper tube per surface area of collector.

3.4. Collector Casing. The purpose of the casing is to provide the collector with some insulation, so as to better prevent heat loss into the environment and also to make the collector more presentable. The collector casing was made of wood since wood is better insulator than any metal.

3.5. Collector Placement: Angle of Inclination. The direct sun power incident on a solar collector depends not only on the power contained in the sunlight, but also on the angle between the solar collector and the sun. When the absorbing surface and the sunlight are perpendicular to each other, the power density on the surface is equal to that of the sunlight (in other words, the power density will always be at its maximum when the solar panel is perpendicular to the sun).

For a collector to function optimally, it should be inclined at a specific angle [9], and large deviations from the optimum angle will result in important losses in energy produced. In tropical regions, where latitude is less than 36 degrees, the duration of effective collector irradiation is constant the whole year [9].

3.6. Power Saved. Through the use of the automated solar water heater, dependency on electric power can be reduced which would lead to savings on power. The power input of the solar water heater can be calculated as

$$\text{Pi} = I\tau\alpha \quad (5)$$

(see [5]), where Pi = power input, τ = cover transmissivity, α = surface black absorptance, and the power output was calculated using Po = Pi $\cdot$ η.

3.7. Carbon Footprint Calculation. Carbon (IV) oxide (CO_2) emission = fuel consumption × fuel emission factor (Tonnes CO_2-emission = litre × kg CO_2 per litre fuel/1000) (Table 6).

Assuming a generator operating for 30 days in a month, consuming 4 litres of fuel per day for water heating.

For gasoline: Tonnes CO_2-emission = (4 litres × 30 days) × 0.0024 = 0.288 tons of CO_2.

For diesel: Tonnes CO_2-emission = (4 litres × 30 days) × 0.0027 = 0.324 tons of Co_2.

3.8. Summary of Design Parameters. These are as presented in Table 7.

4. Construction Methods

Figure 2 shows the heat collector system whilst Figure 3 shows the entire Solar Water Heating System including the storage tank. The copper tube was bent to form a coil shape with the fins attached to it. It was painted black in order to absorb more heat from the sun, basically converting solar energy to heat energy. This copper tube need to be placed in a case. A wooden case was made to hold copper tube. The wooden case was also painted black. Aluminium foil was placed at the back of the wooden case. The reason for this is that the backing should not warm up but only the collector is to absorb heat. The foil will take any sun that was not absorbed by the collector on the first pass and bounce it back over the collector for another try at absorption. The glass is then placed on top of the wooden case. The glass cover will keep all the heat inside the panel for further absorption. This serves to trap the whole infrared radiation from the sun inside our panel where our collector will absorb it. Light can pass through glass, but heat cannot.

4.1. Advantages of the Solar Water Heating System

(i) Full incorporation of solar system as a source of energy for domestic consumption,

(ii) provides a cheaper and more effective way of generating energy for domestic and commercial purposes,

TABLE 5: Table showing the longitude and latitude of towns in Ogun State [3, 6].

Towns	Longitude	Latitude	Postal code
OTA	4.780°	7.950°	112264
ILARO	3.300°	6.530°	111101
IKENNE	3.720°	6.870°	121103
IJEBU ODE	3.580°	6.470°	120252
IJEBU IGBO	4.100°	6.560°	120105
IFO	3.196°	6.814°	112104
ABEOKUTA	3.350°	7.150°	110001
SHAGAMU	3.650°	6.850°	121101

TABLE 6: CO_2 emission factors for combustion of common fossil fuels [7].

Fuel type	Metric tons CO_2 per litre	Metric tons CO_2 per gallon	Metric tons CO_2 per cubic meter	Metric tons CO_2 per therm
Gasoline	0.0024	0.0092	—	—
Diesel	0.0027	0.0104	—	—
Jet fuel	—	0.0100	—	—
Aviation gasoline	0.0024	0.0090	—	—
LPG	0.0016	0.0060	—	—
CNG	—	—	0.0022	0.0054

TABLE 7: Design specification for the Solar Water Heating System.

Component	Specification
Collector area	
Length	9 meters
Collector tube/pipe	Copper
Diameter	0.500 inches = 0.013 meters
Length	9 meters
Inclination of flat plate collector	10°
Absorber	Copper painted black (emissivity = 0.875)
Number of glass cover	One (4 mm thick)
Aluminium foil	
Length	0.840 meters
Breadth	0.450 meters

(iii) evaluates the performance of the solar water heater prototypes ensuring easy construction, repair, and maintenance,

(iv) provides an alternative for the conventional method of water heating for domestic and commercial use,

(v) provides average individuals with ease and accessibility to an alternative source of energy which is affordable, clean and much less dangerous,

(vi) fully optimizes the solar renewable energy source available for domestic and commercial water heating,

(vii) pollution free.

TABLE 8: Inlet temperature and outlet temperature of water in the SWHS.

	Date	Inlet temperature (°C)	Outlet temperature (°C)
Day 1	09/07/2012	28.200	58.300
Day 2	10/07/2012	28.000	60.000
Day 3	11/07/2012	27.800	59.300
Day 4	12/07/2012	28.000	60.700
Day 5	13/07/2012	27.500	59.700
Day 6	14/07/2012	26.500	60.000
Day 7	15/07/2012	27.000	60.700
Day 8	17/07/2012	28.000	60.700
Day 9	18/07/2012	28.500	59.300
Day 10	19/07/2012	28.500	62.300
Average		27.800	60.100

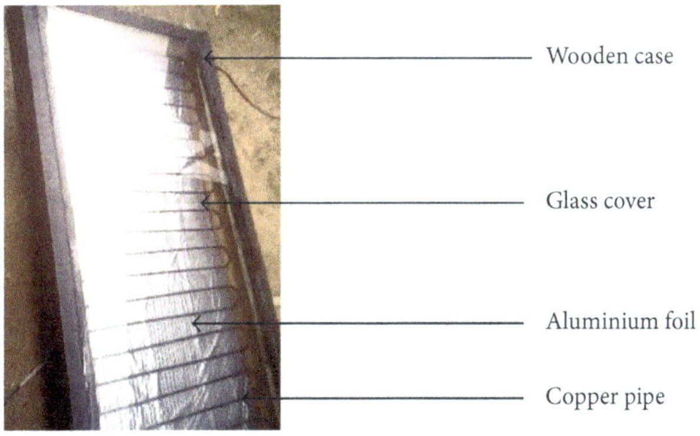

FIGURE 2: Solar water heating (heat collector) system.

4.2. Performance Evaluation/Test and Results. The performance of the Solar Water Heating System was evaluated to determine its reliability in terms of

(i) efficiency,

(ii) power saved.

Testing was carried out on a daily basis between 12 pm and 5 pm for ten days in Camp, Abeokuta, Ogun State. The inlet fluid temperature and outlet fluid temperature was measured using a thermometer. Table 8 shows the average values of the inlet temperature and the outlet temperature per day and the average for the whole test period.

Testing was also carried out for different angles of the solar collector to the horizontal surface. The inlet fluid temperature and outlet fluid temperature ware measured using a thermometer.

Table 9 shows the average values of the inlet temperature and the outlet temperature per day (for different angles) and the average for the whole testing period. A graphical representation of the water heating ability of the developed unit is shown in Figures 4 and 5.

Table 9: Inlet and outlet temperature of water in the SWHS at different angles.

Angle (°)	Temperature (°C)	Day 1 15/08/2012	Day 2 16/08/2012	Day 3 17/08/2012	Day 4 21/08/2012	Average
2.5	Inlet temperature	28.500	28.300	28.000	28.300	**28.300**
	Outlet temperature	58.200	58.300	58.000	58.700	**58.300**
5.0	Inlet temperature	28.500	28.300	28.000	28.300	**28.300**
	Outlet temperature	58.800	59.300	58.300	60.800	**60.000**
7.5	Inlet temperature	28.500	28.300	28.000	28.300	**28.300**
	Outlet temperature	60.300	60.000	59.800	60.800	**60.300**
10.0	Inlet temperature	28.500	28.300	28.000	28.300	**28.300**
	Outlet temperature	60.700	60.000	60.700	62.300	**61.000**
12.5	Inlet temperature	28.500	28.300	28.000	28.300	**28.300**
	Outlet temperature	62.700	62.300	62.300	64.800	**63.000**
15.0	Inlet temperature	28.500	28.300	28.000	28.300	**28.300**
	Outlet temperature	64.300	65.300	63.700	65.000	**64.600**
17.5	Inlet temperature	28.500	28.300	28.000	28.300	**28.300**
	Outlet temperature	60.800	60.700	60.300	60.200	**60.500**
20.0	Inlet temperature	28.500	28.300	28.000	28.300	**28.300**
	Outlet temperature	58.300	59.800	57.800	60.500	**59.100**

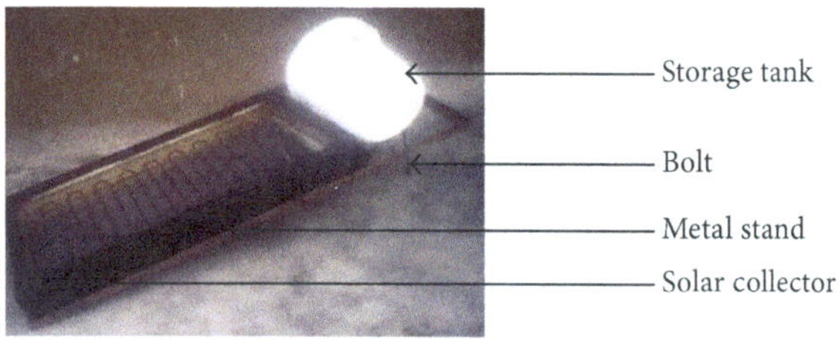

Figure 3: The solar water heating system with the stand.

Efficiency of the Solar Water Heating System

$$\eta = F_R \tau\alpha - F_R U_L \left(\frac{T_i - T_a}{I}\right),$$

$$\eta = (0.900 \times 0.850 \times 0.950) - 0.900 \times 5.500 \times \frac{(300.800 - 298.700)}{4775},$$

$$\eta = 0.72675 - 0.00217696,$$

$$\eta = 0.725 = 72.5\%. \quad (6)$$

Power saved by the Solar Water Heating System

$P_i = 4775 \times 0.85 \times 0.95 = 3.86\,\text{kW}$,
$P_o = P_{i.}\eta$,
$P_o = 3.86\,\text{kW} \times 0.725 = 2.799\,\text{kW}$.

For a 6-hour period of effective sunlight in a day, the total power output of the solar water heater in a day can be calculated as

$$2.799\,\text{kW} \times 6\text{ hours} = 16.794\,\text{kWh}. \quad (7)$$

Table 10: Daily consumption of hot water in a Nigerian home.

Size of family	Morning (litres)	Afternoon (litres)	Evening (litres)	Total (litres)
A family of 5	52	6	5	63
A family of 2	18	3	2	23
A family of 2	20	3	2	25
A family of 4	48	5	4	57

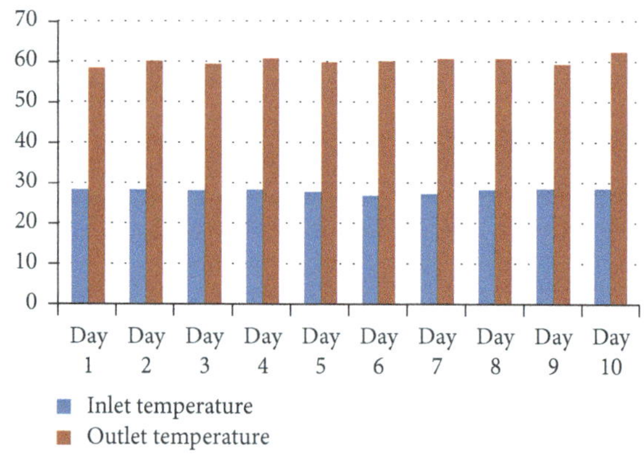

Figure 4: Inlet and outlet temprature of water.

Table 10 shows the amount of hot water used up in different Nigerian homes.

From Table 10, average consumption of hot water can be calculated as.

$$\frac{63 + 23 + 25 + 57}{4} = 42\text{ litres}. \quad (8)$$

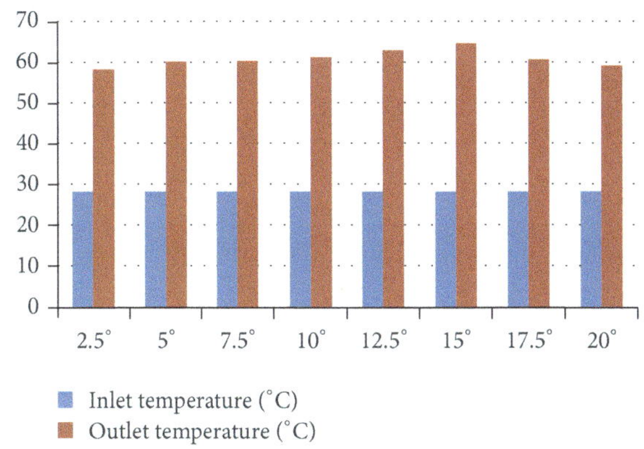

FIGURE 5: Inlet and outlet temperature for the average values at different angles.

TABLE 11: Operation of a 1.5 KW electric heater heating water to 85°C.

Volume (litre)	Duration (hour)	Power (KWh)
0	0	$0 \times 0 = 0$
1	0.057	$0.057 \times 1.5 = 0.086$
10	0.550	$0.550 \times 1.5 = 0.825$
20	1.020	$1.020 \times 1.5 = 1.530$
30	1.490	$1.490 \times 1.5 = 2.235$
40	1.960	$1.960 \times 1.5 = 2.940$
50	2.420	$2.420 \times 1.5 = 3.630$
60	2.880	$2.880 \times 1.5 = 4.320$
70	3.340	$3.340 \times 1.5 = 5.010$

From Table 10, the family of 5 shown consumes about 63 litres of hot water in a day. This is close to 60 litres of hot water. From Table 11, the 1.5 kW water heater consumes energy of 4.320kWh in heating 60 litres of hot water. If the family used the solar hot water heater in place of the electric water heater, the amount saved on electric bills daily would be (₦12.90 is the cost per kWh in Nigeria)

$$4.32\text{kWh} \times ₦12.90\text{k} = ₦55.728\text{k}. \quad (9)$$

The amount saved in a month would be

$$₦55.728\text{k} \times 30 = ₦1671.840\text{k}. \quad (10)$$

The annual savings on electricity bills would be

$$₦1671.840\text{k} \times 12 = ₦20062.080\text{k}. \quad (11)$$

Within a year, the amount saved has summed up to the capital used in purchasing the solar water heater. The work was undertaken and concluded in the month of August 2012. The cost of material used in all aspects of work amounted to ₦21,400 as at august 2012 pricing.

5. Conclusions

A Solar Water Heating System has been developed using locally available materials and was tested. The solar collector was rotated at different angles, and measurements indicate that the water in the tank was heated by the solar energy being absorbed by the solar collector. The developed unit can be deployed for use in the domestic and commercial sectors. The Solar Water Heating System is environmentally friendly and can readily be used as an alternative water heating saving cost, and reducing the amount of power consumed and carbon emission. Consideration can be given to retrofitting the currently used conventional water heaters.

References

[1] NASA, "Nasa Surface meteorology and Solar Energy Data Set," 2012, http://eosweb.larc.nasa.gov/cgi-bin/sse/grid.cgi?uid=3030.

[2] S. Jaisankar, J. Ananth, S. Thulasi, S. T. Jayasuthakar, and K. N. Sheeba, "A comprehensive review on solar water heaters," *Renewable and Sustainable Energy Reviews*, vol. 15, no. 6, pp. 3045–3050, 2011.

[3] D. J. Close, "The performance of solar water heaters with natural circulation," *Solar Energy*, vol. 6, no. 1, pp. 33–40, 1962.

[4] K. S. Ong, "A finite-difference method to evaluate the thermal performance of a solar water heater," *Solar Energy*, vol. 16, no. 3-4, pp. 137–147, 1974.

[5] K. Brown, *Introduction to Solar Water Heating*, Centre for Alternative Technology, London, UK, 2005.

[6] J. A. Duffie and W. A. Beckman, *Solar Engineering of Thermal Processes*, John Wiley and Sons, New York, NY, USA, 2nd edition, 1991.

[7] GHG, "(Green House Gas Protocol) Calculation Tools for Stationary Emission," 2013, http://www.theice.com/publicdocs/ccx/CCX_GHG_Factors.pdf.

[8] B. M. Santos, M. R. Queiroz, and T. P. F. Borges, "A solar collector design procedure for crop drying," *Brazilian Journal of Chemical Engineering*, vol. 22, no. 2, pp. 277–284, 2005.

[9] A. Adell, "Determination of the optimum inclination of a flat solar collector in function of latitude and local climatic data," *Revue de Physique Appliquée*, vol. 17, Article ID 576, p. 569, 1982.

Comparison of Technological Options for Distributed Generation-Combined Heat and Power in Rajasthan State of India

Ram Kumar Agrawal[1] and Kamal Kishore Khatri[2]

[1] *Department of Mechanical Engineering, Kautilya Institute of Technology & Engineering, Jaipur Rajasthan 302022, India*
[2] *Department of Mechanical Engineering, Mody Institute of Technology & Science, Lakshmangarh, Sikar Rajasthan 332311, India*

Correspondence should be addressed to Kamal Kishore Khatri; kamalkishorekhatri@gmail.com

Academic Editor: José L. Míguez

Distributed generation (DG) of electricity is expected to become more important in the future electricity generation system. This paper reviews the different technological options available for DG. DG offers a number of potential benefits. The ability to use the waste heat from fuel-operated DG, known as combined heat and power (CHP), offers both reduced costs and significant reductions of CO_2 emissions. The overall efficiency of DG-CHP system can approach 90 percent, a significant improvement over the 30 to 35 percent electric grid efficiency and 50 to 90 percent industrial boiler efficiency when separate production is used. The costs of generation of electricity from six key DG-CHP technologies; gas engines, diesel engines, biodiesel CI engines, microturbines, gas turbines, and fuel cells, are calculated. The cost of generation is dependent on the load factor and the discount rate. It is found that annualized life cycle cost (ALCC) of the DG-CHP technologies is approximately half that of the DG technologies without CHP. Considering the ALCC of different DG-CHP technologies, the gas I.C. engine CHP is the most effective for most of the cases but biodiesel CI engine CHP seems to be a promising DG-CHP technology in near future for Rajasthan state due to renewable nature of the fuel.

1. Introduction

In centralized generation(CG), electricity is generated in large remote plants. Power must then be transported over long distances at high voltage before it can be used. Distributed generation (DG) can be defined as electric power generation within distribution networks or on the customer side of the network [1]. DG is referred to as the use of small, modular power generation at or near the point of consumption irrespective of size, technology, or fuel used—both off-grid and on-grid as an alternative to large power generation and electricity transport over long distances [2].

The performance of the small power technologies (i.e., reciprocating engine and gas turbine) has improved remarkably over the last decade. In the last decade, technological innovations and a changing economic and regulatory environment have resulted in a renewed interest for distributed generation [3]. This is confirmed by the five major factors that contribute to this evolution; these are developments in distributed generation technologies, constraints on the construction of new transmission lines, increased customer demand for highly reliable electricity, the electricity market liberalization, and concerns about climate change [4]. This has aroused the interest of operators, regulators, and legislators in distributed generation. Three independent trends, first, increasing system capacity needs, second, utility industry restructuring, and third, technology advancements are concurrently laying the groundwork for the possible widespread introduction of DG. DG at the customer's site can also provide benefits to the electric utility that include reduced T&D electric losses, T&D upgrade deferrals, transmission congestion relief, and peak shaving [5]. The potential positive effects of distributed generation for improving tail-end voltages and power factor corrections were mentioned by Dondi et al. [6]. The IEA [7] recognizes the provision of reliable power as the most important future market niche for distributed generation.

Cogeneration, or combined heat and power (CHP), is defined as the simultaneous generation of heat and power in a single process. The power output is usually electricity but may include mechanical power. Heat outputs can include steam, hot water, or hot air for process heating, space heating, or absorption chilling [2]. Distributed (co)generation (DG) represents an alternative paradigm of energy supply and the opportunity for significant CO_2 emission reductions [8]. It is very reasonable to expect that DG will play a role in reducing local, regional, and even global air pollution [9]. One of the keys to the success of fossil-fuelled DG in Europe is the ability to use the waste heat from electricity generation, that is, CHP, raising total system efficiencies up to 90% (higher heating value (HHV)) in the best applications. Overall efficiencies of CHP system can approach 90 percent (considering both power and heating), a significant improvement over the 30 to 35 percent electric grid efficiency and 50 to 90 percent industrial boiler efficiency when separate production is used [5].

The high efficiencies of such applications, that is, CHP, offer both reduced costs and significant reductions of CO_2 emissions. Other factors which are enhanced reliability and security, reduced need for transmission and distribution upgrades, and easier plant siting may also drive increased deployment of DG in the future [10]. DG holds the promise of improved environmental performance of energy sectors due to higher efficiencies (using CHP) and the use of low-carbon fuels, for example, natural gas or carbon-free fuels, for example, hydrogen [11].

DG provides low-cost electricity and gives access to liberalizing generation markets for distribution utilities. DG also offers the potential for improved network management. This network management requires knowledge of projected electricity production and is greatly enhanced by distribution utilities having some control over electricity exports [8]. Distributed energy resources (DER) technologies, such as gas-fired reciprocating engines and microturbines, can be economically beneficial in meeting commercial sector energy loads. Even with a lower electric-only efficiency than that of traditional central station coal steam turbines (CST) and centralized combined cycle gas turbines (CCGT) combined with on-site gas-fired heat boilers (DH), combined heat and power (CHP) applications can increase overall system energy efficiency. From a policy perspective, it is useful to have good estimates of penetration rates of DER under different economic and regulatory scenarios [12]. Banerjee [13] analyzed the different technological options available for DG, their status and evaluated them based on the cost of generation and future potential in India. Narula et al. [14] studied DG technology options motivated by the goal of achieving "universal energy access" by 2030 and looked at electricity access for rural households in the South Asian region. They developed a model which was employed to assess the cost effectiveness of centralized and decentralized distributed generation (DDG) technologies.

There is increased interest in renewable fuels in present time and in Rajasthan, being mainly the dessert land, scope of renewable fuel like biodiesel obtained from the nonedible vegetable oil sources is great. Depleting oil reserves, increasing oil prices, lack of availability of the mineral oil, and the problem of environmental pollutions have prompted research worldwide into alternate fuels for internal combustion (IC) engines. Vegetable-oil-based fuels have been proved as potential alternative greener energy substitute for fossil diesel in compression ignition (CI) engines. The vegetable oils are renewable in nature and have comparable properties with fossil diesel. These are biodegradable, nontoxic and have potential to reduce the harmful emissions [15–18]. Diesel vehicles remain a major cause of street-level air pollution in many cities. For new diesel vehicles, increasingly stringent emissions standards have been imposed to reduce the pollutants emitted by them. Gasoline- and diesel-driven automobiles are the major sources of greenhouse gases (GHG) emission [19–22]. In this century, fuels derived from crude oil have been the major source of the world's energy as well as transportation sector. Future projections indicate that economics and energy needs will increase the focus on the production of synthetic fuels derived from nonpetroleum sources, including biomass and waste products [23, 24].

There is a need for examination of cost effectiveness of various DG-CHP technological options for Rajasthan considering its local requirement and potential within the framework of existing DG technologies. The earlier studies have been related to DG technological options for particular countries without taking the CHP options. But in case of our research study, the comparison for only DG technologies with DG-CHP has been analyzed with selected DG technologies, for example, IC engines fuelled by diesel and biodiesel, microturbines, and so forth for Rajasthan state only. These various DG-CHP technological options are proposed for Rajasthan state where there is mostly desert land and where the population is also scattered. The large distances between the villages and the cities prompt the use of DG technologies. So it is considered that if DG-CHP is used instead of only DG technology, then significant benefits can be achieved. This paper examines the cost effectiveness of the different DG-CHP technological options including biodiesel-fuelled CI engine CHP and evaluates them based on the cost of CHP generation.

2. Power Sector Status in Rajasthan State

Rajasthan state has low population density and people lives in remote places. Mandatory electricity requirement as insured by Rajasthan government for these villages has not been fulfilled by present supply of electricity through transmission lines due to lacking of transmission lines availability and shortage of electricity generation. Rajasthan government is lagging behind their schedule for providing electricity for all villages and hamlets as mentioned in the central government's 11th five-year plan to be implemented by the states. This shortage can be fulfilled by distributed generation of electricity.

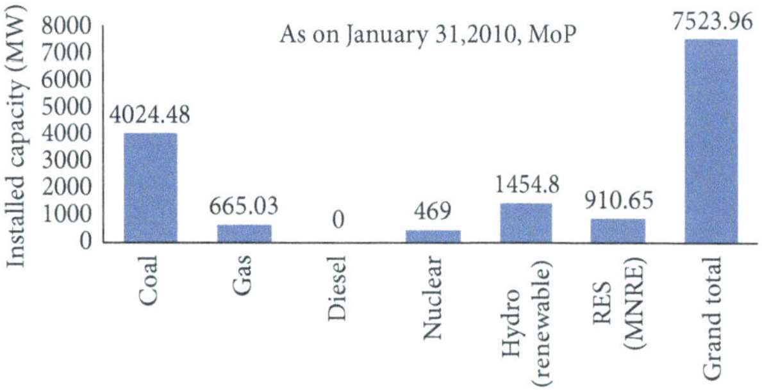

FIGURE 1: Installed capacity of power utilities in Rajasthan.

In Rajasthan, in 1949, the total installed capacity of power houses was 13.27 MW. Total circuit length of 33 kV lines was 137 Km and that of 11 kV was 193 Km. Total consumers were 34518. During first five-year plan in 1951–1956, 15 crores were spent on power development which resulted in an increase in installed capacity to 34.90 MW. The number of villages electrified was 36, wells electrified was 47 and there were 51205 consumers with per capita consumption of energy 2.8 unit.

The Rajasthan State Electricity Board was constituted with effect from July 1, 1957, by government of Rajasthan under the Electricity (Supply) Act, 1948, whose enactment has for its object the coordinate development and rationalization of generation and supply of electricity on a regional basis throughout the country in the most efficient and economical way. Government of Rajasthan on July 19, 2000, issued a gazette notification unbundling Rajasthan State Electricity Board into Rajasthan Rajya Vidyut Utpadan Nigam Ltd. (RVUN) as the generation company; Rajasthan Rajya Vidyut Prasaran Nigam Ltd. (RVPN) as the transmission company, and the three regional distribution companies, namely, Jaipur Vidyut Vitran Nigam Ltd. (JVVNL), Ajmer Vidyut Vitran Nigam Ltd. (AVVNL), and Jodhpur Vidyut Vitran Nigam Ltd. (JdVVNL).

The generation company owns and operates the thermal power stations at Kota and Suratgarh, gas-based power station at Ramgarh, hydel power station at Mahi, and minihydel stations in the state. The transmission company operates all the 400 kV, 220 kV, 132 kV, and 66 kV electricity lines and system in the state. The three distribution companies operate and maintain the electricity system below 66 kV in the state in their respective areas mentioned in their license.

Rajasthan state fulfills its electricity requirement mainly from CG (centralized generation) which includes thermal, hydro, nuclear, and renewable energy sources. Rajasthan had an installed capacity of 7523.96 MW in the centralized power utilities on January 3, 2010. Of this 4689.51 MW is accounted for by thermal power plants, 1454.80 MW by large hydro plants, 910.65 MW by renewable energy sources, and 469.0 MW of nuclear power plants as shown in Table 1 and Figure 1 [25].

Rajasthan state has seen remarkable development in the recent past decade and is in developing stage also. It means energy requirement in the state will be more than that predicted. In this context consumption of diesel and gas will increase in the coming years.

3. Relevance of Distributed Generation in Rajasthan

In Rajasthan, distributed generation has found three distinct markets which are given below where CHP can be used as per requirements:

(i) back-up small power generation systems including diesel generators that are being used in the domestic and small commercial sectors,

(ii) stand-alone off-grid systems or minigrids for electrification of rural and remote areas. The units will be located at dispersed rural locations, which reduces the transmission and distribution (T&D) losses to some extent. The T&D losses in Rajasthan were about 28% in 2002 [26],

(iii) large captive power plants such as those installed by power-intensive industries.

However before going to these three markets, it is necessary to study the government policy about the generation of electricity. A comprehensive electricity bill was drafted in the year 2000 following a wide consultative process. After a number of amendments, it sailed through the legislative process and was enacted on June 10, 2003. It is known as Electricity Act 2003. It replaces the three existing legislations governing the power sector, namely, first, Indian Electricity Act, 1910, second, the Electricity (Supply) Act, 1948, and third, the Electricity Regulatory Commissions Act, 1998. The policy in section 3 (3.3) of the Electricity Act 2003 identifies decentralized distributed generation of electricity by setting up of facilities together with local distribution network based on either conventional or nonconventional resources methods of generation.

Also to promote distributed generation, the government of India has announced many schemes. Two specific schemes of the government of India, the RGGVY (Rajiv Gandhi Grameen Vidyutikaran Yojana) and the RVE (Remote Village Electrification) scheme, provide up to 90% capital subsidy for rural electrification projects using DDG (decentralized distributed generation) options based on conventional and nonconventional fuels, respectively.

Keeping all above factors in mind it seems that DG can be an additional method of electricity generation. Therefore in this paper economic comparison has been made between various technological options of DG. The various technological options selected for this study are reciprocating diesel engine, biodiesel CI engine, reciprocating gas engine, microturbine, small gas turbine, and low-temperature fuel cell.

4. Comparison Methodology

In order to compare the costs of generation of electricity from each of the above technological options, when these options are used as CHP supply, the annualized life cycle cost (ALCC) [13] is used. The annualized life cycle cost represents the annual cost of purchase of the system and its operation. The amount of fuel consumed annually in producing electricity only by DG-CHP option is obtained by subtracting the

TABLE 1: Installed capacity of electricity generation in Rajasthan state.

Sector	Thermal			Total thermal	Nuclear	Hydro (renewable)	RES** (MNRE)	Grand total
	Coal	Gas	Diesel					
State	3240.00	443.80	0.00	3683.80	0.00	987.96	30.25	4702.01
Private	135.00	0.00	0.00	135.00	0.00	0.00	880.4	1015.40
Central	649.48	21.23	0.00	870.71	469.00	466.84	0.00	1806.55
Subtotal	4024.48	665.03	0.00	4689.51	469.00	1454.80	910.65	7523.96

** Renewable energy sources (RES) include SHP, BG, BP, U&I, solar, and wind energy.
Abbreviation: SHP: small hydro project, BG: biomass gasifier, BP: biomass power, U&I: urban and industrial waste power.
Source: [25].

TABLE 2: Heat-to-power ratio (HPR) values.

DG-CHP technology outputs	HPR
Microturbines	2.6
Reciprocating diesel engines	1.6
Reciprocating gas engines	2.1
Low-temperature fuel cells	1.4
Small gas turbines	2.0
Biodiesel CI engine	1.6

Note: heat distribution losses are not included in this table.
Source: [10].

amount of fuel consumed in producing heat from the total amount of fuel consumed annually in DG-CHP technological options. The amount of fuel consumed in producing heat in DG-CHP option is calculated with the help of heat-to-power ratio (HPR) [8] of each technology given in Table 2:

$$\text{HPR} = \frac{\text{Energy produced (or consumed) as heat}}{\text{Energy produced (or consumed) as electricity}}. \tag{1}$$

In this study, it is considered that there is a heating requirement for domestic purposes or for processes in small industries, and so forth. Loads are not proposed for any particular case; it is considered that there is always almost a heating requirement locally and this local requirement can be fulfilled by using DG-CHP technologies (as per the heat-to-power ratio) and a separate boiler will not be needed for heating purpose hence saving fuel cost. The annualized loads are considered in the study mainly concentrating on the variation of load factor and its effect on the life cycle cost of the technologies. In centralized generation of electricity, if heat is produced separately, a boiler will be required. Here in DG-CHP the cost of boiler is saved but it is not subtracted from the cost of the technological option and considered heat will be used locally and heat distribution losses are not taken into account. The cost of generated electricity is obtained by dividing the ALCC by the annual generation of electricity.

The ALCC is computed as

$$\text{ALCC} = C_0\text{CRF}(d, n) + \text{AC}_f + \text{ACO \& M}, \tag{2}$$

where C_0 = the initial capital cost for the technological option, $\text{CRF}(d, n)$ = capital recovery factor for the technological option based on the discount rate "d" and the life of the option "n" in years, AC_f = the annual fuel cost to produce electricity only for the technological option, and ACO & M = the annual operating and maintenance cost for the technological option.

The capital recovery factor (CRF) is computed using the equation

$$\text{CRF}(d, n) = \frac{[d(1+d)^n]}{[(1+d)^n - 1]}. \tag{3}$$

The annual generation of CHP is dependent on the load factor. The cost of generation of CHP is dependent on the size of the equipment and the application load factor. In this paper based on the typical unit size range given in Table 2, a 50 kW peak rating is used as the basis of calculation except for small gas turbine (500 kW). Tables 2, 3, and 4 are used for the ALCC calculations of the selected technological options. The calculations are done with existing Rajasthan state's fuel and equipment prices. Other data used are diesel fuel costs Rs. 42.06 per liter in June 2011 and Rs. 49.30 per liter in January 2013 in Rajasthan. Natural gas price is Rs. $18/\text{sm}^3$ in June 2011 and Rs. $22.5/\text{sm}^3$ in January 2013.

In the case of technologies not commercially available in Rajasthan or India the existing international prices in US $ have been converted to Indian rupees (Rs.) at the prevalent exchange rates (1 US$ = 52.5 Rs. in January 2013). An idea of the comparative costs of technological options, when these are used as only power and as CHP, and impact of the load factor will provide an idea of the viability of the DG-CHP option. The status of each option in Rajasthan is discussed along with some of the issues relevant to its adoption.

5. Cost of Generation

To make the comparison between the various technological options, load factor and ALCC are used. Load factor represents the operation of engine in fraction of hours in one day. ALCC for various technological options is calculated using (2) and (3). Graphs between ALCC and load factor are drawn at two interest rates 10% and 30% with and without heat power ratio obtainable for a particular technological option for the months of June 2011 and January 2013.

Figures 2 and 3 show the annualized life cycle costs of the diesel engine, gas engine, microturbine, small gas turbine, and fuel cell options, as a function of the load factor with a societal discount rate of 10% for the months of June 2011 and

Table 3: Technologies: markets and performance.

	Residential	Commercial	Industrial	Grid distributed	Remote/off-grid distributed	Typical unit size range (installation size can be larger)	2010 Installed capital cost (Rs/kW)	Electric conversion efficiency (%)	Commercial availability	Application
Microturbines[a]		1	1	1	2	25–300 kW	20000	28	Yes	CHP
Reciprocating CI engines	1	1	1	1	1	5 kW–20 MW	7100	35	Yes	Standby services/CHP
Reciprocating gas engines	1	1	1	1	1	5 kW–20 MW	16000	40	Yes	CHP
Low-temperature fuel cells	1	1	2	1	1	2–250 kW	80000	35	Yes	CHP
Small gas turbines[b]			1	1	2	500 kW–20 MW	17000	40	Yes	CHP

[a]Recuperated microturbine.
[b]Forty percent efficiency achieved with advanced turbine cycle.
1: Primary target market.
2: Secondary target market.
Source: table constructed on the basis of information found in [4, 5] and data available in Rajasthan.

Table 4: Distributed generation technologies and characteristics.

DG-CHP technologies	Life in years	*O&M* cost (Rs/kWh)	Fuel used
Microturbines	20	0.30	Generally uses natural gas, but flare, landfill, and biogas can also be used
Reciprocating CI engines	20	0.30	Diesel, also heavy fuel oil and biodiesel
Reciprocating gas engines	20	0.30	Gas, mainly natural gas; biogas and landfill gas can also be used
Small gas turbines	20	0.30	Gas, kerosene
Low-temperature fuel cells	10	0.30	Hydrogen or natural gas. Reforming of CH_4 to H_2 leads to decreased efficiency

Source: table constructed on the basis of information found in [4] and data available in Rajasthan.

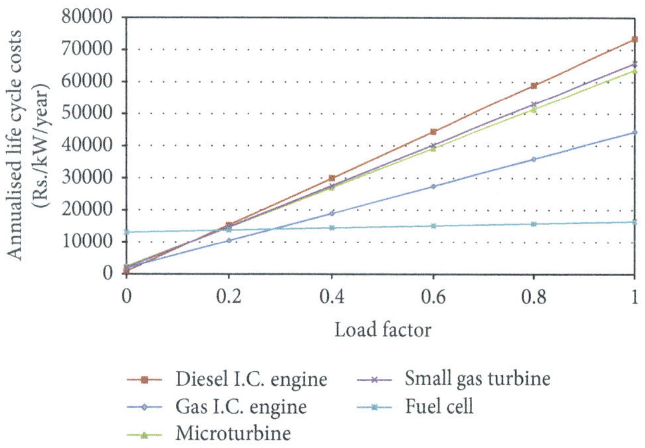

Figure 2: Comparison of annualised life cycle costs for different technological options, when only DG is considered (discount rate (*d*) = 10%) for the month of June 2011.

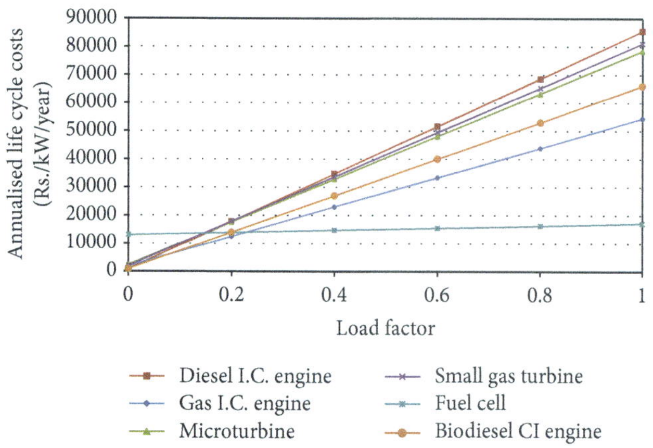

Figure 3: Comparison of annualised life cycle costs for different technological options when only DG is considered (discount rate (*d*) = 10%) for the month of January 2013.

January 2013, respectively, for DG only, that is, not as DG-CHP. Data for biodiesel CI engine is taken in January 2013 only.

It is seen from Figure 2 that fuel cell is the cheapest option at load factor more than 0.3 in June 2011 and from Figure 3 at load factor more than 0.24 in the month of January 2013. It

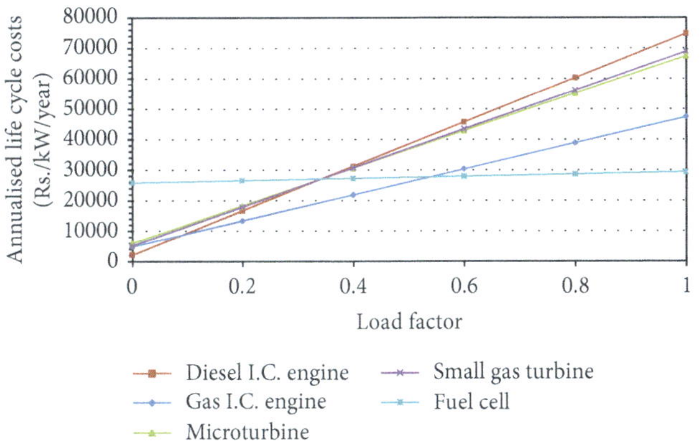

FIGURE 4: Comparison of annualised life cycle costs for different technological options when only DG is considered (discount rate (d) = 30%) for the month of June 2011.

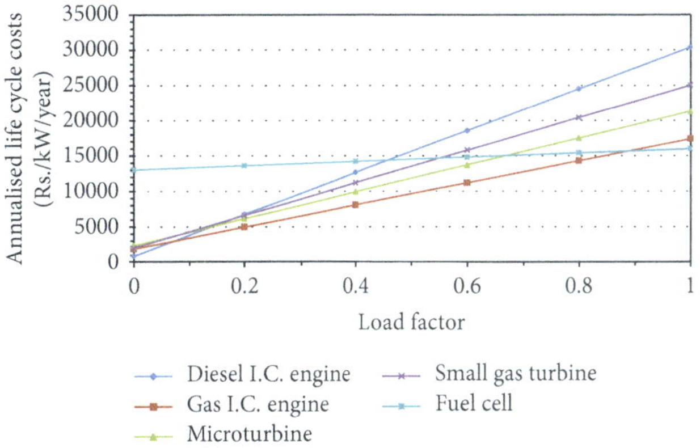

FIGURE 6: Comparison of annualised life cycle costs for different technological options when DG-CHP is considered (discount rate (d) = 10%) for the month of June 2011.

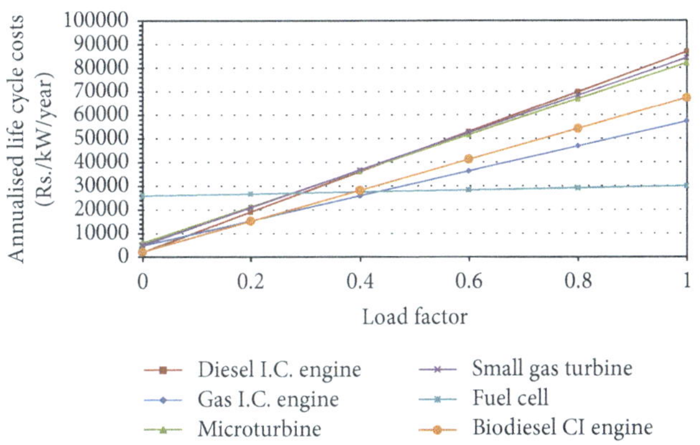

FIGURE 5: Comparison of annualised life cycle costs for different technological options when only DG is considered (discount rate (d) = 30%) for the month of January 2013.

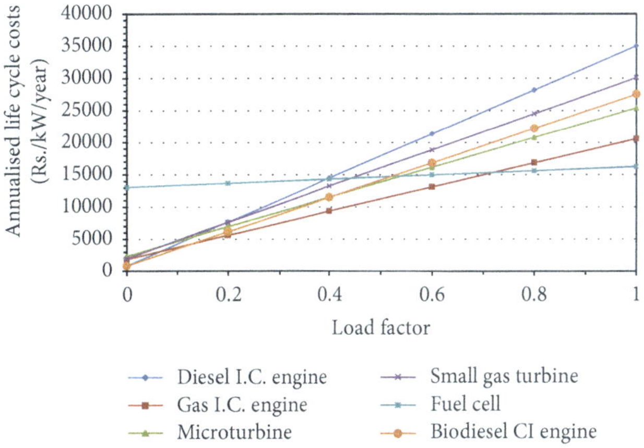

FIGURE 7: Comparison of annualised life cycle costs for different technological options when DG-CHP is considered (discount rate (d) = 10%) for the month of January 2013.

shows that fuel cell is becoming more favorable at low load factor during the passage of time; it is because that cost of diesel is increasing. But technical expertise for fuel cell is not available in the state. Therefore fuel cell is not possible in the present scenario in Rajasthan state but it seems the better option for the future.

From Figures 2 and 3 it can be analyzed that the gas engine is the next cheaper option but the ALCC of gas engine has increased from June 2011 to January 2013 remarkably as compared to fuel cell. The gas engine is less popular in the state due to higher initial capital cost and poor distribution system of gas as a fuel. The improvement in natural gas availability and the presence of gas distribution companies in the state are likely to see an increase in gas engine option. However at low load factor diesel engines are used as a back-up power. Figure 3 shows that at low load factor biodiesel CI engines are better option than diesel engines. Also biodiesel CI engines have better environment-friendly characteristics.

It is seen from Figures 4 and 5 that at high load factors fuel cell is cheaper than other options. Gas engine is the next cheaper option at high load factors. Diesel engine is the costliest at high load factor. At low load factor biodiesel CI engine is the cheapest.

In Figures 4 and 5 (d = 0.3) with comparison to Figures 2 and 3 (d = 0.1), ALCC of fuel cell is increased from Rs. 14000 to Rs. 26000 due to increase in discount rate from 0.1 to 0.3. The possible reason for this increased cost is high initial capital cost of the fuel cell. Because of high initial capital cost, the interest amount will be high which increases the ALCC.

Figures 6 and 8 show the ALCC for societal discount rates of 10% and 30%, respectively, for DG-CHP in the month of June 2011. Here also, except at very low load factors, the gas engine option is cheaper than the diesel engine.

Figures 7 and 9 show the ALCC for societal discount rates of 10% and 30%, respectively, for DG-CHP in the month of January 2013. Here also, except at very low load factors, the gas engine option is cheaper than the diesel engine.

It is noticed from Figures 6–9 that the gas engine is the most favorable option in DG-CHP consideration. That is, ALCC of gas engine is the lowest at almost all the load factors. In case of DG alone, it was seen that ALCC of fuel cell was the lowest but in case of DG-CHP it is more than all other options at low loads and more than gas engine with d = 0.3 at high load factors. With DG-CHP the diesel engine is found as the most expensive option at all load factors. The biodiesel CI engine is always more favorable than diesel engine.

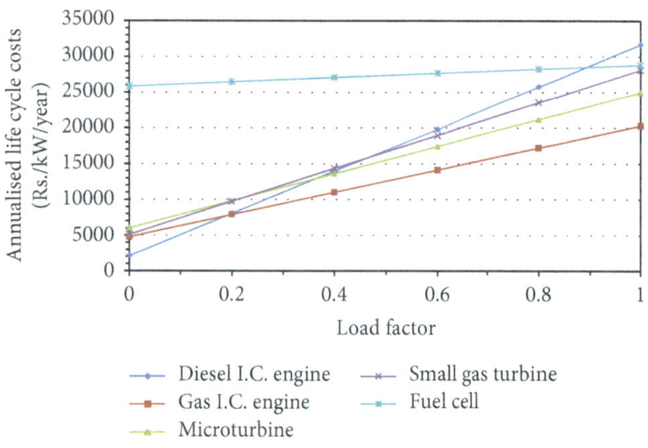

FIGURE 8: Comparison of annualised life cycle costs for different technological options when DG-CHP is considered (discount rate (d) = 30%) for the month of June 2011.

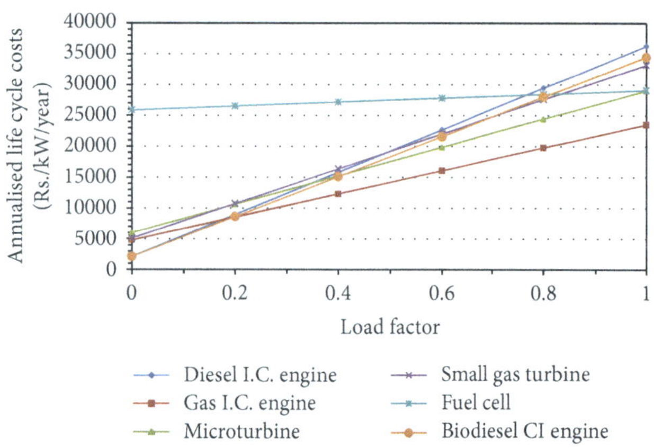

FIGURE 9: Comparison of annualised life cycle costs for different technological options when DG-CHP is considered (discount rate (d) = 30%) for the month of January 2013.

As seen in Figures 8 and 9, except at very low load factor, gas engine is cheaper than diesel engine. Fuel cell is costlier than other options at almost all load factors. Reason for this high cost of fuel cell is that its HPR value is low in comparison to other options. It means useful heat generation is very low which does not contribute towards the cost saving. Initial setup cost of fuel cell is also very high which increases the ALCC.

Figure 9 shows that biodiesel CI engine is more favorable than diesel engine at almost all load factors. It can be seen from the figure for the month of January 2013 and for the month of June 2011 that ALCC of different options has been increased in different manners for the options with the passage of time. Rising trend of ALCC of different options for different years can be observed by comparing Figures 2, 4, 6, and 8 with Figures 3, 5, 7, and 9.

6. Conclusions

DG and DG-CHP are compared for the various technological options at two different discount rates and at two different timings to find out the most suitable option. The following conclusions are drawn from the study.

(1) The fuel cell is found to be very economical but it is not possible to implement the fuel cell due to technical facilities presently in the state.

(2) In case of DG the ALCC for most of the options is found to be more than twice that for the DG-CHP. Therefore DG-CHP may be more beneficial compared to only DG.

(3) In case of DG-CHP the cost for gas engine is found to be the lowest which shows that gas engine is the most effective option in the future when more knowhow about the gas engine will be available in the state.

(4) The biodiesel CI engine-based CHP is found to be very attractive as the ALCC is lower than that for diesel engine CHP and more competitive compared to other options for most of the cases and additionally due to its renewable nature it may emerge as the most effective option considering the local conditions in Rajasthan.

Hence from the above study, it can be concluded that the DG-CHP technologies may be proved as better option for energy sustainability as compared to only DG technologies and the most effective DG-CHP can be selected depending upon the local conditions and availability of fuel sources.

References

[1] T. Ackermann, G. Andersson, and L. Söder, "Distributed generation: a definition," *Electric Power Systems Research*, vol. 57, no. 3, pp. 195–204, 2001.

[2] Fumi Harahap of WADE, "More for less: how decentralized energy can deliver cleaner, cheaper and more efficient energy in Nigeria," A Report by World Alliance for Decentralized Energy (WADE), Christian Aid and the International Centre for Environment and Energy Development (ICEED), 2009.

[3] F. Gullí, "Small distributed generation versus centralised supply: a social cost-benefit analysis in the residential and service sectors," *Energy Policy*, vol. 34, no. 7, pp. 804–832, 2006.

[4] G. Pepermans, J. Driesen, D. Haeseldonckx, R. Belmans, and W. D'haeseleer, "Distributed generation: definition, benefits and issues," *Energy Policy*, vol. 33, no. 6, pp. 787–798, 2005.

[5] D. Arthur, "Distributed generation: system interfaces," Little White Paper, Arthur, D. Little, Inc., Cambridge, Mass, USA, 1999.

[6] P. Dondi, D. Bayoumi, C. Haederli, D. Julian, and M. Suter, "Network integration of distributed power generation," *Journal of Power Sources*, vol. 106, no. 1-2, pp. 1–9, 2002.

[7] IEA, *Distributed Generation in Liberalised Electricity Markets*, vol. 128, International Energy Agency, Paris, France, 2002.

[8] N. Strachan and H. Dowlatabadi, "Distributed generation and distribution utilities," *Energy Policy*, vol. 30, no. 8, pp. 649–661, 2002.

[9] J. E. Allison and J. Lents, "Encouraging distributed generation of power that improves air quality: can we have our cake and eat it too?" *Energy Policy*, vol. 30, no. 9, pp. 737–752, 2002.

[10] N. Strachan and A. Farrell, "Emissions from distributed vs. centralized generation: the importance of system performance," *Energy Policy*, vol. 34, no. 17, pp. 2677–2689, 2006.

[11] E. van der Vleuten and R. Raven, "Lock-in and change: distributed generation in Denmark in a long-term perspective," *Energy Policy*, vol. 34, no. 18, pp. 3739–3748, 2006.

[12] K. M. Maribu, R. M. Firestone, C. Marnay, and A. S. Siddiqui, "Distributed energy resources market diffusion model," *Energy Policy*, vol. 35, no. 9, pp. 4471–4484, 2007.

[13] R. Banerjee, "Comparison of options for distributed generation in India," *Energy Policy*, vol. 34, no. 1, pp. 101–111, 2006.

[14] K. Narula, Y. Nagai, and S. Pachauri, "The role of decentralized distributed generation in achieving universal rural electrification in South Asia by 2030," *Energy Policy*, vol. 47, pp. 345–357, 2012.

[15] A. S. Ramadhas, S. Jayaraj, and C. Muraleedharan, "Use of vegetable oils as I.C. engine fuels—a review," *Renewable Energy*, vol. 29, no. 5, pp. 727–742, 2004.

[16] H. Raheman and A. G. Phadatare, "Diesel engine emissions and performance from blends of karanja methyl ester and diesel," *Biomass and Bioenergy*, vol. 27, no. 4, pp. 393–397, 2004.

[17] H. B. Mathur and L. M. Das, "Utilization of non-edible wild oils as diesel engine fuels," in *Proceedings Bio-Energy Society Second Convention and Symposium*, pp. 198–202, 1985.

[18] N. Hemmerlein, V. Korte, H. Richter, and G. Schoroder, "Performance, exhaust emissions and durability of modern diesel engines running on rapeseed oil," SAE Technical Paper Series 910848, SAE, 1991.

[19] A. K. Agarwal, "Biofuels (alcohols and biodiesel) applications as fuels for internal combustion engines," *Progress in Energy and Combustion Science*, vol. 33, no. 3, pp. 233–271, 2007.

[20] D. G. Kesse, "Global warming-facts, assessment, countermeasures," *Journal of Petroleum Science and Engineering*, vol. 26, pp. 157–168, 2000.

[21] X. Cao, "Climate change and energy development: implications for developing countries," *Resources Policy*, vol. 29, no. 1-2, pp. 61–67, 2003.

[22] T. Johansson and S. McCarthy, "Global warming post-Kyoto: continuing impasse or prospects for progress?" *Energy and Development Report, Energy*, pp. 69–71, 1999.

[23] Z. Utlu and M. S. Koçak, "The effect of biodiesel fuel obtained from waste frying oil on direct injection diesel engine performance and exhaust emissions," *Renewable Energy*, vol. 33, no. 8, pp. 1936–1941, 2008.

[24] G. Tashtoush, M. I. Al-Widyan, and A. O. Al-Shyoukh, "Combustion performance and emissions of ethyl ester of a waste vegetable oil in a water-cooled furnace," *Applied Thermal Engineering*, vol. 23, no. 3, pp. 285–293, 2003.

[25] Ministry of Power, "Annual Report 2009–10 Ministry of Power, Government of India, New Delhi," 2010, http://www.powermin.nic.in.

[26] Ministry of Power, *Discussion Paper on Rural Electrification Policies*, Government of India, New Delhi, India, 2003.

18

Review on the HVAC System Modeling Types and the Shortcomings of Their Application

Raad Z. Homod

Department of Mechanical Engineering, University of Basrah, Qarmat Ali Campus, Basrah 61004, Iraq

Correspondence should be addressed to Raad Z. Homod; raadahmood@yahoo.com

Academic Editor: Mohamed Benghanem

The modeling of the heating, ventilation, and air conditioning (HVAC) system is a prominent topic because of its relationship with energy savings and environmental, economical, and technological issues. The modeling of the HVAC system is concerned with the indoor thermal sensation, which is related to the modeling of building, air handling unit (AHU) equipments, and indoor thermal processes. Until now, many HVAC system modeling approaches are made available, and the techniques have become quite mature. But there are some shortcomings in application and integration methods for the different types of the HVAC model. The application and integration processes will act to accumulate the defective characteristics for both AHU equipments and building models such as nonlinear, pure lag time, high thermal inertia, uncertain disturbance factors, large-scale systems, and constraints. This paper shows types of the HVAC model and the advantages and disadvantages for each application of them, and it finds out that the gray-box type is the best one to represent the indoor thermal comfort. But its application fails at the integration method where its response deviated to unreal behavior.

1. Introduction

The HVAC system modeling implies to the modeling of building, indoor, outdoor, and air handling unit (AHU) equipments. It is normally difficult for one HVAC system model to be completely comprehensive. Therefore, it is possible to divide the comprehensive model into submodels which may be appropriate in some instances. The key to any successful indoor thermal analysis or representation lies in the accuracy of the model of the indoor conditioned space within building and AHU equipments.

The simple hand-calculation methods were available to find out the cooling/heating load until the advent of computer simulation programs for the HVAC systems in the mid 1960s. The first simulation method attempted was to imitate physical conditions by treating the variable time as the independent [1]. Most of the earliest simulation methods were based on the white-box or mathematical (physical) models, which are preferred over other models such as the black-box and the gray-box models because they are easy to analyze even though they are more complex than others [2].

1.1. The Mathematical Model of the HVAC System. There are two types of the white-box or the mathematical model: the lumped and the distributed parameters. The main advantage of the lumped parameter model is that it is much easier to solve than the distributed model.

The mathematical models are very popular for the HVAC systems in representing the processing signal. The processes' signals are constructed based on physical and chemical laws of conservation, such as component, mass, momentum, and energy balance. These laws describe the linking between the input and the output which is transparently represented by a large number of mathematical equations. Furthermore, the mathematical model is a good tool to understand the behavior of the indoor condition by describing the important relationships between the input and the output of the HVAC system. In general, the modeling process of the HVAC systems leads to dynamic, nonlinear, and high-thermal inertia, pure lag time, uncertain disturbance factors, and very high-order models. The whole model can be described by several submodels to alleviate the complexity of the model [3].

These submodel processes are related to fluid flow and heat-and-mass transfers between interfacing submodels, which can be governed by mass, momentum, and energy conservation principles. These principles are usually expressed by differential equations, which may be implemented by time domain or S (frequency) domain, where the S domain can be represented by a transfer function or a state-space function. The limitations of the early building mathematical models are mainly due to the limitation in the computer hardware since the models needed an intensive computational process. But the situation is changing as computational tools capacity has improved by the evolution in software and hardware of computers. The recent mathematical models are being developed to solve the large set of equations that also incorporate submodels mathematical equations.

1.2. The Black-Box Model of the HVAC System. The concept of the black-box model is to fit the transfer function model to the input/output real model data to yield coefficient polynomials that can be factored to provide resonance frequencies and characterization of damping coefficients without knowledge of the internal working. Hence, it does not reflect any specific physical or mathematical structure of the existing behavior in the real model. The mathematical representations of this model include time series such as autoregressive moving average (ARMA), autoregressive model structure with exogenous inputs (ARX), recurrent neural network models, and recurrent fuzzy models. For real-time operation and control, the black-box models are simple enough. On the contrary, physical or mathematical modeling involves detailed analysis of the relationships between all parameters that affect the system. Due to the complex nature of the HVAC systems and the large number of the parameters involved, it is difficult to mathematically model the system [4]. Romero et al. [5] mentioned that the mathematical models require detailed facility information that is sometimes hardly found.

Therefore, the black box is the simplest solution, but at the same time it has to be regularly updated as operation conditions change. Thus, it cannot be used for prediction outside the range of the training data, and such models have poor performance in general.

1.3. The Gray-Box Model of the HVAC System. Some physical processes of the HVAC system are less transparently described where there is much physical insight available, but certain information is lacking. In this case, mathematical models could be combined with black-box models where the resulting model is called a gray-box or a hybrid model. Furthermore, the white-box model of the HVAC system needs modification to provide better performance; this can be accomplished by using a gray-box technique to mimic the output of the white-box analysis. This method is implemented by Leephakpreeda [6] when building a gray-box model based on the white-box model to predict indoor temperature. Some black-box models of the HVAC system are modified to become gray-box models to improve their performance as done by Zhao et al. [7] when they identified the nonlinear link model parameters by a neural network to represent heat exchanger. This means that the main function of a gray box is to improve the performance of the white- or the black-box model.

2. The Background of the HVAC Systems Modeling

The commercial and residential buildings are facing a new era of a growing demand for intelligent buildings worldwide. Intelligent buildings are referred to as energy and water saving, and they provide healthy environment. The first intelligent building was introduced in the late 1970s when buildings were equipped with IT equipments [8]. The developments of the improved building and AHU models are essential to meet the requirements of an intelligent building [9].

The HVAC system modeling evolution of research has been reflected on the representation of the indoor thermal behavior by development and enhancement of identification of buildings and AHU equipments. In general, research on indoor thermal comfort can be divided into two main categories: design-oriented research and research-oriented design as explained by Fallman [10]. This study followed the second category where it depends on the previous research outcomes to develop a design that enhances the indoor thermal comfort.

3. The Evolution of the HVAC System Modeling

Building and AHU modeling has been used for decades to help the HVAC system scientists design, construct, and operate the HVAC systems. The pioneering development in the building and the HVAC system equipments industry is the heat conduction equation model by Joseph Fourier published in 1822, which is the most cited model (see, e.g., [11–14]).

The earlier simulation work in building structure by Stephenson and Mitalas [15, 16] on the *response factor* method significantly improved the modeling of transient heat transfer through the slabs, the opaque fabric, and the heat transfer between internal surfaces and the room air. The heat balance approaches were introduced in the 1970s [17] to enable a more rigorous treatment of building loads. Rather than utilizing weighting factors to characterize the thermal response of the room air due to solar incident, internal gains, and heat transfer through the fabric, instead, the heat balance methodology solves heat balances for the room air and at the surfaces of fabric components.

Since its first prototype was developed over two decades ago, the building model simulation system has been in a constant state of evolution and renewal. Numerical discretization and simultaneous solution techniques were developed as a higher-resolution alternative to the response-factor methods [18]. Essentially, this approach extends the concept of the heat balance methodology to all relevant building and plant components. More complex and rigorous methods for modeling of the HVAC systems were introduced in the 1980s. Transient models and more fundamental approaches were developed [19] as alternatives to the traditional approach which performed mass and energy balances on preconfigured templates of common HVAC systems. The delivery of training and the production of learning materials [20] are also receiving increased attention. Additionally, many validation exercises have been conducted [21] and test procedures developed [22] to assess, improve, and demonstrate the integrity of simulation tools.

The literature presented two types of the HVAC system model: steady-state models, which are extensively presented such as in [23–28], and unsteady-state models presented by [29, 30]. Unsteady-state models can be further categorized into two extreme modeling approaches. The first approach, called *physical or mathematical models*, builds up models entirely based on universal laws, physical laws, and principles [31]. The second approach, called *empirical or black-box models*, constructs models entirely based on experiments or data [32–34].

4. The Mathematical Model

Mathematical models have been widely used in areas as diverse as engineering, economics, medicine, ecology, and agriculture for many kinds of different purposes to satisfy scientific curiosity, prediction, control, fault diagnosis and inadequacies, simulation, and operator training.

In the field of the HVAC system modeling, the most complicated model part is the building model. This is because components that need to be modeled for building are not limited to building construction, such as roof, walls, floor slab, windows, and external shading. Internal loads such as the activity within the space, the number of people, and the heat gain from lighting must also be modeled as well. The subdivision of the building model is an extensive scope of the HVAC models' field, and there are rarely studies that include the entirely scope of a building model. For example, Lü [35] studied the transmission of heat and moisture throughout the walls, roofs, and ceilings to estimate the indoor air temperature and humidity. He did not consider the transmission of heat and moisture throughout ventilation, filtration (doors and windows), and internal load. Furthermore, the moisture conservation equation used assumed the temperature to be the same in all of the phases. He constructed his model by applying the conservation of mass and energy theory based on the fundamental thermodynamic relations. For mass conservation law, he implemented this in moisture transmission by applying Darcy's law and Fick's law. For energy, he implemented Fourier's law. He used the distributed white-box model where partial differential equations are discretized in space by using finite element with time marching scheme and Crank-Nicolson scheme.

On the other hand, Ghiaus and Hazyuk [36] used the mathematical model to estimate the heating load in dynamic simulation by using steady-state heat balance for normalized outdoor conditions. They applied the superposition theorem for electrical circuits to obtain their model's parameters. And they assumed that the thermal model of the building is linear, thermal capacity of the wall and the indoor air is lumped and considered that the time series of disturbances (such as weather and internal loads) and occupational programs are known because they used model predictive control (MPC) which proposed an unconstrained optimal control algorithm to solve the load estimation problem. They obviously have imposed many assumptions to facilitate the calculations of heating load, which leads to lack of accuracy in the results. In addition, they used a single-input single-output- (SISO-) type model that does not consider the moisture transmission, an important element in deciding thermal comfort.

For the air-handling unit (AHU) mathematical model, Wang et al. [37–40] built models of heat exchanger for air-handling unit based on the conservation of energy and applied thermal balance equation on control volume for heat exchanger. This model is characterized as a SISO model since it does not take into account the effect of the mixing air chamber and assumed that the temperature of air supplied to the conditioned space is equal to the surface temperature of heat exchanger. Furthermore, they neglected the humidity of the moisture air supplied to the conditioned space because they do not want to include the effectiveness of humidity variation on thermal comfort. Therefore, they supposed that the type of cooling coil is of a dry type and that there is no indoor latent load.

5. The Black-Box Model

The physical model involves detailed study of the relationships between all parameters that affect the hygrothermal (the variation of humidity and temperature) system. Due to the complex nature of hygrothermal systems and the large number of parameters involved, physical modeling has become more complicated in application. Usage of the black-box model is sometimes preferred because it is straightforward to construct and there is no need to have knowledge of the system's internal structure.

Mustafaraj et al. [4] identified the humidity and thermal behavior models of an office in a modern commercial building by using different methods of the black-box model such as Box-Jenkins (BJ), autoregressive with external inputs (ARX), autoregressive moving average exogenous (ARMAX) structure, and output error (OE) models. They adopted linear parametric models to predict room temperature and relative humidity for different time scales. The linear model is adopted to obtain a simple and low number of model parameters, but this caused downbeat on the accuracy, especially in the representation of the heat storage or flywheel effect on the instantaneous load. In this group of models, they found out that the BJ model is suitable for the winter season where the ARMAX and ARX models give good results for the summer and autumn seasons and that the OE model is appropriate for the summer season. This means that there is no specific model that can represent indoor temperature and humidity for all four seasons. Furthermore, Mustafaraj et al. [41] created the neural network-based nonlinear autoregressive model with external inputs (NNARX) model, which is suitable to predict indoor office temperature and relative humidity for the summer season. The learning of NNARX model is done off-line because this type of model is well known for having a sluggish learning process. On top of this, they used the optimal brain surgeon (OBS) strategy which made the learning much slower, so it is not suitable to apply the online learning process [42].

Yiu and Wang [43] created a generic SISO and MIMO black-box model for AHU. The ARX and ARMAX structures are used, where their parameters are identified by using the recursive extended least squares (RELS) method. In general, the selection of model structure, between SISO and MIMO as well as between ARX and ARMAX, is a compromise between model simplicity and accuracy. The accuracy of the anticipated model outputs is in contrary with the simplicity and the time period of the updating weight. Furthermore, the more the accurate the ARMAX structure is, the more complex it becomes, which will also yield more residual white noise.

Barbosa and Mendes [44] integrated the works of a group of researchers in order to obtain a comprehensive model, for the chiller model is quoted from CA [45] by applying empirical equations based on regression functions. The cooling tower model used is based on Merkel's theory for the mass and sensitive heat transfer between the air and water in a counter flow cooling tower. The pump and fan models are quoted from Brandemuehl et al. [46] where power for variable flow is calculated from a regression of part-load power consumption as a function of part-load flow with the assumption that motor efficiency is constant. For the cooling and dehumidification coil model, there are three possible conditions for the coil: completely dry, partially wet, or completely wet. The model for all three conditions is quoted from Elmahdy and Biggs [47] based on coil outside surface temperature and air dew-point temperature. The room building model for heat and moisture transfer is based on the Philip and DeVries theory, which solves the partial differential governing equations for room control volume within the porous building element, which is quoted from Mendes et al. [48], where it is assumed that the water vapor behaves like a perfect gas and the vapor exchanged between the wall and the air is in a linear function of the differences between the temperature and moisture content.

6. The Gray-Box Model

The Gray-box model, sometimes called the semiphysical or hybrid model, is created by a combination of *physical and empirical models*, which is to compensate for their deficiencies as individual approaches.

In some gray-box modeling, the model structures are derived mathematically from physical or thermodynamic principles, while their parameters are determined from catalog, commissioning, or operating data. This is what Braun et al. [49] and Wang et al. [50] did when they developed an effective model through introducing the idea of air saturation specific heat. Catalog data at an operating condition are used to obtain the number of transfer units, which is then employed to obtain the performance at other operating conditions.

Based on the same concept of Braun et al., Wang et al. [51] built their gray-box model for predicting the performance of chilled water cooling coils in a static state. The mathematical part they built was based on the heat transfer mechanism and the energy balance principle. A model with no more than three characteristic parameters that represent the lumped geometric terms was developed. Procedures for determining the unknown parameters using commissioning or catalog information by linear or nonlinear least squares methods are used. Using this method, the model captures the inherent nonlinear characters of the AHU. Both Braun's and Wang's models have high levels of uncertainties because they evaluated models parameters depending on catalog and operation data where most of these data are estimated from ideal operation conditions. Some data are measured from the real operation, but these parameters values will eventually change due to the aging of the HVAC system.

Meanwhile, Ghiaus et al. [52] used a gray-box model to identify the AHU by imposing in the mathematical part that air temperature difference occurs in cooling coils and that the humidity ratio difference occurs in the humidifier only, meaning that the cooling coil is of dry type and that there is no change in the air temperature through the humidifier. This is to separate the transfer functions for each element in order to overcome the coupling between the temperature and humidity, where the parameters of the discrete form of these models are then experimentally identified. It is obvious that the assumptions made by the authors are too unrealistic and cannot be achieved except in some rare cases. This led to avoidable inaccuracy in the model's output data.

7. The Shortcomings in the Previous Works

From the reading through the literature on the topics of this research's objectives which are summarized in the previous sections, some shortcomings that prevent the possibility to be implemented in the simulation environment are faced. Thus, it can be concluded that the most important discouraging gaps to implement the analysis simulation properly and accurately for each objective are listed as follows.

There are a lot of deficiencies for each of the models studied that need to be addressed. These deficiencies resulted from various simplifying assumptions to reduce the complexity of thermal interactions, unmeasured disturbances, uncertainty in thermal properties of structural elements, and other parameters, which makes it quite a challenge to obtain reliable analytical models.

The most prominent shortcoming in this area of studies is the fact that there is no model that includes both building and AHU with all of the details. The other shortcoming is that the existing building models do not represent the lag time cooling load and solar gains incident on the surfaces of wall, roof, and window. In addition, the AHU model represented only the cooling coil without precooling coil and neglected the effectiveness of air mixing chamber. Furthermore, there are some studies where the models are simplified, eliminating many of the important features such as the humidity transmission, the change of air temperature through humidifier.

Besides, most of the literature on previous works presented the SISO type-model which is easy to manipulate using linear controller [53]. And the most important of all is that a lot of studies assume that the cooling coil is dry, which is contrary to the reality.

8. The Alternative Methods

In addition to the shortcomings due to the inaccurate application of the HVAC system modeling, there are difficulties in the characterstics of the HVAC systems which do not allow its modeling to deploy efficiently; these characteristics are listed as follows.

(1) For the building and AHU model, consider the following.

   (a) It is difficult to represent solar gain incident on the building envelops.

   (b) Large number of variables is required to describe the HVAC system model.

(2) For the indoor thermal comfort model, consider the following.

   (a) It is difficult to represent indoor thermal comfort model as white-box to analyze.

   (b) The indoor thermal comfort is identified as an implicit nonlinear model, and this is difficult to implement in the real time.

To address the aforementioned shortcomings and the difficulties in the existing model, the following building and AHU model procedures are suggested for this study.

(1) Use a physical-empirical hybrid modeling to describe the HVAC with its various thermal inertia parts.

(2) Systematize the HVAC system into five subsystems to reduce the complexity of the modeling process.

(3) Use the empirical methods for thermal load's calculations which are friendly with variable air volume (VAV) to enhance energy savings.

(4) Use precooling coil to control indoor humidity, and such method results in reducing energy waste as presented by Homod et al. [54].

The alternative method for this issue is proposing a hybrid model combing building structure with the equipments of the HVAC system in one model as adopted by Homod et al. [55, 56]. The hybrid identification is built with physical and empirical functions of thermal inertia quantity. The empirical residential load factor (RLF), which is modeled by the residential heat balance (RHB), is required to calculate a building cooling/heating load. The model parameters can be calculated differently from room to room and are appropriate for the variable air volume (VAV) factor [57]. In this work, a precooling coil is added to humidify the incoming air, which controls the humidity more efficiently inside conditioned space.

The modifications of the HVAC system modeling recommended for the alternative method are based on the reorganizing of the subsystem models. The model for the HVAC system is divided into two parts: building and AHU model and indoor thermal comfort model. The building and AHU model adapted is based on hybridization between two methods: physical and empirical methods, and it depends on the thermal inertia quantity. Physical laws are used to build a submodel for subsystems that have low thermal inertia, while the empirical method is used to build a submodel for subsystems with high thermal inertia. The empirical method used is the residential load factor (RLF). Furthermore, a precooling coil is added to the AHU to humidify the incoming air, which controls the humidity more efficiently inside conditioned space. The indoor thermal comfort model is to evaluate indoor comfort situations using predicted mean vote (PMV) and predicted percentage of dissatisfaction (PPD) indicators. These indicators are identified by a Takagi-Sugeno (TS) fuzzy model and tuned by the Gauss-Newton method for nonlinear regression (GNMNR) algorithm. This type of model is regarded as a white-box model, which is useful for analytical processes such as prediction and extrapolation [58]. The variation behavior of PPD versus PMV is imperative for the HVAC system to control indoor desired conditions as adopted by many researchers [59–64].

The space physical behavior for the building and AHU model for the two main output components is given by combining thermal model equation with moisture model equation deriving the whole subsystem state-space equation of conditioned space as presented by Ghiaus and Hazyuk [36]. Then, eliminating the states vector x, we follow a similar method in the precooling coil by taking the Laplace transformation on both sides of the state-space equation to get the following:

$$\begin{bmatrix} T_r(s) \\ \omega_r(s) \end{bmatrix} = \begin{bmatrix} G_{1,19} & G_{1,20} & G_{1,21} & G_{1,22} & G_{1,23} & G_{1,24} & G_{1,25} & G_{1,26} & G_{1,27} \\ G_{2,19} & G_{2,20} & G_{2,21} & G_{2,22} & G_{2,23} & G_{2,24} & G_{2,25} & G_{2,26} & G_{2,27} \end{bmatrix} = \begin{bmatrix} T_{\mathrm{wl_{in}}}(s) \\ T_{g_{\mathrm{in}}}(s) \\ T_{\mathrm{slb_{in}}}(s) \\ T_o(s) \\ T_s(s) \\ f_4 \\ \omega_s(s) \\ \omega_o(s) \\ Q_{\mathrm{ig},l} \end{bmatrix}, \tag{1}$$

where $G_{1,19}(s)$ $G_{1,20}(s), \ldots, G_{1,27}(s)$ and $G_{2,19}(s)$ $G_{2,20}(s), \ldots,$ $G_{2,27}(s)$ are the input-output transfer factors that are presented in the Appendix.

The combined model block diagram for building and AHU model and indoor thermal comfort model represents a good overall picture of the relationships among transfer function variables of a subsystem model. It is possible to arrange the final subsystem's transfer functions in a way to reflect reality where the output of the first subsystem is the input to the next subsystem and so on. This is illustrated by Figure 1. Note here that it is difficult to arrange and derive the overall mathematical model that represents the system's general equation by only looking at these equations. Therefore, we sought the help of graphics.

9. Conclusion

This paper has reviewed the previous works related to the HVAC systems modeling to guide the researcher to develop the best and rational solutions for the problem statements of the current HVAC system modeling. Through the literature review for different types of modeling for the HVAC system, a lot of advantages, disadvantages, and shortcomings of the current models have been discovered. The review of the models' literatures are aimed at figuring out which type of model is the most suitable to represent the behavior of the real HVAC system. By displaying the advantages and disadvantages of each of the three model types, it became evident that the gray-box model has many features which discriminate it from the other models by closely representing the real behavior of a HVAC system. The hybrid or the gray-box method is employing energy and mass conservation law to obtain the overall system model, where this model is built with physical and empirical functions of thermal inertia quantity. The building and HVAC system structures are including both types of high and low thermal inertia, and this study recommended the hybridization between the two modeling approaches, physical and empirical, to arrive at an accurate model of the overall system. However, to do that for such a system with various thermal inertia subsystems, care must be given to the heat storage capacity of the subsystem and its relation to the difference in the temperature (input and output temperatures of control volume) and the difference in the humidity ratio. If heat storage is a function of these two properties only, then we can apply physical laws directly. The integration between the building and the HVAC system equipments model usually found in the previous studies incorporated two subsystems: heat exchanger for heating or cooling of supply air to the building and a conditioned space, which included the building structure and indoor and outdoor heating/cooling loads. This integration method leads to the trimming of some influencing subsystems in order to reduce a multiple subsystem into two subsystems, and this leads to inaccurate representation of the real model. To represent identical real model, the trimmed subsystems must be incorporated inside the model. So, a real model consists of an enormous number of variables and parameters that complicate it since it is difficult to deal with these variables and parameters. To reduce the complexity of the HVAC model, some studies divided it into parts and adopted a hybrid method that uses both physical and empirical modeling schemes to arrive at a model that can accurately represent a building and AHU model with its various thermal inertia subsystems. An important finding of this study is that one of the major unpredicted disturbances to the system is the variation of solar radiation, which is very hard to model correctly. For these reasons, the empirical analyses that were employed on those parts of the system are recommended to follow.

Highlights

(i) The advantages, disadvantages, and shortcomings of the HVAC systems modeling application are shown.

(ii) The entire literature modeling aspects that related to indoor thermal sensation are summarized.

(iii) It is found out that the gray-box model is the best one to represent the indoor thermal comfort.

(iv) An alternative method to combing building structure with AHU for the gray-box model is show.

Appendix

Parameters, Factors, and Origin of (1) (Source: Homod et al. [55])

The definitions of system parameters and variables are given as follows.

M_{He} is the mass of heat exchanger (kg), cp_{He} is the specific heat of heat exchanger and (J/kg·°C), $\dot{m}_{w,t}$ is the mass flow rate of chilled water at time t (kg/sec.), $T_{h,t}$, $T_{\mathrm{os},t}$, $T_{o,t}$ are the temperatures of heat exchanger, outside supply air, and outside air, respectively, at time t (°C), T_{wo} and T_{win} are the water out/in heat exchanger temperature (°C), $\dot{m}_{o,t}$ is the mass flow rate of outside air at time t (kg/sec.), M_{ahe} is the mass of air in heat exchanger (kg), $T_{\mathrm{os},t}$ and $\omega_{\mathrm{os},t}$ are the temperature and humidity ratio of outside air supplied, respectively, M_m is the mass of air in control volume of mixing air chamber (kg), cp_a is the specific heat of moist air (J/kg·°C), $T_{m,t}$, $T_{\mathrm{os},t}$, and $T_{r,t}$ are the mixing, outside supply, and return temperature, respectively, at time t (°C), $\dot{m}_{\mathrm{os},t}$, $\dot{m}_{r,t}$, and $\dot{m}_{m,t}$ are the mass flow rate of ventilation, return, and mixing air at time t (kg/sec.), $M_m cp_a$ is the heat capacitance of air for mixing air chamber (J/°C), M_{mHe} is the mass of main heat exchanger (kg), cp_{He} is the specific heat of heat exchanger J/(kg·°C), $\dot{m}_{\mathrm{mw},t}$ is the mass flow rate of main cooling coil chilled water at time t kg/(sec.), $T_{h,t}$, $T_{s,t}$, and $T_{m,t}$ are the heat exchanger, supply air, and mixing air temperature, respectively, at time t (°C), T_{wo}, and T_{win} are the water out/in heat exchanger temperature (°C), ($T_{\mathrm{wo}} - T_{\mathrm{win}}$) = Δt_{mw} is the cooling design temperature difference (°C), $M_{\mathrm{wl}} cp_{\mathrm{wl}}$ is the heat capacitance of walls, ceilings, roofs, and doors (J/K), $\sum_i \dot{Q}_{opq_{\mathrm{in}}}$ and $\sum_i \dot{Q}_{opq_{\mathrm{out}}}$ are the heat gains and losses through walls, ceilings, roofs, and doors, U is the construction U-factor W/(m^2·K), Δt is the cooling design temperature difference, °C, DR is the cooling daily range (K), OF_t, OF_b, and OF_r are the opaque-surface cooling factors, the k parameters are k_1 which is the function of thermal resistant and outside temperature, k_2 which is the function of thermal resistant and solar radiation incident on the surfaces (°C), and k_3 which is the function of thermal resistant and convection heat transfer, $\mathrm{CF}_{\mathrm{fen}_j}$ is given by equation, $\dot{Q}_{\mathrm{fen}}$ is the fenestration cooling load (W), A_{fen} is the fenestration

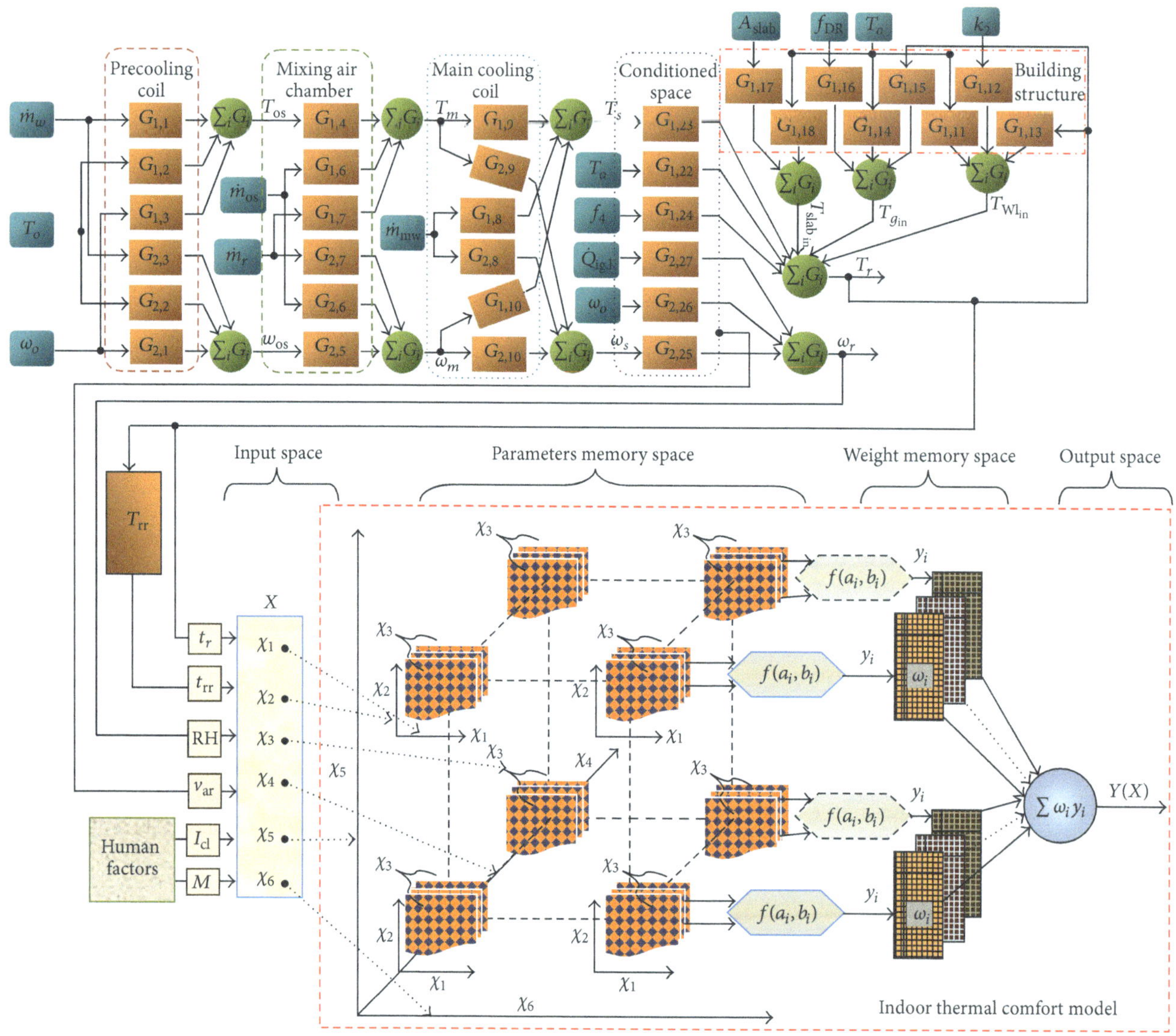

FIGURE 1: Subsystems model block diagram.

area (including frame) (m^2), CF_{fen} is the surface cooling factor (W/m^2), u_{NFRC} is the fenestration NFRC heating U-factor, $W/(m^2{\cdot}K)$, NFRC is the National Fenestration Rating Council, Δt is the cooling design temperature difference (K), DR is the cooling daily range (K), SHGC is the fenestration rated or estimated NFRC solar heat gain coefficient, IAC is the interior shading attenuation coefficient, FF_s is the fenestration solar load factor, PXI is a peak exterior irradiance (W/m^2), E_t, E_d, and ED are peak total, diffuse, and direct irradiance, respectively (W/m^2), T_X is a transmission of exterior attachment (insect screen or shade screen), F_{shd} is a fraction of fenestration shaded by permanent overhangs, fins, or environmental obstacles, L = site latitude, °N, ψ = exposure (surface azimuth), ° from south (−180 to +180), SLF is the shade line factor, D_{oh} is the depth of overhang (from plane of fenestration) (m), X_{oh} is the vertical distance from top of fenestration to overhang (m), and h is the height of fenestration (m), IAC is the interior attenuation coefficient of fenestration with partially closed shade, F_{cl} is the shade fraction closed (0 to 1), IAC_{cl} is the interior attenuation coefficient of fully closed configuration, $\sum_i \dot{Q}_{slab_{in}}$ and $\sum_i \dot{Q}_{slab_{out}}$ are the heat gain and loss through slab floors, respectively (W), $M_{wl}cp_{wl}$ is the heat capacitance of slab (J/K), $\dot{Q}_{slab_{out}}$ is the heat loss through slab floors (W), f_t is the heat loss coefficient per meter of perimeter W/(m·K), P is the perimeter or exposed edge of floor (m), $T_{slab_{in}}$ is the inside slab floor temperature or indoor temperature (°C), T_o is the outdoor temperature (°C), A_{slab} is the area of slab (m^2), Cf_{slab} is the slab cooling factor (W/m^2), h_{srf} is the effective surface conductance, including resistance of slab covering material (R_{cvr}), C_{slab} is the heat capacitance of slab floors (J/k), C_s is the air sensible heat factor (w/(L·S·K)), A_L is the building effective leakage area, cm^2, IDF is the infiltration driving force ($L/(s{\cdot}cm^2)$), C_{af} is the heat capacitance of indoor air and furniture, $\dot{m}_{inf}$ is the infiltration air mass flow rate (kg/s), f_{fen} is the direct radiation (W), ω_s and ω_o are the humidity ratios of outdoor and supply air, respectively, and $\dot{Q}_{ig,l}$ is the latent cooling load from internal gains, where $G_{1,19} = k_{wl}/f_2(\tau_6 S + 1)$, $G_{1,20} = 1/f_2 R_g(\tau_6 S + 1)$, $G_{1,21} = k_{slb}/f_2(\tau_6 S + 1)$, $G_{1,22} = f_3/f_2(\tau_6 S + 1)$, $G_{1,23} = \dot{m}_m cp_a/f_2(\tau_6 S + 1)$, $G_{1,24} = 1/f_2(\tau_6 S + 1)$, $G_{1,25} = 0$, $G_{1,26} = 0$, $G_{1,27} = 0$, $G_{2,19} = 0$, $G_{2,20} = 0$, $G_{2,21} = 0$, $G_{2,23} = 0$, $G_{2,24} = 0$, $G_{2,25} = \dot{m}_s/\dot{m}_r(\tau_r S + 1)$, $G_{2,26} = \dot{m}_{inf}/\dot{m}_r(\tau_r S + 1)$, $G_{2,27} = 1/h_{fg}\dot{m}_r(\tau_r S + 1)$, $k_{wl} = \sum_j A_{w_j} h_{i_j}$, $k_{slb} = \sum_j A_{slb_j} h_{i_j}$, and $f_3 = C_s \times A_L \times \text{IDF}$ (W/k); C_s is the air sensible heat factor (w/(L·S·K)), A_L is the building effective leakage area cm^2, IDF is the infiltration driving force (L/(s·cm^2)), $f_2 = \sum_j A_{w_j} h_{i_j} + 1/R_g + \sum_j A_{slb_j} h_{i_j} + C_s \times A_L \times \text{IDF} + \dot{m}_m cp_a$, (W/k), $\tau_6 = C_{af}/f_2$, (sec.), C_{af} is the heat capacitance of indoor air and furniture, $\dot{m}_{inf}$ is the infiltration air mass flow rate, (kg/s), $f_4 = f_{fen} + 136 + 2.2A_{cf} + 22N_{oc}$ (W), and $f_{fen} = \sum_j A_{fen_j} \text{PXI}_j \times \text{SHGC}_j \times \text{IAC}_j \times \text{FF}_{sj}$ is the direct radiation (W). ω_s, ω_o is the humidity ratio of outdoor and supply air respectively, and $\dot{Q}_{ig,l}$ is the latent cooling load from internal gains.

References

[1] GATC, *Computer Program For Analysis of Energy Utilization in Postal Facilities*, vol. 1, General American Transportation Corporation, 1967.

[2] J. V. Candy, *Model-Based Signal Processing*, John Wiley & Sons, New York, NY, USA, 2006.

[3] P. S. Agachi, Z. K. Nagy, M. V. Cristea, and A. I. Lucaci, *Model Based Control*, John Wiley & Sons, New York, NY, USA, 2006.

[4] G. Mustafaraj, J. Chen, and G. Lowry, "Development of room temperature and relative humidity linear parametric models for an open office using BMS data," *Energy and Buildings*, vol. 42, no. 3, pp. 348–356, 2010.

[5] J. A. Romero, J. Navarro-Esbrí, and J. M. Belman-Flores, "A simplified black-box model oriented to chilled water temperature control in a variable speed vapour compression system," *Applied Thermal Engineering*, vol. 31, no. 2-3, pp. 329–335, 2011.

[6] T. Leephakpreeda, "Grey prediction on indoor comfort temperature for HVAC systems," *Expert Systems with Applications*, vol. 34, no. 4, pp. 2284–2289, 2008.

[7] Y. Zhao, S. Zhou, and L. Li, "Dynamic characteristics modelling of a heat exchanger using neural network," in *Proceedings of the IEEE International Conference on Digital Object Identifier*, pp. 13–18, November 2008.

[8] R. Caffrey, "The intelligent building: an ASHRAE opportunity intelligent buildings," *ASHRAE Technical Data Bulletin*, vol. 4, no. 1, pp. 1–9, 1998.

[9] J. Wong, H. Li, and J. Lai, "Evaluating the system intelligence of the intelligent building systems. Part 1: development of key intelligent indicators and conceptual analytical framework," *Automation in Construction*, vol. 17, no. 3, pp. 284–302, 2008.

[10] D. Fallman, "Design-oriented human-computer interaction," in *Proceedings of the SIGCHI Conference on Human Factors in Computing Systems ACM*, pp. 225–232, New York, NY, USA, April 2003.

[11] X. Lu and P. Tervola, "Transient heat conduction in the composite slab: analytical method," *Journal of Physics A*, vol. 38, no. 1, pp. 81–96, 2005.

[12] X. Lu, P. Tervola, and M. Viljanen, "An efficient analytical solution to transient heat conduction in a one-dimensional hollow composite cylinder," *Journal of Physics A*, vol. 38, no. 47, pp. 10145–10155, 2005.

[13] X. Lu, P. Tervola, and M. Viljanen, "A novel and efficient analytical method for calculation of the transient temperature field in a multi-dimensional composite slab," *Journal of Physics A*, vol. 38, no. 39, pp. 8337–8351, 2005.

[14] X. Lu, P. Tervola, and M. Viljanen, "A new analytical method to solve the heat equation for a multi-dimensional composite slab," *Journal of Physics A*, vol. 38, no. 13, pp. 2873–2890, 2005.

[15] D. G. Stephenson and G. P. Mitalas, "Cooling load calculations by thermal response factor method," *ASHRAE Transactions*, vol. 73, no. 1, pp. 508–515, 1967.

[16] D. G. Stephenson and G. P. Mitalas, "Calculation of heat conduction transfer functions for multilayer slabs," *ASHRAE Transactions*, vol. 77, no. 2, pp. 117–126, 1971.

[17] T. Kusuda, *The Computer Program For Heating and Cooling Loads in Buildings*, NBS Building Science Series No. 69, National Bureau of Standards, Washington, DC, USA, 1976.

[18] J. A. Clarke, *Environmental Systems Performance [Ph.D. thesis]*, University of Strathclyde, Glasgow, UK, 1977.

[19] J. Lebrun, *Proceedings International Conference on System Simulation in Buildings*, Commission of the European Communities, Liège, Belgium, 1982.

[20] S. O. Jensen, "Validation of building energy simulation programs," Part I and II, Research Report PASSYS Subgroup Model Validation and Development, CEC, Brussels, EUR, 15115 EN, 1993.

[21] R. Judkoff and J. Neymark, International Energy Agency Building Energy Simulation Test (BESTEST) and Diagnostic Method, IEA Energy Conservation in Buildings and Community Systems Programme Annex 21 Subtask C and IEA Solar Heating and Cooling Programme Task 12 Subtask B, 1995.

[22] ASHRAE, "Standard method of test for the evaluation of building energy analysis computer programs," ASHRAE Standard 140P, The American Society of Heating, Refrigerating, and Air-Conditioning Engineers, Atlanta, Ga, USA, 1998.

[23] G. Knabe and H. Le, "Building simulation by application of a HVAC system considering the thermal and moisture behaviors of the perimeter walls," in *Proceedings of the International Conference on Building Performance Simulation (IBPSA '01)*, vol. 1, pp. 965–972, 2001.

[24] J. L. M. Hensen, *On the Thermal Interaction of Building Structure and Heating and Ventilating System [Ph.D. thesis]*, Eindhoven University of Technology, 1991.

[25] T. T. Chow, J. A. Clarke, and A. Dunn, "Primitive parts: an approach to air-conditioning component modelling," *Energy and Buildings*, vol. 26, no. 2, pp. 165–173, 1997.

[26] J. C. Lam, S. C. M. Hui, and A. L. S. Chan, "Regression analysis of high-rise fully air-conditioned office buildings," *Energy and Buildings*, vol. 26, no. 2, pp. 189–197, 1997.

[27] J. Cui, T. Watanabe, Y. Ryu, Y. Akashi, and N. Nishiyama, "Numerical simulation on simultaneous control process of indoor air temperature and humidity," in *Proceedings of the International Conference on Building Performance Simulation (IBPSA '99)*, vol. 2, pp. 10–12, 1999.

[28] J. P. Bourdouxhe and P. André, "Simulation of a centralized cooling plant under different control strategies," in *Proceedings of the International Conference on Building Performance Simulation (IBPSA '97)*, vol. 1, pp. 95–102, 1997.

[29] R. M. Barbosa and N. Mendes, "Dynamic simulation of fan-coil systems," in *Proceedings of the 17th International Congress of Mechanical Engineering (COBEM '03)*, Sao Paulo, Brazil, 2003.

[30] P. R. Novak, N. Mendes, and G. H. C. Oliveira, "Simulation and analysis of a secondary HVAC system using MATLAB/SIMULINK platform," in *Proceedings of the ASME International Mechanical Engineering Congress and Exposition*, pp. 387–392, November 2004.

[31] J. A. Orosa, "A new modelling methodology to control HVAC systems," *Expert Systems with Applications*, vol. 38, no. 4, pp. 4505–4513, 2011.

[32] X. Lü and M. Viljanen, "Controlling building indoor temperature and reducing heating cost through night heating electric stove," *Energy and Buildings*, vol. 33, no. 8, pp. 865–873, 2001.

[33] X. Lü, "Modelling of heat and moisture transfer in buildings: II. Applications to indoor thermal and moisture control," *Energy and Buildings*, vol. 34, no. 10, pp. 1045–1054, 2002.

[34] S. Ogonowski, "Modeling of the heating system in small building for control," *Energy and Buildings*, vol. 42, no. 9, pp. 1510–1516, 2010.

[35] X. Lü, "Modelling of heat and moisture transfer in buildings: I. Model program," *Energy and Buildings*, vol. 34, no. 10, pp. 1033–1043, 2002.

[36] C. Ghiaus and I. Hazyuk, "Calculation of optimal thermal load of intermittently heated buildings," *Energy and Buildings*, vol. 42, no. 8, pp. 1248–1258, 2010.

[37] J. Wang, C. Zhang, and Y. Jing, "Hybrid CMAC-PID controller in heating ventilating and air-conditioning system," in *Proceedings of the IEEE International Conference on Mechatronics and Automation (ICMA '07)*, pp. 3706–3711, August 2007.

[38] J. Wang, C. Zhang, and Y. Jing, "Research of cascade control with an application to central air-conditioning system," in *Proceedings of the IEEE Conference on Digital Object Identifier*, pp. 498–503, 2007.

[39] J. Wang, C. Zhang, and Y. Jing, "Analytical design of decoupling control for variable-air-volume air-conditioning system," in *Proceedings of the IEEE Conference on Digital Object Identifier*, pp. 630–635, 2008.

[40] J. Wang, Y. Jing, and C. Zhang, "Robust cascade control system design for central air-conditioning system," in *Proceedings of the World Congress on Intelligent Control and Automation (WCICA '08)*, pp. 1506–1511, June 2008.

[41] G. Mustafaraj, G. Lowry, and J. Chen, "Prediction of room temperature and relative humidity by autoregressive linear and nonlinear neural network models for an open office," *Energy and Buildings*, vol. 43, no. 6, pp. 1452–1460, 2011.

[42] M. J. Nazemosadat, P. Setoodeh, and A. A. Safavi, "Improving neural network models for forecasting seasonal precipitation in south-western Iran: the evaluation of oceanic-atmospheric indices," *Advances in Geosciences, Atmospheric Science*, vol. 16, 2008.

[43] J. C.-M. Yiu and S. Wang, "Multiple ARMAX modeling scheme for forecasting air conditioning system performance," *Energy Conversion and Management*, vol. 48, no. 8, pp. 2276–2285, 2007.

[44] R. M. Barbosa and N. Mendes, "Combined simulation of central HVAC systems with a whole-building hygrothermal model," *Energy and Buildings*, vol. 40, no. 3, pp. 276–288, 2008.

[45] "Pacific Gas and Electric," in *Proceedings of the PG and E Energy Center DOE-2 Lunch Series: Chiller Plant Performance Curves*, San Francisco, Calif, USA, 1996.

[46] M. J. Brandemuehl, S. Gabel, and I. Andresen, *HVAC2 Toolkit: A Toolkit For Secondary HVAC System Energy Calculation*, ASHRAE, Atlanta, Ga, USA, 1993.

[47] A. H. Elmahdy and R. C. Biggs, "Finned tube heat exchanger: correlation of dry surface heat transfer data," *ASHRAE Transactions*, vol. 85, no. 2, pp. 262–273, 1985.

[48] N. Mendes, P. C. Philippi, and R. Lamberts, "A new mathematical method to solve highly coupled equations of heat and mass transfer in porous media," *International Journal of Heat and Mass Transfer*, vol. 45, no. 3, pp. 509–518, 2001.

[49] J. E. Braun, S. A. Klein, and J. W. Mitchell, "Effectiveness models for cooling towers and cooling coils," *ASHRAE Transactions*, vol. 95, no. 2, pp. 164–174, 1989.

[50] X. Wang, J. Xia, X. Zhang, S. Shiochi, C. Peng, and Y. Jiang, "Modelling and experiment analysis of variable refrigerant flow air-conditioning systems," in *Proceedings of the IBPSA Conference on Building Simulation*, pp. 361–368, 2009.

[51] Y.-W. Wang, W.-J. Cai, Y.-C. Soh, S.-J. Li, L. Lu, and L. Xie, "A simplified modeling of cooling coils for control and optimization of HVAC systems," *Energy Conversion and Management*, vol. 45, no. 18-19, pp. 2915–2930, 2004.

[52] C. Ghiaus, A. Chicinas, and C. Inard, "Grey-box identification of air-handling unit elements," *Control Engineering Practice*, vol. 15, no. 4, pp. 421–433, 2007.

[53] R. Z. Homod, K. S. M. Sahari, H. A. Almurib, and F. H. Nagi, "Hybrid PID-cascade control for HVAC system," *International Journal of Systems Control*, vol. 1, no. 4, pp. 170–175, 2010.

[54] R. Z. Homod, K. S. M. Sahari, H. A. F. Almurib, and F. H. Nagi, "Gradient auto-tuned Takagi-Sugeno Fuzzy Forward control of a HVAC system using predicted mean vote index," *Energy and Buildings*, vol. 49, pp. 254–267, 2012.

[55] R. Z. Homod, K. S. M. Sahari, H. A. F. Almurib, and F. H. Nagi, "Double cooling coil model for non-linear HVAC system using RLF method," *Energy and Buildings*, vol. 43, no. 9, pp. 2043–2054, 2011.

[56] R. Z. Homod, K. S. M. Sahari, F. Nagi, H. A. F. Mohamed, and F. H. Nagi, "Modeling of heat and moisture transfer in building using RLF method," in *Proceedings of the 8th IEEE Student Conference on Research and Development*, pp. 287–292, December 2010.

[57] R. Z. Homod and K. S. M. Sahari, "Energy savings by smart utilization of mechanical and natural ventilation for hybrid residential building model in passive climate," *Energy and Buildings*, vol. 60, pp. 310–329, 2013.

[58] R. Z. Homod, K. S. Mohamed Sahari, H. A. F. Almurib, and F. H. Nagi, "RLF and TS fuzzy model identification of indoor thermal comfort based on PMV/PPD," *Building and Environment*, vol. 49, no. 1, pp. 141–153, 2012.

[59] J. Liang and R. Du, "Design of intelligent comfort control system with human learning and minimum power control strategies," *Energy Conversion and Management*, vol. 49, no. 4, pp. 517–528, 2008.

[60] F. Calvino, M. L. Gennusa, M. Morale, G. Rizzo, and G. Scaccianoce, "Comparing different control strategies for indoor thermal comfort aimed at the evaluation of the energy cost of quality of building," *Applied Thermal Engineering*, vol. 30, no. 16, pp. 2386–2395, 2010.

[61] Y. Farzaneh and A. A. Tootoonchi, "Controlling automobile thermal comfort using optimized fuzzy controller," *Applied Thermal Engineering*, vol. 28, no. 14-15, pp. 1906–1917, 2008.

[62] J. Liang and R. Du, "Thermal comfort control based on neural network for HVAC application," in *Proceedings of the IEEE Conference on Control Applications*, pp. 819–824, 2005.

[63] G. Ye, C. Yang, Y. Chen, and Y. Li, "A new approach for measuring predicted mean vote (PMV) and standard effective temperature (SET)," *Building and Environment*, vol. 38, no. 1, pp. 33–44, 2003.

[64] J. A. Orosa, "A new modelling methodology to control HVAC systems," *Expert Systems with Applications*, vol. 38, no. 4, pp. 4505–4513, 2011.

Preparation and Characterization of Malaysian Dolomites as a Tar Cracking Catalyst in Biomass Gasification Process

M. A. A. Mohammed,[1] A. Salmiaton,[1] W. A. K. G. Wan Azlina,[1] M. S. Mohamad Amran,[1] and Y. H. Taufiq-Yap[2]

[1] *Department of Chemical & Environmental Engineering, Faculty of Engineering, Universiti Putra Malaysia, 43400 UPM Serdang, Selangor, Malaysia*
[2] *Department of Chemistry, Centre of Excellence for Catalysis Science and Technology, Faculty of Science, Universiti Putra Malaysia, 43400 Serdang, Selangor, Malaysia*

Correspondence should be addressed to A. Salmiaton; mie@eng.upm.edu.my

Academic Editor: Johan E. Hustad

Three types of local Malaysian dolomites were characterized to investigate their suitability for use as tar-cracking catalysts in the biomass gasification process. The dolomites were calcined to examine the effect of the calcination process on dolomite's catalytic activity and properties. The modifications undergone by dolomites consequent to thermal treatment were investigated using various analytical methods. Thermogravimetric and differential thermal analyses indicated that the dolomites underwent two stages of decomposition during the calcination process. The X-ray diffraction and Fourier-transform infrared spectra analyses showed that thermal treatment of dolomite played a significant role in the disappearance of the $CaMg(CO_3)_2$ phase, producing the MgO-CaO form of dolomite. The scanning electron microscopy microphotographs of dolomite indicated that the morphological properties were profoundly affected by the calcination process, which led to the formation of a highly porous surface with small spherical particles. In addition, the calcination of dolomite led to the elimination of carbon dioxide and increases in the values of the specific surface area and average pore diameter, as indicated by surface area analysis. The results showed that calcined Malaysian dolomites have great potential to be applied as tar-cracking catalysts in the biomass gasification process based on their favorable physical properties.

1. Introduction

Natural dolomite is the mineral form of the double carbonate of calcium and magnesium, $CaMg(CO_3)_2$. It occurs as the major constituent of sedimentary formations in association with calcite, $CaCO_3$. Dolomite was first described in 1791 as a rock by the French engineer and mineralogist Deodat Gratet de Dolomieu (1750–1801) when he observed exposures in the Dolomite Alps of Northern Italy [1]. In Malaysia, dolomite is abundantly found in Perlis, the northern part of Peninsular Malaysia, and it is called "Batu Reput" by the local people. The main quarries of dolomite can be found in the surroundings of Chuping City in Perlis. Malaysian dolomite is relatively soft and easily crushed to fine powder. It has a grayish color and when crushed to form the powder, it becomes yellowish [2]. Dolomite in Malaysia is mainly used in the float glass and fertilizer industries, for soil conditioning, for use as building products in construction applications, and to prepare roads before paving with asphalt. Dolomite also has the potential to be used as a catalyst in certain chemical processes such as gasification and pyrolysis.

Gasification and pyrolysis of biomass for the production of fuel gas constitute one of the most promising processes for clean energy based on the substitution of fossil fuel with renewable energy; moreover, it decreases the global emission levels of carbon dioxide (CO_2) [3]. Tar formation is a major drawback when biomass is converted into syngas or fuel gas during the gasification process. The most common method that can effectively reduce the tar content in raw fuel gas is thermal and catalytic cracking. These methods are preferred

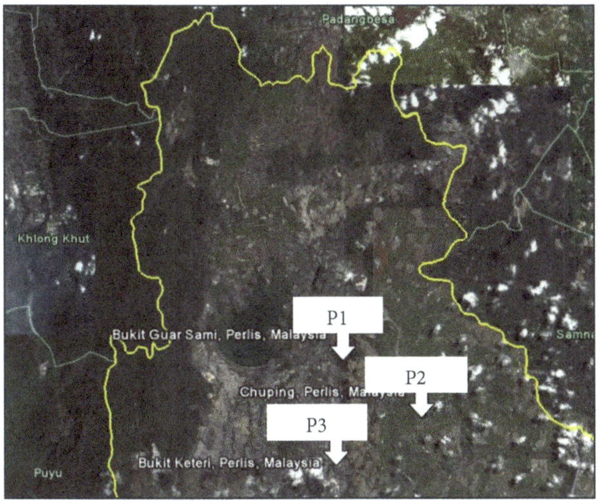

Figure 1: Dolomite quarries locations in Perlis State.

because the gas is maintained at a high temperature. This approach has been intensively studied for more than two decades because of its advantages including (1) conversion of tar into useful gases and (2) adjustability of the composition of the gas product. In a review of catalytic tar cracking by Sutton et al. [4], catalysts were divided into three distinct groups, namely, (1) dolomite catalysts, (2) alkali metal and other metal catalysts, and (3) nickel catalysts. Han and Kim [5] added a fourth group, that is, novel metal catalysts, which includes Rh, Pd, Pt, and Ru.

The application of dolomite as a catalyst in biomass gasification has attracted the interest of many researchers, and dolomite has been proved to have excellent tar-cracking capacity because it is cheap, easily replaceable, and can increase gas yield at the expense of the liquid products. However, the main limitations of using dolomite catalysts are their low activity levels during reforming of the methane present in the product gas and their easy erosion and breaking [4, 10]. The chemical composition and surface area of dolomite varies with its source of supply. Several research groups [6, 8, 9, 11–13] have investigated a system of raw gas cleaning involving different types of dolomites, such as Spanish, Swedish, Finnish, and Chinese dolomites, as catalysts for tar removal.

Dolomite seems to have a significant potential to solve the problems of high tar content in producer gas and will be of significant benefit to the Malaysian economy and environment. Thus, this work was aimed at studying the suitability of Malaysian dolomite for use as a tar-cracking catalyst in the biomass gasification process. To achieve this objective, three types of local dolomites from different locations in Perlis State were prepared and characterized in terms of their properties using different analytical techniques, such as X-ray fluorescence (XRF), thermogravimetric analysis (TGA), X-ray diffraction (XRD), Fourier-transform infrared absorption spectra (FTIR), nitrogen adsorption-desorption, and scanning electron microscopy (SEM).

2. Experimental

2.1. Materials. Three types of natural Malaysian dolomite (P1, P2, and P3) were used and characterized in this study. Dolomite samples were obtained from dolomite quarries located in Perlis State on the northern part of Peninsular Malaysia (as illustrated in Figure 1). These samples were provided by Northern Dolomite Sdn. Bhd., Perlis; UMPAN Global Sdn. Bhd., Penang; and Kangar Dolomite Sdn. Bhd., Kedah, respectively. The original samples were in the form of stones and had to be crushed and ground until they changed to the powder form; then, they were subjected to size classification using a Retsch Test Sieve (ASTM E11, 200 μ). The dolomites were then calcined under nitrogen in a furnace at 1000°C for four hours. The dolomite powders with particle sizes below 0.2 mm for both the natural and the calcined forms were used for characterization.

2.2. Analysis

2.2.1. XRF Study. The chemical composition of dolomites was estimated using wavelength-dispersive XXRF (WDXRF; model BRUKER, S8 TIGER). The fused-bead technique was used to prepare the dolomite samples for XRF analysis. Specifically, 0.5 g of sample was mixed with 5.0 g of spectraflux LT-110 ($Li_2B_4O_7$ = 66.5%, $LiBO_2$ = 33.5%) and placed in a platinum crucible. The mixture was then fused at temperatures between 900 and 1250°C and cast in a platinum casting dish. The samples were subjected to precise control of the cooling process to ensure the reproducibility of the flat glass discs and to avoid cracking or crystallization. The samples were then placed in an XRF instrument within nine minutes for analysis.

2.2.2. Thermal Analysis. To evaluate the thermal decomposition behavior of dolomites during the calcination process, thermal analysis was carried out in a simultaneous thermogravimetric analyzer (model Mettler Toledo, TGA/SDTA851) for both TGA and differential thermal analysis (DTG). Approximately 50 mg of dolomite powder was placed in an alumina crucible and weighed exactly. The sample was then placed inside the furnace and programmed for continuous heating from room temperature to 1000°C at a heating rate of 10°C/min under nitrogen flow at the rate of 100 mL/min.

2.2.3. XRD Analysis. To examine the structural changes induced by the calcination process, the dolomite samples were characterized using a diffractometer-XRD system (model Shimadzu XRD6000), using Cu-Kα (at wavelength 1.5406 Å) radiation, fitted with a Cu filter on the secondary optics, and operated under 40 kV power and 30 mA current. The scanning 2θ range was from 20° to 90°, at a scan rate of 3°/min.

2.2.4. FTIR Spectral Analysis. FTIR spectra (model Perkin-Elmer, 100 series) were used to confirm the structural transformation of the dolomites during the calcination process.

TABLE 1: Analysis results of the compositions of dolomite (%).

Compounds	Malaysian dolomite (P1) This study	Malaysian dolomite (P2) This study	Malaysian dolomite (P3) This study	Chinese dolomite (Zhejiang) [6]	Swedish dolomite (Sala) [6]	Indian dolomite (D_{01}) [7]	Spanish dolomite (Malaga) [8]	Finnish dolomite [9]
SiO_2	0.07	0.09	15.37	1.35	2.21	0.18	—	2.5
Al_2O_3	0.04	0.08	1.69	0.14	0.11	0.25	0.4	0.6
Fe_2O_3	0.07	0.12	0.51	0.03	0.54	0.63	0.01	2.1
CaO	30.0	32.0	23.0	30.72	30.5	30.24	30.6	26.6
MgO	21.0	21.0	17.2	20.12	20.2	21.33	21.2	18.3
K_2O	0.014	0.026	0.195	0.02	0.04	0.03	—	—
Na_2O	0.013	0.013	0.013	0.01	0.03	0.23	—	—
P_2O_5	0.013	0.017	0.019	0.01	0.01	—	—	—
MnO	0.01	0.011	0.013	0.002	—	21.33	—	—
SrO	0.008	0.009	0.009	—	—	—	—	—
TiO_2	0.006	0.007	0.015	0.00	0.01	0.18	—	0.3

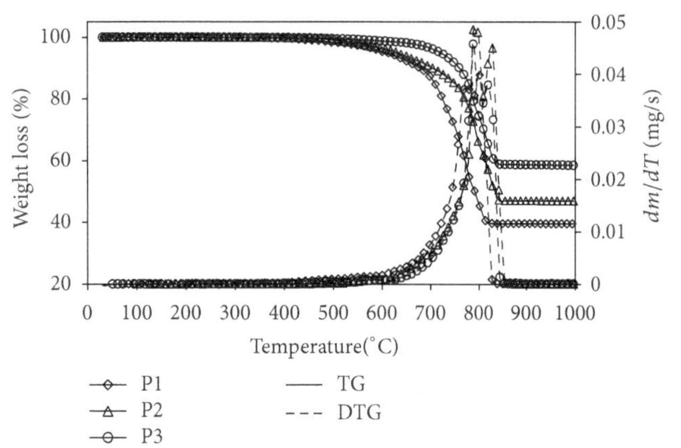

FIGURE 2: TG and DTG curves of Malaysian dolomites.

The spectral data were used to identify the sample compounds by matching the fingerprints of the samples with those in the database. In addition, the functional groups and structural characteristics of a compound enable us to elucidate the possible structure types. The spectrum range of the FTIR study was the mid-infrared region, which covers the frequency ranges of 500–4000 cm^{-1}, at a resolution of 8 cm^{-1}.

2.2.5. SEM Analysis. The surface and microstructure analysis of dolomites was carried out using an SEM instrument (model HITACHI, S-3400N). The dolomite powder was mounted onto the SEM stubs (layered with sticky carbon tape). The stub was then placed in a sputter coater (EMITECH, K550X) for five minutes for coating with gold to provide high reflectivity during the scanning process. The samples were then placed at 40°C in an oven before SEM analysis.

2.2.6. Physical Properties Analysis. The physical properties of the dolomites, such as specific surface area, total pore volume, and pore size distribution, were obtained by measuring their nitrogen adsorption-desorption isotherm at 77 K using a Quantachrome instrument (model Autosorb-1). The Brunauer-Emmet-Teller (BET) surface area was estimated from the adsorption-isotherm data using the relative pressure (p/p_o) range of 10^{-6} to 1. The total pore volume was assessed by converting the amount of nitrogen gas adsorbed at relative pressure to the volume of liquid adsorbate. The analytical method consisted of three steps, including dehydration, degassing under low vacuum pressure, and nitrogen gas adsorption at 77 K. The resulting isotherm was analyzed using the BET adsorption method, and the pore-size distributions were derived by the Barrett-Joyner-Halenda (BJH) desorption method.

3. Results and Discussion

3.1. Chemical Composition of Dolomite. The chemical compositions of the various Malaysian dolomite samples, which were determined using XRF analysis, are listed in Table 1. The main contents of dolomite are CaO and MgO, followed by low levels of undesirable impurities, such as SiO_2, K_2O, Na_2O, and Al_2O_3. The chemical composition of P3 dolomite differed significantly from that of P1 and P2. As shown in Table 1, the chemical composition analysis indicated that the CaO and MgO contents of the P1 and P2 dolomites were very close to those of other types of dolomites, whereas in the case of P3, these contents were relatively low. Furthermore, P1 and P2 had a much lower content of impurities than the other types of dolomites, whereas P3 had a higher level of impurities, especially silica (SiO_2, 15.37%) and alumina (Al_2O_3, 1.69%). These impurities combine with CaO at elevated temperatures to form a slag, reducing the pore volume and the amount of available active lime [7]. Many research groups have reported that the chemical composition of dolomites play an important role when used as a catalyst for tar reforming in gasification processes [6, 8, 9, 11–13]. Myren and others reported that a mixture of silica and dolomite was not as effective as the same

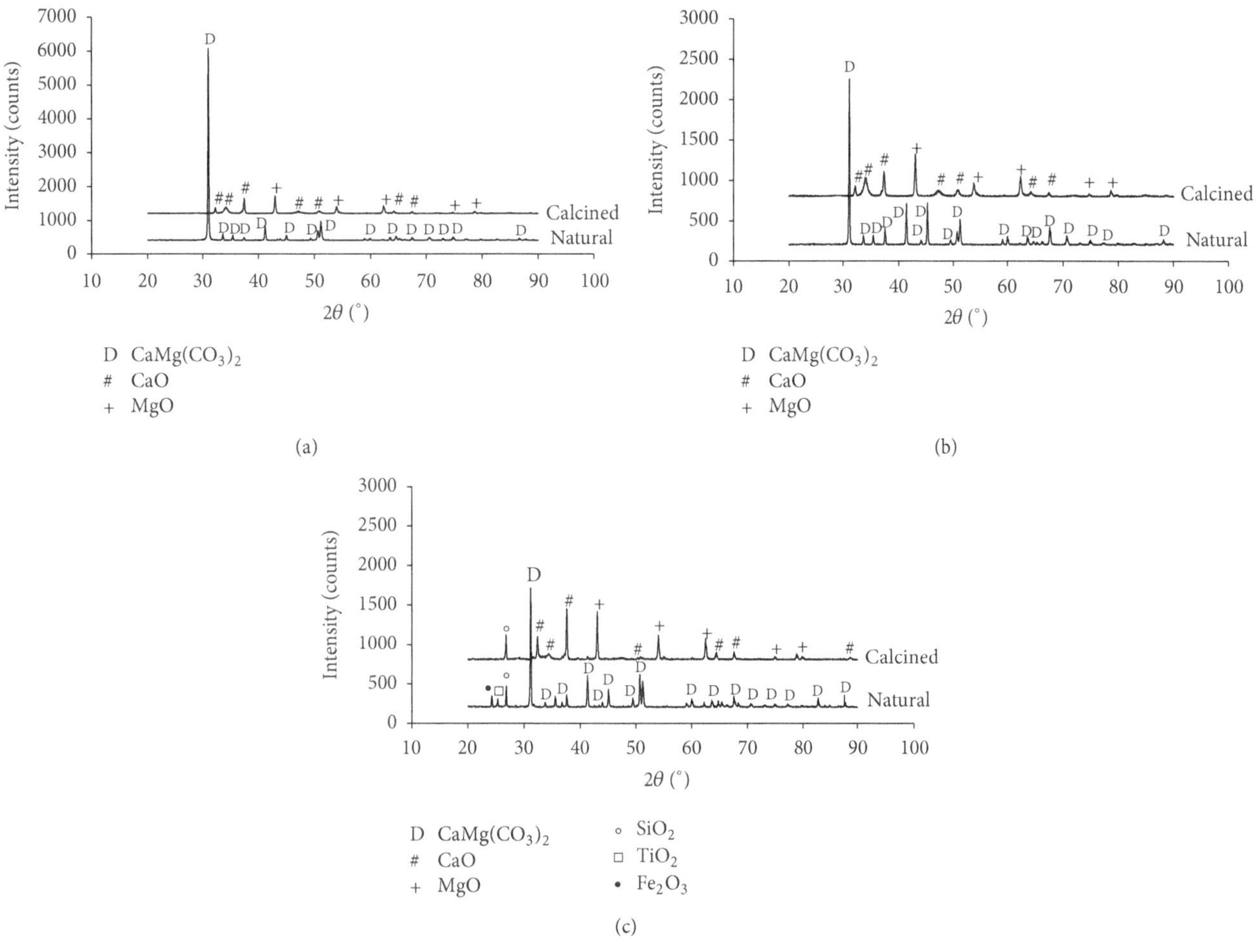

FIGURE 3: XRD analysis before and after calcination (a) P1 dolomite, (b) P2 dolomite, and (c) P3 dolomite.

amount of pure dolomite for tar cracking [14]. In general, dolomites with the lowest content of CaO and MgO show the lowest tar cracking efficiency [6].

3.2. Thermal Decomposition of Dolomite. To evaluate the thermal decomposition behavior of dolomites under calcination temperatures, thermal analysis was conducted in a simultaneous TG-DTA analyzer. The thermal decomposition characteristics of dolomites with reference to the TGA and DTG curves are displayed in Figure 2. Dolomite is the mineral form of the double carbonate of calcium and magnesium, $CaMg(CO_3)_2$. Under heating, dolomite is often assumed to be decomposed in two separate stages as follows [2, 15].

First stage:

$$CaMg(CO_3)_{2(s)} \longrightarrow CaCO_{3(s)} + MgO_{(s)} + CO_{2(g)} \quad (1)$$

Second stage:

$$CaCO_{3(s)} \longrightarrow CaO_{(s)} + CO_{2(g)} \quad (2)$$

The first thermal decomposition stage occurs after 600°C and is followed by the second stage after 800°C. From Figure 2, the observed weight loss below 600°C for P1, P2, and P3 was 4.59%, 3.42% and 1.24% respectively, whereas, the weight loss between 600 and 830°C was 55.77%, 49.26% and 39.84% respectively. The main weight loss in the 600–830°C range can be attributed to the decomposition of the carbonates mineral that exists in dolomite and the release of carbon dioxide as explained by (1). Therefore, the weight loss for P1 and P2 in this temperature range was higher compared to that of P3 due to the higher content of carbonates. In Figure 2, the DTG curves of the three types of dolomites show two peaks. For P1, the onset of the first peak begins at 600°C, reaches a maximum at 767.7°C, and ends at 790.3°C, whereas the second peak begins at 790.3°C, reaches a maximum at 801.3°C, and ends at 830°C. For P2, the first peak begins at 600°C, reaches a maximum at 787.5°C, and ends at 809.5°C, whereas the second peak begins at 809.5°C, reaches a maximum at 829.5°C, and ends at 830°C. For P3 also, the first peak begins at 600°C, reaches a maximum at 787.5°C, but ends at 799.3°C, whereas the second peak begins at 799.3°C, reaches a maximum at 819.6°C, and ends at 830°C. As shown in (1) and (2), the peak at the lower temperature represents the decomposition of the dolomite structure, with the release of carbon dioxide from the carbonate ion associated with the magnesium part of the structure, accompanied by the formation of calcite

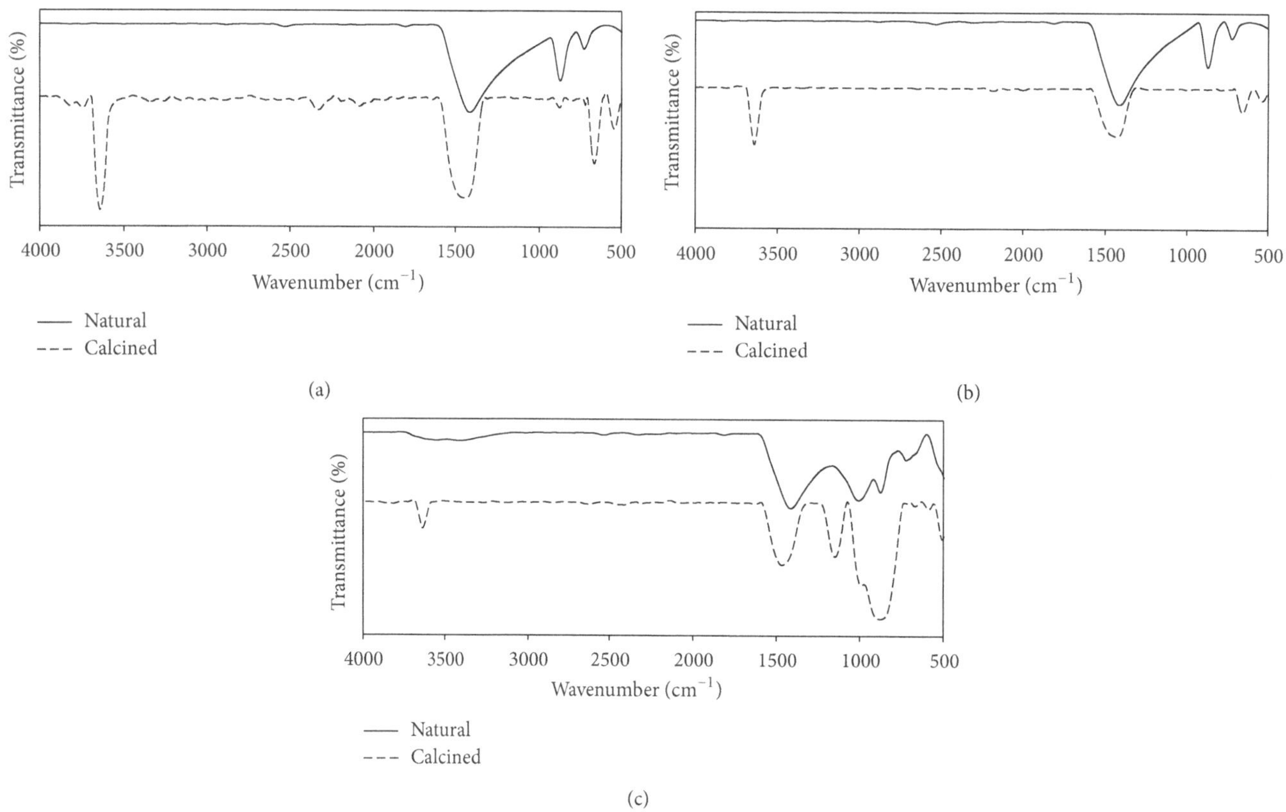

FIGURE 4: FTIR spectra before and after calcination (a) P1 dolomite, (b) P2 dolomite, and (c) P3 dolomite.

and magnesium oxide. The peak at the higher temperature corresponds to the decomposition of calcite, with the release of carbon dioxide. As shown in Figure 2, the decomposition process for P1 is initiated at lower temperatures compared to the other types of Malaysian dolomites. This probably is due to its higher content of CaO, as presented in Table 1. These results are in agreement with the previous works by McIntosh et al. [15] and Gunasekaran and Anbalagan [7]. In addition, Farizul et al. [2] have also found that the DTG plot of Malaysian dolomite shows two peaks at 750.0 and 930.0°C. The temperature of the second peak is higher compared to that in the present study, which can be attributed to the partial pressure of carbon dioxide in the dolomite samples, whereby the temperature of the second peak increases with increasing partial pressure of carbon dioxide. The peak temperature also depends on other procedural variables, such as the heating rate and the sample mass, because these, in turn, influence the actual atmosphere around the decomposing sample [15]. As a result, the thermal treatment process (calcination) of dolomite produced a catalyst with very high activity, due to disappearance of the $CaMg(CO_3)_2$ phase and producing of the MgO-CaO form of dolomite.

3.3. XRD Analysis. The dolomites were characterized before and after calcination using XRD to analyze the effect of calcination on the dolomite structure. The results are shown in Figure 3. The XRD patterns for natural dolomites show that all peaks correspond to those of $CaMg(CO_3)_2$ (JCPDS 36-0426). All these peaks also have been observed in other research works [3, 7], indicating that the Malaysian dolomite samples have almost the same properties as the other dolomites. The XRD pattern for calcined dolomites, however, changes drastically compared to that of natural dolomites. New dominant peaks which belong to CaO (JCPDS 04-0777) and MgO (JCPDS 87-0652) are observed. Natural P3 sample (Figure 3(c)), on the contrary, shows three additional peaks between $2\theta = 20^\circ$ and $2\theta = 30^\circ$, which are assigned to Fe_2O_3 (JCPDS 26-1319), TiO_2 (JCPDS 71-1169), and SiO_2 (JCPDS 83-0539), occurring at $2\theta = 25.2^\circ$, 26.7°, and 28°, respectively. The changes in the peaks illuminate the fact that calcination of Malaysian dolomites results in the formation of the MgO-CaO form, which is the active catalytic component in tar conversion.

3.4. FTIR Spectral Analysis. The typical transmittance FTIR spectra of the Malaysian dolomites before and after the calcination process are shown in Figure 4. The FTIR spectra for all natural dolomite samples indicate that the most prevalent component present in dolomite is carbonate. The carbonate mineral group (magnesium and calcium carbonates) shows a strong and broad band near 1412 cm^{-1}, whereas the other main components of dolomite are clearly differentiated by characteristic bands near 872 and 724 cm^{-1} [16]. In the case of P3 (Figure 4(c)), in addition to the band corresponding to the carbonate group, the FTIR spectra show a band at 1005 cm^{-1},

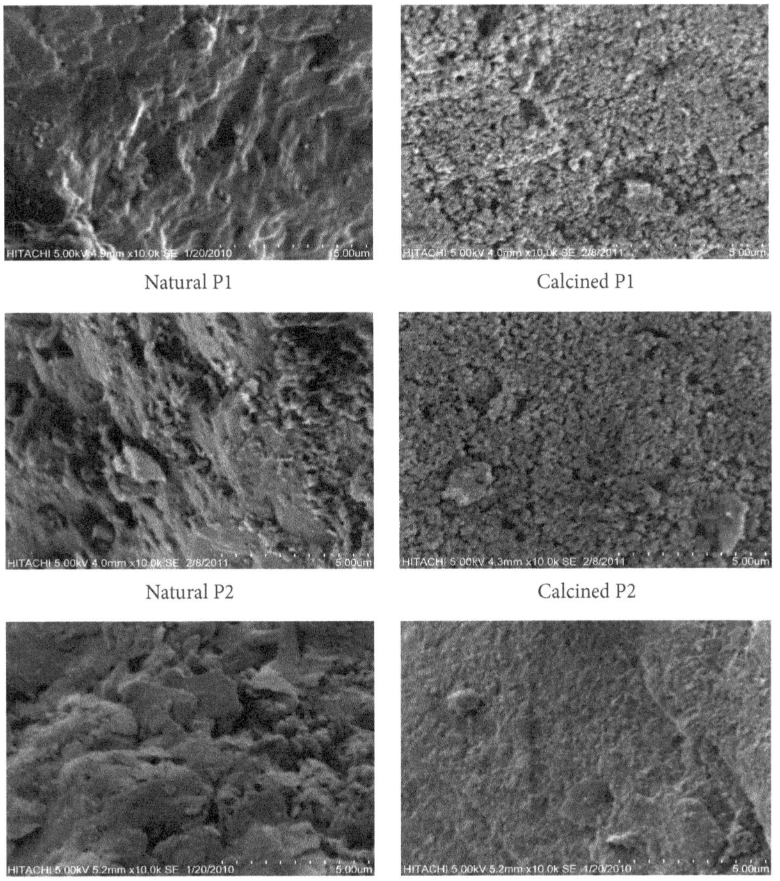

Natural P1 — Calcined P1

Natural P2 — Calcined P2

Natural P3 — Calcined P3

FIGURE 5: SEM microphotograph of natural and calcined Malaysian dolomites.

TABLE 2: Pore characteristics of the calcined dolomites.

	P1 (this study)	P2 (this study)	P3 (this study)	Sala [6]	Zhejiang [6]	Malaga [8]
BET surface area (m^2/g)	15.25	16.85	6.16	7.3	5.1	12
Pore volume (cm^3/g)	0.32	0.104	0.09	0.09	0.03	0.07
Average pore diameter (Å)	416	233	215	496	269	144

referring to the presence of Si–O vibration of the silicate phase, which is in agreement with the chemical analysis data.

The FTIR spectra for dolomites after calcination clearly show the structural transformation of natural dolomite to calcium and magnesium oxides. At this stage, the main broad band at 1412 cm^{-1} shifts to 1430 cm^{-1}, and the intensity of the band decreases, showing the presence of calcium and magnesium oxides [17]. Moreover, strong and intense bands are observed at 3641 and 660 cm^{-1} due to calcium oxide, in addition to the band at 540 cm^{-1} attributable to magnesium oxide. The weak bands at 875 and 1005 cm^{-1} for natural P3 (Figure 4(c)) combine together to form a strong and broad band after calcination, which can be attributed to the effect of impurities such as silicates on the dolomite structure at elevated calcination temperatures. Moreover, another strong band is observed at 1144 cm^{-1} for calcined P3 dolomite due to the presence of silica, whereas the intensity of the calcium oxide bond at 3641 cm^{-1} is weak compared to the bands of the calcined samples of P1 and P2. These results are in complete agreement with the results obtained from XRD analysis.

3.5. Scanning Electron Microscopy (SEM) Analysis. As illustrated in Figure 5, the SEM micrographs of the calcined dolomites are significantly different from those of the dolomites before calcination. The SEM micrographs of natural dolomites show a rough and disordered surface with low-porosity grains, whereas the surfaces of calcined dolomites show clusters of tidy and porous grains, confirming the thermal decomposition.

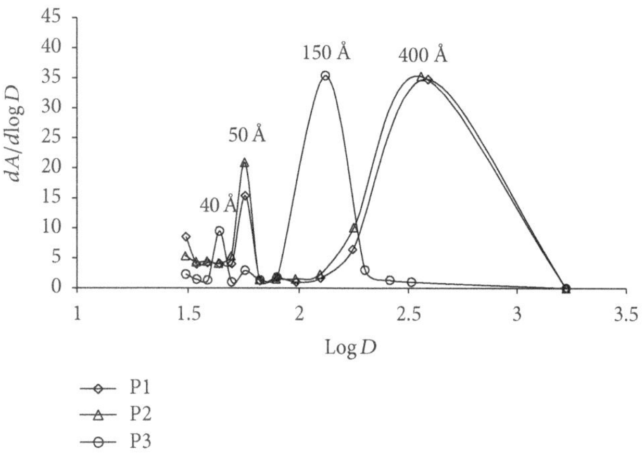

FIGURE 6: Pore size distribution of calcined Malaysian dolomites: D: pore diameter (Å), A: BET surface area (m^2/g).

3.6. Surface Area and Total Pore Volume Analysis. The BET surface area, the BJH pore volume, and the average pore size for calcined Malaysian dolomites are summarized in Table 2. A comparison of the surface area and pore volume (17–3000 Å) shows that P1 and P2 dolomites have the highest surface area and pore volume compared to P3 and other published dolomites. This is possibly due to the high CaO and MgO contents in P1 and P2 dolomites and the lower contents of impurities, such as SiO_2, K_2O, Na_2O, and Al_2O_3. Slag is produced when these impurities combine with CaO at high temperatures, leading to a reduction in both the pore volume and the amount of available active lime [18]. The surface properties, apart from the chemical composition of dolomite, are the most important factors for selecting dolomite as the cracking catalyst in many thermochemical processes.

The pore size-distribution curves of the calcined Malaysian dolomites are shown in Figure 6. The pore-size distribution of P1 and P2 dolomites differs from that of P3. The pore sizes of P1 and P2 dolomites are mainly in the 400 Å region, whereas the major pore size of P3 is in the 150 Å region.

4. Conclusions

The natural Malaysian dolomites, after undergoing thermal treatments at high temperatures (1000°C), produced activated dolomite, which could be used as a cracking catalyst in the biomass gasification process. Malaysian dolomites P1 and P2 contained higher amounts of CaO and MgO, with lower amounts of SiO_2, K_2O, Na_2O, and Al_2O_3, compared to P3 and other dolomites (Table 1). The DTG curves showed two stages of decomposition during the calcination process. The first thermal decomposition occurred after 600°C, followed by the second step after 800°C. The XRD and FTIR results confirmed the structural transformation of natural dolomite to the calcium and magnesium oxide form, and the results of the BET and BJH analyses showed that P1 and P2 dolomites had higher surface areas and pore volumes compared to P3 and other types of dolomites. This suggests that P1 and P2 dolomites have great potential for use as tar-cracking catalysts. The most important factors that contribute to the selection of dolomites with highly efficient tar-cracking capacity are calcination balance and chemical composition in additional to geometrical surface area and average pore diameter.

Acknowledgments

The authors would like to thank Northern Dolomite Sdn. Bhd., UMPAN Global Sdn. Bhd., and Kangar Dolomite Sdn. Bhd., for their support in providing the dolomites samples. The work was financed by the Department of Chemical and Environmental Engineering, Faculty of Engineering, Universiti Putra Malaysia.

References

[1] W. L. Roberts, T. J. Cambell, and G. E. Rapp, *Encyclopedia of Minerals*, Van Nostrand-Reinhold, New York, NY, USA, 2nd edition, 1990.

[2] H. K. Farizul, P. C. Goh, D. M. S. N. Suhardy, H. M. D. Irfan, and S. Saiful Azhar, "Dolomite as a raw material in fertilizer production," in *Proceedings of the AEESEAP International Conference*, N. A. Rahim, Ed., pp. 4–8, Kuala Lumpur, Malaysia, 2005.

[3] G. Hu, S. Xu, S. Li, C. Xiao, and S. Liu, "Steam gasification of apricot stones with olivine and dolomite as downstream catalysts," *Fuel Processing Technology*, vol. 87, no. 5, pp. 375–382, 2006.

[4] D. Sutton, B. Kelleher, and J. R. H. Ross, "Review of literature on catalysts for biomass gasification," *Fuel Processing Technology*, vol. 73, no. 3, pp. 155–173, 2001.

[5] J. Han and H. Kim, "The reduction and control technology of tar during biomass gasification/pyrolysis: an overview," *Renewable and Sustainable Energy Reviews*, vol. 12, no. 2, pp. 397–416, 2008.

[6] Q. Z. Yu, C. Brage, T. Nordgreen, and K. Sjostrom, "Effects of Chinese dolomites on tar cracking in gasification of birch," *Fuel*, vol. 88, no. 10, pp. 1922–1926, 2009.

[7] S. Gunasekaran and G. Anbalagan, "Thermal decomposition of natural dolomite," *Bulletin of Materials Science*, vol. 30, no. 4, pp. 339–344, 2007.

[8] P. Pérez, P. M. Aznar, M. A. Caballero, J. Gil, J. A. Martín, and J. Corella, "Hot gas cleaning and upgrading with a calcined dolomite located downstream a biomass fluidized bed gasifier operating with steam-oxygen mixtures," *Energy and Fuels*, vol. 11, no. 6, pp. 1194–1197, 1997.

[9] P. Simell, E. Kurkela, P. Ståhlberg, and J. Hepola, "Catalytic hot gas cleaning of gasification gas," *Catalysis Today*, vol. 27, no. 1-2, pp. 55–62, 1996.

[10] L. Wang, C. L. Weller, D. D. Jones, and M. A. Hanna, "Contemporary issues in thermal gasification of biomass and its application to electricity and fuel production," *Biomass and Bioenergy*, vol. 32, no. 7, pp. 573–581, 2008.

[11] A. Orio, J. Corella, and I. Narvaez, "Performance of difference dolomites on hot raw gas cleaning from biomass gasification with air," in *Proceedings of Conference on Developments in Thermochemical Biomass Conversion*, pp. 1144–1150, Banff, Canada, 1996.

[12] J. Delgado, M. P. Aznar, and J. Corella, "Biomass gasification with steam in fluidized bed: effectiveness of CaO, MgO, and CaO–MgO for hot raw gas cleaning," *Industrial & Engineering Chemistry Research*, vol. 36, pp. 1535–1543, 1997.

[13] V. Vassilatos, G. Taralas, K. Sjostrom, and E. Bjornbom, "Catalytic cracking of tar in biomass pyrolysis gas in the presence of calcined dolomite," *Canadian Journal of Chemical Engineering*, vol. 70, no. 5, pp. 1008–1013, 1992.

[14] C. Myren, C. Hornell, E. Bjornbom, and K. Sjoström, "Catalytic tar decomposition of biomass pyrolysis gas with a combination of dolomite and silica," *Biomass and Bioenergy*, vol. 23, no. 3, pp. 217–237, 2002.

[15] R. M. McIntosh, J. H. Sharp, and F. W. Wilburn, "The thermal decomposition of dolomite," *Thermochimica Acta*, vol. 165, no. 2, pp. 281–296, 1990.

[16] V. Ramasamy, V. Ponnusamy, S. Sabari, S. R. Anishia, and S. S. Gomathi, "Condensed matter: structural, mechanical and thermal properties," *Indian Journal of Pure and Applied Physics*, vol. 47, pp. 586–591, 2009.

[17] D. O. Kevin and T. A. R. R. C. Jackson, "A guide to identifying common inorganic fillers and activators using vibrational spectroscopy," *Internet Journal of Vibrational Spectroscopy*, vol. 2, pp. 1–11, 1998.

[18] R. K. Chan, K. S. Murthi, and D. Harrison, "Thermogravimetric analysis of Ontario limestones and dolomites I. Calcination, surface area, and porosity," *Canadian Journal of Chemistry*, vol. 48, no. 19, pp. 2972–2978, 1970.

The Current Situation of Wind Energy in Turkey

Raşit Ata

Department of Electrical & Electronic Engineering, Celal Bayar University, 45140 Manisa, Turkey

Correspondence should be addressed to Raşit Ata; rasit.ata@cbu.edu.tr

Academic Editor: Mattheos Santamouris

Wind energy applications and turbine installations at different scales have increased since the beginning of this century. As wind energy is an alternative clean energy source compared to the fossil fuels that pollute the atmosphere, systems that convert wind energy to electricity have developed rapidly. Turkey's domestic fossil fuel resources are extremely limited. In addition, Turkey's geographical location has several advantages for extensive use of wind power. In this context, renewable energy resources appear to be one of the most efficient and effective solutions for sustainable energy development and environmental pollution prevention in Turkey. Among the renewable sources, Turkey has very high wind energy potential. According to the Organization for Economic Cooperation and Development (OECD) Turkey theoretically has 166 TWh a year of wind potential. However the installed wind power capacity is approximately 14% of total economical wind potential. In this study, Turkey's installed electric power capacity and electric energy production are investigated and also the current situation of wind energy in Turkey is examined. The wind data used in this study were taken from Turkish Wind Energy Association (TUREB) for the year 2012. This paper reviews the assessment of wind energy in Turkey as of the end of July 2012 including wind energy applications.

1. Introduction

During recent years according to global environmental pollutions, trends towards the sustainable energy and green power sources such as solar, wind, biomass, and geothermal energy were largely increased. It is now widely accepted that the renewable energy sources are very important for the future of the countries. Wind energy is one of the economic renewable sources and a valuable supplement to conventional energy sources. The wind technology was gradually improved since the early 1970s. By the end of the 1990s, wind energy has re-emerged as one of the most important renewable energy resources [1]. The cost of wind electricity production cost has been gradually decreasing with improving technology. At present, wind energy has been widely used to produce electricity in many countries in America, Asia, and especially Europe. For instance, while the world established wind power is 24,322 MW in 2001, it has increased to 237,016 MW in the end of 2011 [2].

Our country, just as the majority of all world countries, faces some short and long term problems in energy procurement. Turkey should at first attempt to increase the amount of energy acquired from new and renewable energy sources to cover basic requirements of society and realize economic expansion, refrain as much as possible from local and global environmental problems caused by energy consumption and production, and especially reduce foreign-dependency in energy. This is valid for not only our country but also other countries of the world [3].

Especially in the last decade, many scientific studies were presented about the development of wind energy. A general review about wind energy development in Turkey is considered by Sahin [4]. Some authors have researched the developments in wind engineering and turbines, for example, Ackerman and Söder [5], Sahin [6], Thresher et al. [7], and Joselin Herbert et al. [8]. In [9–11] authors presented control techniques for composite wind turbine blades. Many authors have studied the status of wind energy, the wind energy potential of Turkey's different geographical regions [12–33]. A general review about wind energy status, renewable energy potential, and development in Turkey is considered by researchers in [34–41]. In these studies, Turkey's installed wind power capacity and electric energy production are investigated and also Turkey current wind energy status is examined by researchers. Ozgener and Hepbasli [42]

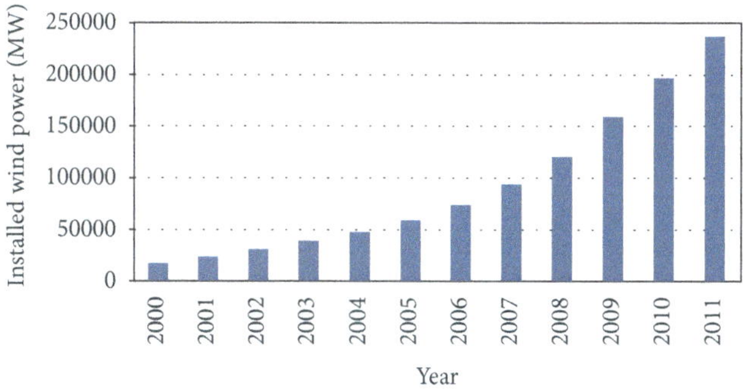

FIGURE 1: Installed wind power capacity of the world.

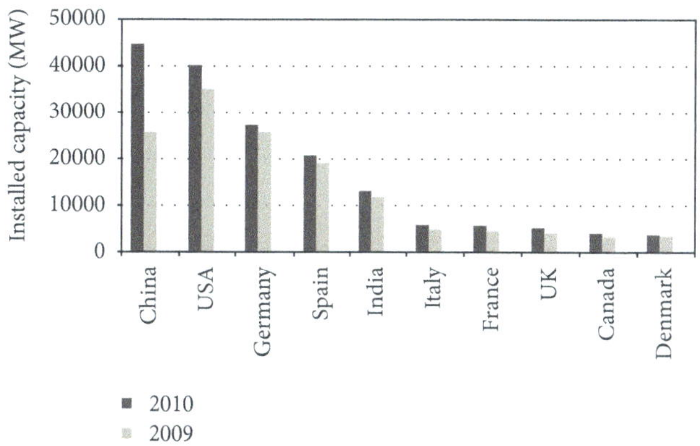

FIGURE 2: Top ten countries in terms of total installed wind power capacity in the world.

presented future directions of wind energy applications in Turkey. Many authors have researched the status, potential, utilization, environmental pollution, and future perspectives of renewable energy sourches, for example, Evrendilek and Ertekin [43], Ocak et al. [44], Yüksel [45], Oğulata [46], Keleş and Bilgen [47], Çapik et al. [48], Yuksel and Kaygusuz [49], Volkan and Ediger [50], Kaya [51], Kaygusuz and Sari [52], and Akpinar et al. [53]. In addition, Kenisarin et al. [54] has discussed wind power engineering in the world and its development in Turkey.

In this paper, wind energy and its importance are discussed firstly, and then the current status and development of Turkey's wind power plants are investigated in detail.

2. Situation of Wind Energy in the World

Wind energy is the most advanced and widespread renewable energy source being the most convenient in commercial terms. Being a clean energy source, wind energy is environment friendly having no possibility of extinction as long as the sun exists. It is an ever growing energy source despite being continuous and despite the fact that it is not exactly known whether available amount will be at hand when required. For instance, while the world established wind power is 24,322 MW in 2001, it has increased to 215,000 MW in June 2011 [2]. Figure 1 shows installed wind power capacity in the world between 2000 and 1st half 2011 [2, 55]. There is an increasing trend in installed wind energy and average increasing rate is 25% over this period. It is estimated that installed wind power will be reached in 24,500 MW by the end of 2011 [2].

Approximately 43.7% of the installed wind capacity of the world is in Europe, 22.5% in North America, and 31.1% in Asia. However, China has the highest installed wind capacity with 44,733 MW which is equal to 52% of Europe's and 22.7% of world installed capacity. The top ten wind energy markets are shown in Figure 2 between 2009 and 2010 [56].

As it is seen from Figure 2 China shows the highest development in installed wind energy capacity with 73% between 2009 and 2010. USA has also high development with 14.2%. The increasing rate of installed wind capacity in this term for Germany, Spain, and India is 5.5, 7.9, 10.6%, respectively. Although the established power of wind energy has recently increased in the countries in the Asian continent, 43.7% of the established power of world wind energy is in the European continent, according to 2010 data.

According to 2010 data of Balkan countries being under the influence of almost similar wind systems to Turkey, our bordering neighbor Greece has 1208 MW of wind energy established power, while Bulgaria has above 375 MW of established wind power. Based on a comparison with our neighbors, the significance of the potential of our country is blatant, only when we consider the great surface area of our country.

3. Turkey's Electrical Energy Status

The main purpose of energy policy in Turkey is to supply the sufficient energy to the utilization taking environmental and economic aspects into account by supporting the economical growing and social development [57]. The trend of growing has been calculated to be continuing in future. Turkey's total installed power capacity obtained from hydraulic, thermal (natural gas, coal, lignite, fuel-oil, LPG, etc.), and wind sources is 49,524 MW at the end of the 2010, an increase of 10.6% over 2009. From 2000 to 2010 total capacity increased by 74.6%. This was one of the largest increases in IEA member countries, reflecting a significant build-up in CCGTs but also coal and hydro capacity [58]. Together, natural gas, hydropower, oil, and coal plants account for 97%. Distribution of the installed power capacity of Turkey according to the sources is given in Figure 3 [59]. As it is seen from Figure 3, the rate of thermal source power plants is very high and approximately 75% of these are natural gas plants.

The total gross electrical energy production in 2010 was 211,207.7 GWh and changes in the production rate depended on the economic situations and technological developments. Average increasing rate was approximately 8.4% in this term [59]. Turkey's electrical energy production from 2005 to 2010 is shown in Figure 4 [59]. Although electrical energy production has increased gradually, there was a decrease in electrical energy production in 2009. Electricity production was realized as 194,812 GWh with a decrease about 2% in 2009 when compared to the values of 2008, that is, 198,418 GWh.

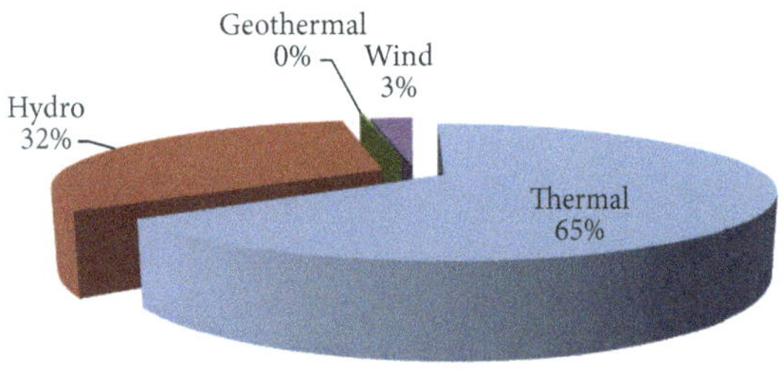

FIGURE 3: Distribution of installed power capacity of Turkey.

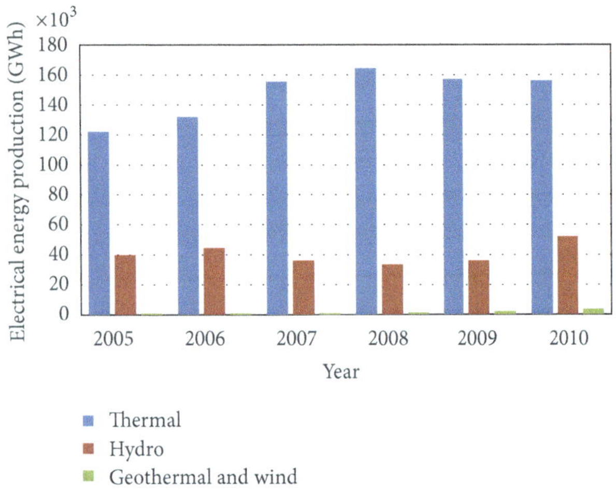

FIGURE 4: Turkey's gross electrical energy production for the years 2005–2010.

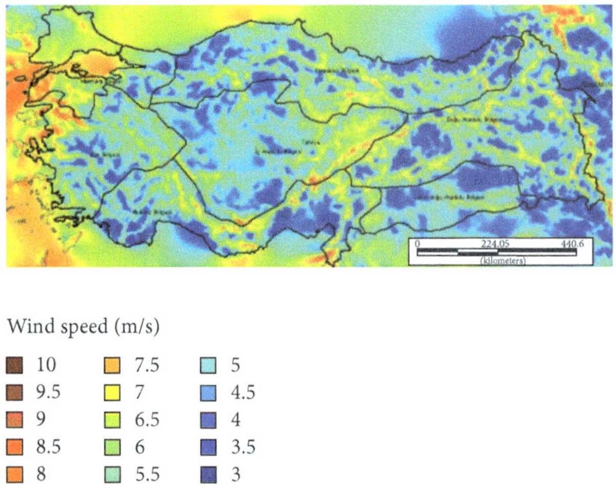

FIGURE 5: Scattering of average wind speed at 50 m high in Turkey.

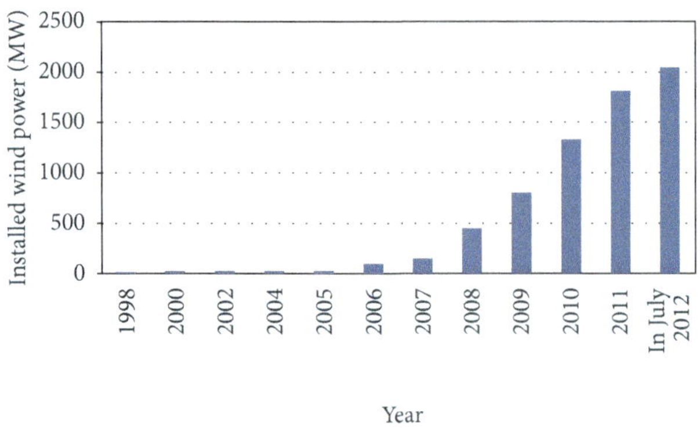

FIGURE 6: Installed wind power capacity in Turkey.

Annual developments of the gross electrical energy production according to sources between 2005 and 2010 are given in Table 1 [59]. As it is seen from Table 1, 73% of the electrical energy is supplied from thermal sources. It can be seen that the proportion of natural gas in the electrical energy production increased approximately 33.6% in this term. In electrical energy production, consumption of natural gas reached 46% in 2010 whereas it was 37% in 2000 [39]. However, the natural gas share of the world in production of electrical energy is 21.4% [60]. Since Turkey imports almost all required natural gas, this high rate usage in producing electricity than the world is an important point to be examined economically [39]. On the other hand, electrical energy production from the wind has grown rapidly in recent years.

4. The Current Situation of Wind Energy in Turkey and Its Future

A wind atlas of Turkey published by the Turkish Energy Market Regulatory Agency (EPDK) in May 2002 indicates that the regions with the highest potential for wind speeds at height of 50 m are the Aegean, Marmara, and Eastern Mediterranean regions of Turkey, as well as some mountainous regions of central Anatolia [61]. Figure 5 shows scattering of average wind speed in 50 m high in Turkey [3]. In addition to this, meteorological data by the USA space studies have shown that Turkey has high wind capacity [39].

Power intensity in 50 m of elevation above ground, which is significant to establish turbines, in places with 4-5 m/s of average annual wind speed at 50 m of elevation above ground mostly exceed annual average of 500 w/m^2. Estimated figures resulting from the researcher conducted in the field, technical wind energy potential of Turkey, established power, and average efficiencies are available in Table 2 [3]. In this table, the land of Turkey has been classified by means of wind energy resource degree. It can be seen from this table that approximately 37% of the land of Turkey has capacity above medium.

First small-scale application to generate electrical energy in Turkey was started with a plant that has 55 kW installed power in Izmir-Çeşme in the Aegean region in 1986. The first power plant in large-scale was also installed in 1998, Çeşme-Germiyan with 1.74 MW capacity. In 1998, the ARES wind farm was built in Çeşme-Alaçatı and included 12 × 600 kW wind turbines. The biggest wind energy power plant in Turkey has 140.1 MW capacity constructed in Manisa-Soma in 2012. Current wind power plants under operation in Turkey are listed in Table 3 [62]. Installed wind power capacity for electrical production is shown in Figure 6 [62].

TABLE 1: Distribution of resources of the Turkey's gross electrical energy production (GWh).

Resources	2005	2006	2007	2008	2009	2010
Hard coal + Imp. coal	13246.2	14216.6	15236.2	15857.5	16595.6	19104.3
Lignite	29946.3	32432.9	38294.7	41858.1	39089.5	35942.1
Fuel-oil	5120.7	4232.4	6469.6	7208.6	4439.8	2143.8
Diesel	2.5	57.7	13.3	266.3	345.3	4.3
Natural gas	73444.9	80691.2	95024.8	98685.3	96094.7	98143.7
Renew. and wastes	122.4	154.0	213.7	219.9	340.1	457.5
Others	359.3	50.3	43.9	43.6	18.0	31.9
Total thermal	122242.3	131835.1	155196.2	164139.3	156923.4	155827.6
Hydro	39560.5	44244.2	35850.8	33269.8	35958.4	51795.5
Geothermal + wind	153.4	220.5	511.1	1008.9	1931.1	3584.6
Total	161956.2	176299.8	191558.1	198418.0	194812.9	211207.7

TABLE 2: Areal distribution of wind speed, power and potential energy amount in Turkey.

Wind source degree	Wind class	Wind power at 50 m. (W/m^2)	Wind speed at 50 m. (m/s)	Overall area (km^2)	Windy land (%)	Potential capacity (MW)
Medium	3	300–400	6.8–7.5	16781.39	2.27	83906.96
Good	4	400–500	7.5–8.1	5851.87	0.79	29259.36
Perfect	5	500–600	8.1–8.6	2598.86	0.35	12994.32
Perfect	6	600–800	8.6–9.5	1079.98	0.15	5399.92
Perfect	7	>800	>9.5	39.17	0.01	195.84
Total				26351.28	3.57	131756.40

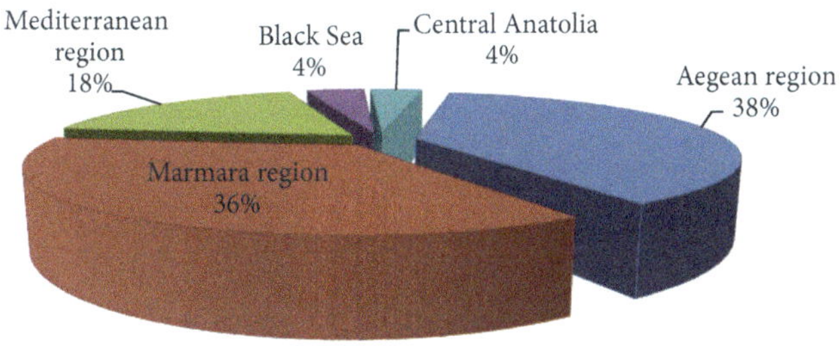

FIGURE 7: Distribution of wind energy power stations according to the regions in Turkey.

Having a look at Table 3, the existing wind power stations are active only in the five of the seven geographical regions of the country, being Aegean, Marmara, Mediterranean, Black Sea, and Central Anatolia regions. While Aegean ranks first with 786.2 MW of established power, Marmara region with 730.65 MW of established power ranks second to be followed by the Mediterranean region with 372.5 MW of established wind power energy station capacity. Distribution of wind energy power stations according to the regions in Turkey is shown in Figure 7.

Considering the distribution of installed power according to provinces, Balıkesir ranks first, as is seen in Table 4, having almost 20% of the overall capacity. Manisa ranks second with 17% of share, İzmir, again a city from the Aegean region, ranks third with 16% of the total capacity, then the other provinces listed. In the near future, the number of provinces with active wind power stations is expected to be further augmented.

In 2010, 528 MW of new wind energy capacity was added in Turkey, bringing the total up to 1329 MW. This represents a year-on-year growth rate of 66%. According to TEIAS (the state-owned transmission company and system operator) it is projected that up to 415.8 MW of wind projects might be added in 2011 [61].

Installed wind capacity is expected to grow between 500–1,000 MW per year reaching more than 5 GW by 2015. Turkey hopes to install up to 20 GW by 2023, helping the country to source 30% of its electricity generation from renewable sources by that date. In order to reach this target, however, the transmission infrastructure will require substantial upgrades to allow such large scale developments to be connected to the power grid. This issue will need to be addressed in the near future [61].

The predicted wind power development capacity in Turkey can be seen in Table 5 [54]. Two different assumptions have been made in this table. According to both predictions, the power obtained from wind will increase and exceed 10,000 MWs by 2030.

The new renewable energy law introduced fairly attractive incentives for wind power plants in Turkey. In early 2007, the Turkish government updated the renewable energy law that guarantees wind generators with 10 year agreements involving a fixed tariff between 50/MWh and 55/MWh. Although somewhat lower than tariffs in other developing markets, for example, 73/MWh in Brazil or the 66/MWh set in Portugal, the tariff, coupled with the high wind potential,

TABLE 3: Wind power plants under operation in Turkey (July 2012).

Name of windfarm	Installed capacity (MW)	Turbine power	Turbine brand	Location	Production date
Çeşme	1.50	0.5 MW	Enercon	İzmir-Çeşme	1998
Ares	7.20	0.6 MW	Vestas	İzmir-Çeşme	1998
Bozcaada	10.20	0.6 MW	Enercon	Çanakkale-Bozcaada	2000
Intepe	30.40	0.8 MW	Enercon	Çanakkale-İntepe	2007
Karakurt	10.80	1.8 MW	Vestas	Manisa-Akhisar	2007
Burgaz	14.90	0.8 MW + 0.9 MW	Enercon	Çanakkale-Gelibolu	2007
Sayalar	34.20	0.9 MW	Enercon	Manisa-Sayalar	2008
Çatalca	60.00	3 MW	Vestas	İstanbul-Çatalca	2008
Yuntdağ	57.50	2.5 MW	Nordex	İzmir-Aliağa	2008
Kemerburgaz	24.00	2 MW	Enercon	İstanbul-Gaziosmanpaşa	2008
Mare manastır	39.20	0.9 MW + 0.8 MW	Enercon	İzmir-Çeşme	2006/2007
Sunjut	1.20	0.6 MW	Enercon	İstanbul-Hadımköy	2003
Teperes	0.85	0.85 MW	Vestas	İstanbul-Silivri	2006
Bandırma	35.00	1.5 MW	GE + Nordex	Balıkesir-Bandırma	2006
Şamlı	114.00	3 MW	Vestas	Balıkesir-Şamlı	2008
Datça	29.60	0.8 MW + 0.9 MW	Enercon	Muğla-Datça	2008
Sebenoba	30.00	2 MW	Vestas	Hatay-Samandağ	2008
Akbuk	31.50	2.1 MW	Suzlon	Aydın-Didim	2009
Çamseki	20.80	2 MW + 0.8 MW	Enercon	Çanakkale-Ezine	2009
Keltepe	20.70	0.9 MW	Enercon	Balıkesir-Susurluk	2009
Gökçedağ	135.00	2.5 MW	GE	Osmaniye-Bahçe	2009/2010
Düzova	30.00	2.5 MW	GE	İzmir-Bergama	2009/2010
Mazı-3	30.00	2.5 MW	Nordex	İzmir-Çeşme	2009/2010
Ayyıldız	15.00	3 MW	Vestas	Balıkesir-Bandırma	2009
Bandırma	60.00	3 MW	Vestas	Balıkesir-Bandırma	2009/2010
Soma	140.10	2 MW + 0.9 MW	Enercon	Manisa-Soma	2011/2012
Belen	36.00	3 MW	Vestas	Hatay-Belen	2009/2010
Sarıkaya	28.80	2 MW + 0.8 MW	Enercon	Tekirdağ-Şarköy	2009
Kocadağ-2	17.50	2.5 MW	Nordex	İzmir-Urla	2010
Bandırma-3	25.00	2.5 MW	Nordex	Balıkesir-Bandırma	2010
Mersin	33.00	3 MW	Vestas	Mersin-Mut	2010
Boreas-1	15.00	2.5 MW	Nordex	Edirne-Enez	2010
Aliağa	90.00	2.5 MW	Nordex	İzmir-Aliağa	2010
Şenbuk	15.00	3 MW	Vestas	Hatay-Belen	2010
Ziyaret	57.50	2.5 MW	GE	Hatay-Samandağ	2010/2011
Soma	90.00	2.5 MW	Nordex	Manisa-Soma	2010
Kuyucak	25.60	2 MW + 0.8 MW	Enercon	Manisa-Kırkağaç	2010
Sares	22.50	2.5 MW	GE	Çanakkale-Ezine	2010/2011
Turguttepe	24.00	2 MW	Vestas	Aydın-Çine	2010
Çataltepe res	16.00	2 MW	Enercon	Balıkesir	2010
Çanakkale	29.90	2.3 MW	Siemens	Çanakkale-Ezine	2011
Susurluk	45.00	2.5 MW	Nordex	Balıkesir-Susurluk	2011
Seyitali res	30.00	2 MW	Enercon	İzmir	2011
Söke-Çatalbük	30.00	2 MW	Gamesa	Aydın	2012
Aksu	72.00	2 MW	Vestas	Kayseri	2012
Amasya	40.00	2.5 MW	Nordex	Amasya	2012
Metristepe	40.00	2.5 MW	Nordex	Bilecik	2012
Dağpazarı	39.00	3 MW	Siemems	Mersin	2012
Şenköy	27.00	3 MW	Alstom W.	Hatay	2012
Şahres	93.00	3 MW	Vestas	Balıkesir	2011

TABLE 3: Continued.

Name of windfarm	Installed capacity (MW)	Turbine power	Turbine brand	Location	Production date
Ayres	5.00	1.8 MW	Vestas	Çanakkale	2011
Akres	45.00	2.5 MW	Nordex	Manisa	2011
Karadağ	10.00	2.5 MW	GE	İzmir	2012
Bozyaka	12.50	2.5 MW	Nordex	İzmir	2011
Killik	40.00	2.5 MW	Nordex	Tokat	2012
Saray	4.00	2 MW	Enercon	İstanbul	2012
Total	2041.35				

TABLE 4: Distribution of installed capacity according to Provinces of Turkey.

Provinces	Installed capacity (MW)
Balıkesir	423.10
Manisa	345.70
İzmir	325.40
Hatay	165.50
Osmaniye	135.00
Çanakkale	133.70
İstanbul	90.05
Aydın	85.50
Mersin	72.00
Kayseri	72.00
Tokat	40.00
Amasya	40.00
Bilecik	40.00
Muğla	29.60
Tekirdağ	28.80
Edirne	15.00

TABLE 5: Prediction of wind power development capacity in Turkey.

Years	Wind energy, MW* [35]	Wind energy, MW* [54]
2000**	19	19
2010**	1329	1329
2015	5142	7230
2020	7849	10620
2025	9733	13160
2030	11200	15885

*Calculation of the authors assuming a wind turbine utilization factor.
**Statistical data.

TABLE 6: Electricity generation costs by fuel type (cent/kWh).

Power source	Minimum	Maximum
Large hydro	3.0	13.0
Small hydro	4.0	14.0
Municipal solid wastes	4.2	6.3
Bio mass	4.2	7.9
Natural gas	4.3	5.4
Coal	4.5	7.0
Agricultural residues	4.5	9.8
Wind	4.7	7.2
Geothermal	4.7	7.8
Hydraulic	5.2	18.9
Nuclear	5.3	9.3
Solar thermal hybrid	6.0	7.8
Wave/tidal	6.7	17.2
Energy crops	10.0	20.0
Solar PV	28.7	31.0

TABLE 7: Cost structure of a typical 2 MW wind turbine installed in Europe (€2006).

	Investment (€1,000/MW)	Share of total cost %
Turbine (ex works)	928	75.6
Grid connection	109	8.9
Foundation	80	6.5
Land rent	48	3.9
Electric installation	18	1.5
Consultancy	15	1.2

provides good market prospects. The law also guarantees reduced costs for land access, generation, and licenses, as well as no VAT or custom taxes for wind equipment [4]. It is seen that this law has encouraged many investors for making investments given below regarding renewable energies in Turkey and especially wind energy.

5. Wind Energy Economics

The economics of wind power plants is influenced by a number of factors. These include the quality of the wind resource, technology efficiency and reliability, the availability of long-term power contracts, and the ability to forecast at least several hours ahead. By far the most significant factors that contribute to wind energy value are related to the wind resource and the characteristics of the grid and the evolving market rules. As additional wind capacity is developed, these variables will be quantified more precisely. A cost comparison between wind energy and other energy production methods was shown in Table 6 [40]. It can be seen that wind energy is as economically usable as other common energy sources.

Approximately 75% of the total cost of energy for a wind turbine is related to upfront costs such as the cost of the turbine, foundation, electrical equipment, and grid-connection.

Obviously, fluctuating fuel costs have no impact on power generation costs. Thus a wind turbine is *capital-intensive* compared to conventional fossil fuel fired technologies such as a natural gas power plant, where as much as 40–70% of costs are related to fuel and operation and maintenance. Table 7 gives the price structure of a typical 2 MW wind turbine [63].

6. Conclusions

Under global climate change conditions, all countries have trended towards renewable energy sources to reduce carbon dioxide emissions into the atmosphere. Wind energy is the most suitable energy source among renewable sources because wind energy has great potential throughout the world, including Turkey, and it is sustainable and does not pollute the environment. So, it has become crucial for electricity production. In general, potential wind energy areas in Turkey lie in northern and the north-western parts, at locations along the Aegean Sea and Marmara Sea coast. Aegean, Marmara, East-Mediterranean, and South East Anatolia regions of Turkey are generally seen as promising of higher wind power potential compared to other part of Turkey. In Turkey the available wind power was 801 MW by the end of the year 2009. This capacity reached 1329 MW at the end of 2010. This capacity became 2041.35 MW at the end of July 2012. The installed wind capacity of Turkey is approximately 14% of Turkey's total economical wind potential. However this rate will be increased after installing the licensed projects.

Turkey is facing serious challenges in satisfying its growing energy demand. To fuel a rapidly growing economy, the country's electricity consumption is increasing by an average of 8-9% every year, and significant investments are needed in generation, transmission, and distribution facilities to balance the power system's supply and demand. Finally, Turkey is an energy-importing country. In order to be less dependent on other countries, Turkey needs to use its sustainable sources. From this point of view, wind power is a very attractive choice, since it is economical, sustainable, environment friendly, and a familiar energy source in Turkey.

References

[1] T. Burton, D. Sharpe, N. Jenkins, and E. Bossanyi, *Wind Energy Handbook*, John Wiley & Sons, Chichester, UK, 2001.

[2] The World Wind Energy Association, Annual report, 2011, http://www.wwindea.org/.

[3] I. Akova, "Development potential of wind energy in Turkey," *Echos Géographiques de Turquie*, vol. 16, 2011.

[4] A. D. Sahin, "A review of research and development of wind energy in Turkey," *Clean Soil Air Water*, vol. 36, no. 9, pp. 734–742, 2008.

[5] T. Ackerman and L. Söder, "Wind energy technology and current status: a review," *Renewable & Sustainable Energy Reviews*, vol. 4, pp. 315–374, 2000.

[6] A. D. Sahin, "Progress and recent trends in wind energy," *Progress in Energy and Combustion Science*, vol. 30, pp. 501–543, 2004.

[7] R. Thresher, M. Robinson, and P. Veers, "Wind energy technology: current status and R&D future," in *Proceedings of the Physics of Sustainable Energy Conference*, University of California at Berkeley, 2008.

[8] G. M. Joselin Herbert, S. Iniyan, E. Sreevalsan, and S. Rajapandian, "A review of wind energy technologies," *Renewable & Sustainable Energy Reviews*, vol. 11, no. 6, pp. 1117–1145, 2007.

[9] U. K. Galappaththi, A. K. M. De Silva, M. MacDonald, and O. R. Adewale, "Review of inspection and quality control techniques for composite wind turbine blades," *Insight: Non-Destructive Testing and Condition Monitoring*, vol. 54, pp. 82–85, 2012.

[10] Z. Qi and E. Lin, "Integrated power control for small wind power system," *Journal of Power Sources*, vol. 217, pp. 322–328, 2012.

[11] J. Yang, C. Peng, J. Xiao, J. Zeng, and Y. Yuan, "Application of videometric technique to deformation measurement for large-scale composite wind turbine blade," *Applied Energy*, vol. 98, pp. 292–300, 2012.

[12] A. Öztopal, A. D. Sahin, Z. Sen, and N. Akgün, "On the regional wind energy potential of Turkey," *Energy*, vol. 25, pp. 189–200, 2000.

[13] H. Balat, "Wind energy potential in Turkey," *Energy*, vol. 23, pp. 51–56, 2005.

[14] R. T. Ogulata, "Energy sector and wind energy potential in Turkey," *Renewable & Sustainable Energy Reviews*, vol. 7, pp. 469–484, 2003.

[15] R. K. Köse, "An evaluation of wind energy potential as a power generation source in Kütahya Turkey," *Energy Conversion and Management*, vol. 45, pp. 1631–1641, 2004.

[16] M. G. Gökçek, A. Bayülken, and S. Bekdemir, "Investigation of wind characteristics and wind energy potential in Kirklareli Turkey," *Renewable Energy*, vol. 32, pp. 1739–1752, 2007.

[17] M. Durak and Z. Şen, "Wind power potential in Turkey and Akhisar case study," *Renewable Energy*, vol. 25, no. 3, pp. 463–472, 2002.

[18] E. K. Akpinar and S. Akpinar, "Determination of the wind energy potential for Maden-Elazig, Turkey," *Energy Conversion and Management*, vol. 45, no. 18-19, pp. 2901–2914, 2004.

[19] R. Köse, M. Arif Ozgur, O. Erbas, and A. Tugcu, "The analysis of wind data and wind energy potential in Kutahya, Turkey," *Renewable & Sustainable Energy Reviews*, vol. 8, no. 3, pp. 277–288, 2004.

[20] M. Bilgili, B. Sahin, and A. Kahraman, "Wind energy potential in Antakya and Iskenderun regions, Turkey," *Renewable Energy*, vol. 29, pp. 1733–1745, 2004.

[21] N. Eskin, H. Artar, and S. Tolun, "Wind energy potential of Gökçeada Island in Turkey," *Renewable & Sustainable Energy Reviews*, vol. 12, pp. 839–851, 2008.

[22] A. Ucar and F. Balo, "Evaluation of wind energy potential and electricity generation at six locations in Turkey," *Applied Energy*, vol. 86, no. 10, pp. 1864–1872, 2009.

[23] Y. Oner, S. Ozcira, and N. Bekiroglu, "Prediction wind energy potential using by wind data analysis in bababurnu-Turkey," in *Proceedings of the International Conference on Clean Electrical Power (ICCEP '09)*, pp. 232–235, June 2009.

[24] B. Sahin and M. Bilgili, "Wind characteristics and energy potential in Belen-Hatay, Turkey," *International Journal of Green Energy*, vol. 6, no. 2, pp. 157–172, 2009.

[25] B. Sahin, M. Bilgili, and H. Akilli, "The wind power potential of the eastern Mediterranean region of Turkey," *Journal of Wind Engineering and Industrial Aerodynamics*, vol. 93, no. 2, pp. 171–183, 2005.

[26] I. T. Togrul and C. Ertekin, "A statistical investigation on the wind energy potential of Turkey's geographical regions," *Energy Sources Part A*, vol. 33, no. 15, pp. 1399–1421, 2011.
[27] M. Bilgili and E. Simsek, "Wind energy potential and turbine installations in Turkey," *Energy Sources Part B*, vol. 7, pp. 140–151, 2012.
[28] E. Camadan, "An assessment on the current status and future of wind energy in Turkish electricity industry," *Renewable & Sustainable Energy Reviews*, vol. 15, pp. 4994–5002, 2011.
[29] C. Ilkılıç, "Wind energy and assessment of wind energy potential in Turkey," *Renewable & Sustainable Energy Reviews*, vol. 16, pp. 1165–1173, 2012.
[30] C. Ilkılıç and I. Türkbay, "Determination and utilization of wind energy potential for Turkey," *Renewable & Sustainable Energy Reviews*, vol. 14, pp. 2202–2207, 2010.
[31] A. A. Coskun and Y. O. Türker, "Wind energy and Turkey," *Environmental Monitoring and Assessment*, vol. 184, pp. 1265–1273, 2012.
[32] M. S. Celiktas and G. Kocar, "Foresight analysis of wind power in Turkey," *International Journal of Energy Research*, vol. 36, pp. 737–748, 2012.
[33] A. N. Celik, "Review of Turkey's current energy status: a case study for wind energy potential of Canakkale Province," *Renewable & Sustainable Energy Reviews*, vol. 15, pp. 2743–2749, 2011.
[34] A. Hepbasli and O. Ozgener, "A review on the development of wind energy in Turkey," *Renewable & Sustainable Energy Reviews*, vol. 8, no. 3, pp. 257–276, 2004.
[35] H. Aras, "Wind energy status and its assessment in Turkey," *Renewable Energy*, vol. 28, pp. 2213–2220, 2003.
[36] M. Bilgili, "A global review of wind power installations and their development in Turkey," *Clean Soil Air Water*, vol. 37, no. 3, pp. 195–202, 2009.
[37] M. Gökçek, "Developing Wind Energy in Turkey," 2012, http://www.intechopen.com/.
[38] K. Kaygusuz, "Wind energy status in renewable electrical energy production in Turkey," *Renewable & Sustainable Energy Reviews*, vol. 14, pp. 2104–2112, 2010.
[39] O. Guler, "Wind energy status in electrical energy production of Turkey," *Renewable & Sustainable Energy Reviews*, vol. 13, pp. 473–478, 2009.
[40] E. Erdogdu, "On the wind energy in Turkey," *Renewable & Sustainable Energy Reviews*, vol. 13, pp. 361–371, 2009.
[41] C. Ilkılıç, H. Aydın, and R. Behçet, "The current status of wind energy in Turkey and in the world," *Energy Policy*, vol. 39, pp. 961–967, 2011.
[42] O. Ozgener and A. Hepbasli, "Current status and future directions of wind energy applications in Turkey," *Energy Sources*, vol. 24, no. 12, pp. 1117–1129, 2002.
[43] F. Evrendilek and C. Ertekin, "Assessing the potential of renewable energy sources in Turkey," *Renewable Energy*, vol. 28, no. 15, pp. 2303–2315, 2003.
[44] M. Ocak, Z. Ocak, S. Bilgen, S. Keleş, and K. Kaygusuz, "Energy utilization, environmental pollution and renewable energy sources in Turkey," *Energy Conversion and Management*, vol. 45, no. 6, pp. 845–864, 2004.
[45] I. Yüksel, "Global warming and renewable energy sources for sustainable development in Turkey," *Renewable Energy*, vol. 3, pp. 802–812, 2008.
[46] R. Oğulata, "Potential of renewable energies in Turkey," *Journal of Energy Engineering*, vol. 133, pp. 63–68, 2007.
[47] S. Keleş and S. Bilgen, "Renewable energy sources in Turkey for climate change mitigation and energy sustainability," *Renewable & Sustainable Energy Reviewsiews*, vol. 16, pp. 5199–5206, 2012.
[48] M. Çapik, A. O. Yilmaz, and T. Çavuşoğlu, "Present situation and potential role of renewable energy in Turkey," *Renewable Energy*, vol. 46, pp. 1–13, 2012.
[49] I. Yuksel and K. Kaygusuz, "Renewable energy sources for clean and sustainable energy policies in Turkey," *Renewable & Sustainable Energy Reviews*, vol. 15, pp. 4132–4144, 2011.
[50] S. Volkan and E. K. Ediger, "Renewable energy potential as an alternative to fossil fuels in Turkey," *Energy Conversion & Management*, vol. 40, pp. 743–755, 1999.
[51] D. Kaya, "Renewable energy policies in Turkey," *Renewable & Sustainable Energy Reviewsiews*, vol. 10, pp. 152–163, 2006.
[52] K. Kaygusuz and A. Sari, "Renewable energy potential and utilization in Turkey," *Energy Conversion Management*, vol. 44, pp. 459–478, 2003.
[53] A. Akpinar, M. I. Komurcu, M. Kankal, I. H. Ozolcer, and K. Kaygusuz, "Energy situation and renewables in Turkey and environmental effects of energy use," *Renewable & Sustainable Energy Reviews*, vol. 12, pp. 2013–2039, 2007.
[54] M. Kenisarin, M. V. Karsli, and M. Caglar, "Wind power engineering in the world and perspectives of its development in Turkey," *Renewable & Sustainable Energy Reviews*, vol. 10, pp. 341–369, 2006.
[55] Global cumulative installed wind capacity 1996–2010, http://www.gwec.net/.
[56] Wind Energy International 2011/2012 yearbook, http://www.sonnenseite.com/.
[57] EÜAŞ Sector Report, "Electrical Energy Production Sector Report," Turkey, 2008.
[58] Energy Policies of IEA Countries, 2010, http://www.iea.org/.
[59] Electricity Generation & Transmission Statistics of Turkey, 2010, http://www.teias.gov.tr/.
[60] Key World Energy Statistic, International Energy Agency; 2011.
[61] Global Wind Report-Annual market update, 2010, http://www.gwec.net/.
[62] http://www.tureb.com.tr/.
[63] A report by the European Wind Energy Association, The economics of wind power, March 2009, http://www.wind-energy-the-facts.org/.

21

Practical Coupled Resonators in Domino Arrangements for Power Transmission and Distribution: Replacing Step-Down Power Transformers and Their Branches across the Power Grid

Athanasios G. Lazaropoulos

School of Electrical and Computer Engineering, National Technical University of Athens, 9 Iroon Polytechniou Street, Zografou, 15780 Athens, Greece

Correspondence should be addressed to Athanasios G. Lazaropoulos; aglazaropoulos@gmail.com

Academic Editor: Kamaruzzaman Sopian

This paper considers the potential of replacing step-down power transformers of the entire power grid as well as part of their transmission line branches with wireless power transfer (WPT) technology components. Exploiting the state-of-the-art evolutions in the fields of WPT technology, coupled resonators in domino arrangements—domino coupled resonator (DCR) configurations—are proposed as suitable technological substitute for step-down power transformers and are investigated in terms of performance metrics such as power transfer efficiency (PTE) and transformation ratio (TR). The contribution of this paper is fivefold. First, an analytical theoretical analysis appropriate to the study of practical DCR configurations is demonstrated. In order to support the DCR configuration replacement venture, a detailed set of assumptions regarding efficient mid- and long-range high-power WPTs as well as related technical issues is first presented. The validity of the theoretical analysis is verified through experimental measurements. Second, applying the proposed theoretical analysis, a wealth of system parameters that mainly influences the PTE and TR of DCR configurations is identified. Their quantitative effect as well as corresponding DCR configuration adjustments are first presented. Third, an approximate method, denoted as approximate chain scattering matrix (CSM) method, is first introduced. Based on the scattering matrix theory formalism, the approximate CSM method is suitable for mid- and long-range DCR configurations when the theoretical analysis becomes computationally slow. The numerical results of approximate CSM method are compared with the respective ones of theoretical analysis validating the extent and the accuracy of approximate CSM method. Fourth, the potential of power transformer replacement with practical DCR configurations is thoroughly investigated in terms of their TRs. A plethora of high-voltage/medium-voltage (HV/MV), MV/low-voltage (MV/LV), and HV/LV power transformers used across the world is investigated verifying their replacement potential with practical DCR configurations in all the cases examined. Fifth, based on a detailed collection of dimensions concerning power transformers and transmission line branches, it is first verified that practical DCR configurations cannot only substitute all step-down power transformers of the today's power grid but also replace entire transmission line branches too. Finally, it is obvious that there is a long journey ahead for WPT technology and its ultramodern DCR configurations to be affordably, widely, reliably, sustainably, and safely adopted in the human society. During these first steps of WPT development for power transmission and distribution, theoretical analyses and visions are necessary. The last cable problem, that is, the seamless power delivery as easily as information is now transmitted through the air, is one of the major technological challenges of the 21st century, and, thus, WPT technology will certainly play key role.

1. Introduction

The fundamentals of wireless power transfer (WPT) have been discovered in the works of M. Faraday, H. C. Ørsted, and N. Tesla in the 19th and 20th centuries. WPT technology describes the possibility of cutting the last cable, allowing utilities to seamlessly deliver power as easily as information is now transmitted through the air [1–3].

Nowadays, WPT technology attracts much academic and business interest due to its user-friendly nature. Many academic studies have investigated the WPT application potential from small electronic devices, such as medical implants [4–11], irons, vacuum cleaners, televisions [12], and chip/transistor power supply [13], to larger high-power ones, such as power distribution [14–18], electric vehicles [19, 20], wireless-powered lighting systems [21], industrial robots [22, 23], space vehicles, and satellites [24–27]. Recently, new products have been launched focusing on contactless battery charging for mobile appliances (charging pads, phones, and laptops) [28–33] and, generally, on cord-free desks [34, 35]. The common ground of all these cases is that WPT technology will be more practical if its WPT configurations become more efficient, more high-power and capable of transferring power over significantly larger distances through the air [36]. Yet, there is no such WPT configuration to simultaneously satisfy the previous requirements.

More specifically, to configure WPT configurations, there are two implementation methods: the resonant inductive coupling method and the magnetic resonance coupling (MRC) method. The resonant inductive coupling method uses the electromagnetic wave. It is first recognized and exploited by Tesla [1, 32, 37–42]. Although it is an efficient nonradiative WPT method that can charge various electronic devices, the ambitious Tesla's goal of global power transfer through this method is utopian. As it has widely been verified, the resonant inductive coupling method is not suitable for realizing long-range high-power WPT configurations [36, 43]. Conversely, the MRC method uses electromagnetic near-field waves combined with the coupled resonance phenomenon [3, 36, 44–53]. This method was essentially rediscovered and proposed in [1, 44, 45, 54]. Actually, MRC method is characterized by its high power transfer efficiency (PTE) of approximately 90% within 1 m and, today, is suitable for practical short- and mid-range WPT configurations [36, 44, 45, 52, 54–56].

In fact, MRC method becomes the epoch-making WPT technology establishing efficient WPT channels [57, 58]. The first-generation WPT configurations, which are widely used in daily activities, that are based on MRC method consists of two low-loss coupled resonators (i.e. the source and the load coupled resonators). Their source coupled resonator fills its surrounding environment with low-loss nonradiative magnetic fields oscillating at kHz or MHz frequencies [58]. The occurred nonradiative fields through their resonant nature allow high PTEs between the source and the load coupled resonators ensuring that the interaction with nonresonant objects of the surrounding environment is minimal. However, the coverage of these first-generation WPT configurations still remains limited. To increase the effectiveness of WPT configurations in long-range WPT applications, multiple coupled resonators are deployed between source and load coupled resonators extending their effective transfer distance [40, 43, 46, 52, 59–66].

The multiple coupled resonators can be deployed in various domino arrangements forming respective Domino Coupled Resonators (DCR) configurations. Among them, straight DCR configurations, which are studied in this paper, have the unique feature that their coupled resonators centers lie along the same transfer distance axis forming one main power flow path [40, 63]. These DCR configurations enjoy higher PTEs for a given long-range transfer distance and higher application potential for power levels that range from few watts to many kilowatts in comparison with the limited uses of first-generation WPT configurations [40, 46, 52, 60, 62–65].

In assessing DCR configuration performance, the equivalent circuit approach is widely used in the literature [3, 31, 52, 67–69]. In this paper, a theoretical analysis, which is based on the standard coupled circuit theory [70, 71] and the magnetoinductive resonator system analysis [44, 72], is analytically presented. On the basis of a small set of input parameters and a reduced number of WPT assumptions, the proposed theoretical analysis gives accurate results in terms of PTE when different DCR configurations are considered. Actually, the theoretical analysis is applied in order to unveil the DCR configuration key parameters and investigate their influence on PTE. The numerical results derived from the theoretical analysis are validated through a set of experimental measurements.

Although the proposed theoretical analysis accurately copes with the PTE behavior of DCR configurations, its main deficiency is its slow response when DCR configurations with a great number of coupled resonators are deployed. In order to facilitate the handling and to optimize the required computational time of these mid- and long-range DCR configurations, except for the accurate theoretical analysis, an approximate method is first proposed. This approximate method is denoted as chain scattering matrix (CSM) method and gives quick and reliable numerical results for practical DCR configurations. Its formulation is based on the scattering matrix [73–75] and ABCD matrix [24, 26] formalism while its high fidelity and accuracy is validated through the comparison of its numerical results with the respective experimentally validated results of the theoretical analysis.

Already mentioned in the literature [52, 64], a DCR configuration can be interpreted as an expansion of a power transformer, which consists of two coils. Recognizing the physical similarities between these two different systems [76–78] and by exploiting the newly developed tools (i.e., the theoretical analysis and the approximate CSM method), the potential of replacing all step-down power transformers of power grid (i.e., power transformers that step down the supply voltage to a level suitable for the lower voltage level they feed) with practical DCR configurations is first proposed and investigated in this paper. Based on performance metrics such as PTE and transformation ratio (TR), that is, the percentage ratio of output to input voltage, it is reported that power transformers can comfortably be replaced with practical DCR configurations regardless of the power transformer type—either high-voltage (HV)/medium-voltage (MV) or MV/low-voltage (LV) or HV/LV power transformers.

DCR configurations and their potential of substituting power transformers can establish a new vision for future's smart grids. In fact, DCR configurations define a state-of-the-art and convenient solution that cannot only substitute power transformers of the power grid but also can replace entire transmission line branches too. Therefore, DCR configurations constitute an important step towards the solution of last cable problem; say, the power delivery to either large or small customers eliminating the presence of power transformers and transmission line branches.

Anyway, WPT is a newly born technology that requires time, patience, and attention to become fully operational. Hence, apart from some technical issues concerning practical mid- and long-range high-power DCR configuration requirements for the power transformer replacement issue, a brief section is devoted to a research road map and some future steps that must be followed so that WPT technology and its promising DCR configurations become economically viable and technically feasible power grid solution. Finally, it is evident that the best performance results of future's smart grids concerning power transmission and distribution are ensured through the cooperative operation (cooperative power delivery) and further integration of DCR configurations with today's well-proven power grid infrastructure.

The rest of this paper is organized as follows. In Section 2, the transmission and distribution power grids are analytically presented. Special emphasis is given on the transmission lines and power transformers of power grid as well as the main advantages for adopting DCR configurations in order to replace these today's power grid components. Section 3 provides the entire analysis framework concerning the operation of DCR configurations: theoretical analysis, DCR configuration assumptions, approximate CSM method, and various approximations concerning WPT transmission and distribution. In Section 4, numerical results and discussion identify and assess the influence of specific DCR configuration design parameters on PTE and TR performance as well as the accuracy of theoretical analysis and approximate CSM method. The replacement potential of step-down power transformers and their transmission line branches with practical DCR configurations is thoroughly investigated. Section 5 includes background to ideas for future WPT works. Section 6 concludes this paper.

2. Transmission and Distribution Power Grids

The general principles of power generation, transmission, and distribution have been established more than a century ago. Incremental updates have been introduced in terms of renewable sources and other complex power electronic components. Generated electric power is transmitted to the load centres and consumers through this sophisticated network mainly consisting of transmission lines and power transformers [79–81].

2.1. Transmission Lines of Power Grid. Depending on the amount of carried power and its voltage level, a power grid can be further divided into two main interconnected power networks.

(i) *Transmission Power Network.* Its main purpose is to connect all major generator stations and critical load points in the energy system via loop structure. It comprises overhead HV transmission lines. This bulk supply energy system handles large power blocks allocating them among its available electrical energy corridors so that the best overall operating economy and/or other technical objectives are accomplished.

(ii) *Distribution Power Network.* It has the largest share in power grid since it comprises the finest meshes of power grid. Implemented in radial structure, it covers the power demands of medium large, residential, and commercial consumers. Distribution power network utilizes its overhead and underground infrastructure that consists of MV and LV transmission lines.

Although transmission lines determine a complete infrastructure network that encompasses power generation, transmission, and distribution capabilities, the required interconnection among different power networks is achieved through the combined use of transmission line branches and power transformers [79–85].

2.2. Power Transformers of Power Grid. Power transformers provide the required interconnections among the different functions of the power grid; more specifically, power grid performs three major functions.

(i) *Power Generation.* It is the first step in delivering electricity and is performed at power stations. Through the generating transformers, the generating voltages are stepped up and connected to the transmission power network. Generating (step-up) transformers are out of the scope of this paper.

(ii) *Power Transmission.* It is the second step in carrying electricity and involves the power transfer from generating transformers to HV/MV substations, which comprise HV/MV transformers. At HV/MV transformers, the voltage is stepped down and power is fed into the distribution power network.

(iii) *Power Distribution.* It completes the power grid through delivering power to consumers. It involves the power transfer from HV/MV transformers to the primary and secondary distribution power networks.

(a) The primary distribution power network comprises MV/LV transformers and feeds power to medium-large consumers. In general, these MV/LV transformers are owned by the fed medium large consumers.

(b) The secondary distribution feeders consist of MV/LV transformers that supply residential and commercial customers. These MV/LV transformers are owned by power utilities.

(c) In rare and remote cases of power distribution, either medium-large or residential or commercial customers can be straightforwardly supplied by HV/LV power transformers.

Note that HV/MV, MV/LV, and HV/LV power transformers are known as step-down power transformers and constitute the main interest of this paper [24, 25, 43, 62, 86, 87].

2.3. From Traditional Power Grid to WPT Systems. Due to their design, procurement, and manufacturing process, power transformers are custom-oriented equipment, entail significant capital expenditures and demand long lead times. In fact, the cost of a transformer is in the order of thousands of euros while its weight ranges from 100 to 400 tons. In addition, lead times could stretch beyond 20 months due to certain key parts and/or materials [86].

The impact of WPT systems on the way that power is transformed and delivered is immense. Instead of using power transformers, one can bring electrical power that is efficiently stepped down without the constraints of expensive, weighty and permanent electrical equipment. Inspired by relevant recent breakthroughs in WPT technology [6, 8, 19, 20, 29, 32, 36, 47, 87–91], this paper tries to redefine the traditional power grid structure through the use of efficient WPT configurations that can replace step-down power transformers.

3. Theoretical Analysis of DCR Configurations

Until now, the main reason for the lack of WPT applications for mid- and long-range power transfer was their low PTE because all WPT efforts have focused essentially on Tesla's experiments and near-field coupling approaches—that is, one pair of coupled resonators—[1]. Recently, new interest has arisen in the field of WPT technology due to the fact that, instead of using one pair of coupled resonators, overall PTE can significantly be improved by using several coupled resonators placed in or near the short-range region where the PTE of coupled resonators is high [40, 52, 62–64, 88–91]. Furthermore, the use of a great number of coupled resonators can drastically increase the overall transfer distance between the excitation power source and the load achieving the required TR.

3.1. General DCR Configurations and Respective Theoretical Analysis. A DCR configuration may be considered as a cascade of magnetically coupled resonators, which are shown in Figure 1(a). These coupled resonators are placed in close proximity and they can be deployed in various domino arrangements. A practical DCR configuration is depicted in Figure 1(b).

Due to their design, DCR configurations are characterized by their high PTE efficiencies, permitting the delivery of a great number of WPT applications for power levels that range from few watts to several megawatts. As it has already been reported in [60, 62–64], the mathematical formulation that is usually employed to analyze these DCR configurations is based on the standard coupled circuit theory [70, 71] and the magnetoinductive resonator system analysis [44, 72].

More specifically, the general theoretical DCR configuration, which is used across this paper, is presented in Figure 2(a). This overall circuit configuration comprises the concatenation of n coupled resonators. Each coupled resonator i, $i = 1, \ldots, n$ consists of the serial connection of three subcomponents: its winding self-inductance L_i, its resistance R_i, and its winding compensating capacitance C_i.

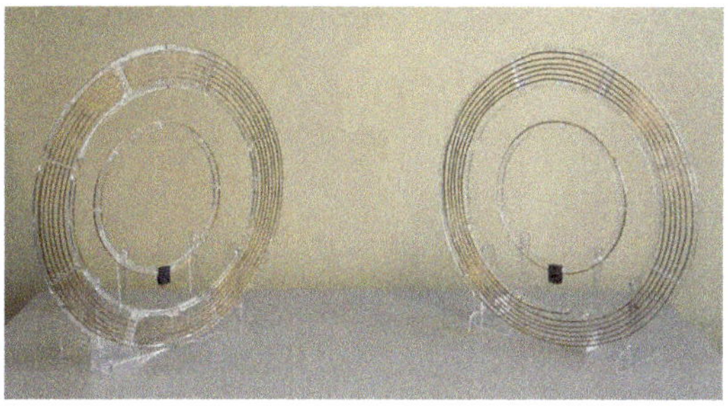

(a)

(b)

FIGURE 1: (a) Practical coupled resonators [3]. (b) Practical DCR configuration with ten coupled resonators [62].

In particular, the first ($i = 1$) and last ($i = n$) coupled resonators are denoted as source and load coupled resonators, respectively. Apart from their typical subcomponents, source coupled resonator also includes the source—that is, excitation voltage source V_S with its impedance Z_S—whereas load coupled resonator includes the load—that is, impedance Z_L. To determine the aforementioned subcomponent parameters of coupled resonators, a plethora of analytical models has been proposed in the literature [4, 92–103]. Across this paper, the mathematical formulations of [4, 42, 53, 92, 103] have been adopted due to their theoretical simplicity and their well-proven validity.

As it is clearly shown in Figure 2(a), each coupled resonator i, $i = 1, \ldots, n$ provides path for a current I_i to flow. The coupled resonator currents $\mathbf{I}^{\mathrm{DCR}} = \left[I_1 \ \cdots \ I_i \ \cdots \ I_n\right]^T$ are determined through the following general circuit equation:

$$\mathbf{A} \cdot \mathbf{I}^{\mathrm{DCR}} = \left[-V_S \ \ \mathbf{0}_{1\times(n-1)}\right]^T, \quad (1)$$

where $[\cdot]^T$ denotes the transpose of a matrix, $\mathbf{0}_{m\times n}$ is an $m \times n$ matrix with zero elements, and $\mathbf{A}$ is $n\times n$ matrix depending on the frequency, coupled resonator characteristics, source and load impedances, and mutual inductances between coupled resonators M_{ij} $(i, j = 1, \ldots, n, \ i \neq j)$ [40, 52, 60, 62–64]. It should be noted that (1) describes a DCR configuration that is not restricted to operate in its (self-) resonance frequency (for more details, see also in Section 4.1.5).

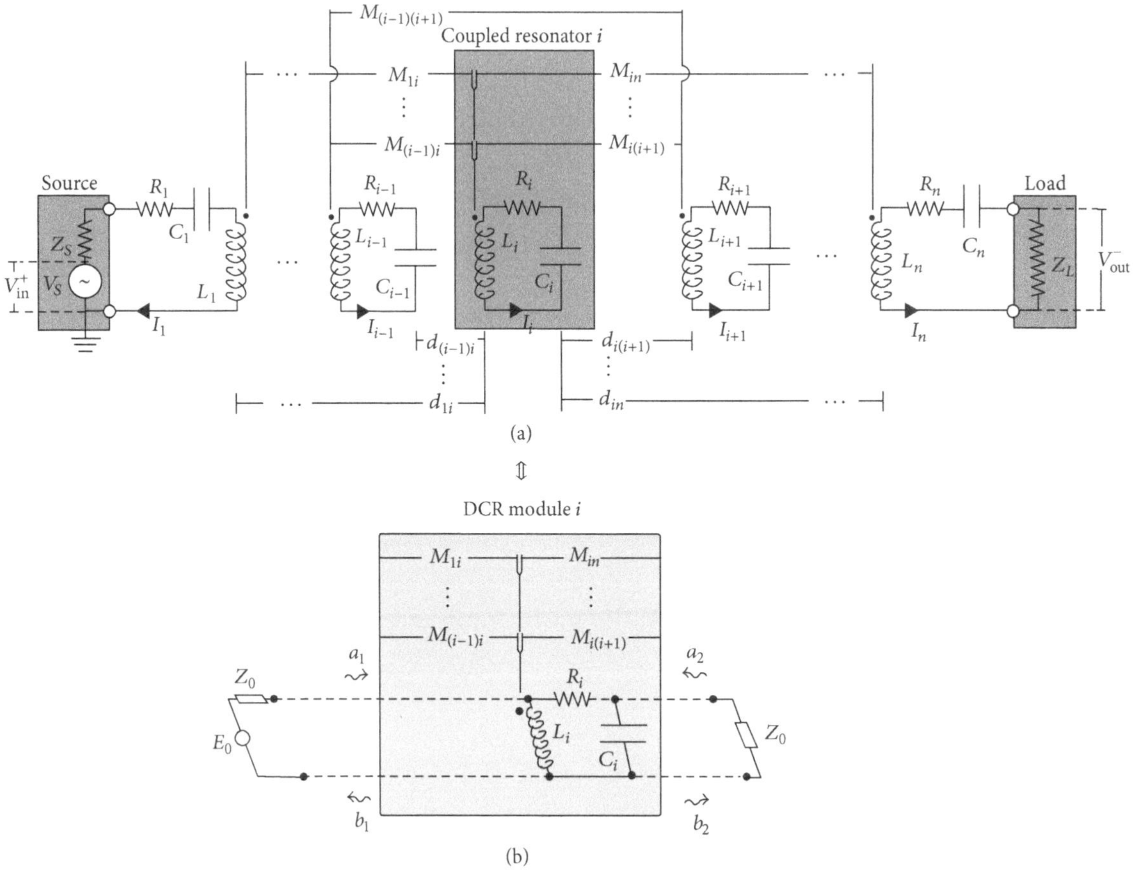

FIGURE 2: Typical DCR configuration with n coupled resonators. (a) Equivalent circuit. (b) DCR module of approximate CSM method.

Under three assumptions concerning WPT to DCR configurations—see Section 3.2—PTE of the DCR configuration depicted in Figure 2(a) is given by

$$\text{PTE} = \left\langle \left\{ \sum_{i=1}^{n-1} \left[\left| \frac{I_i}{I_n} \right|^2 \cdot \left| \frac{R_i}{Z_L} \right| \right] \right\} + \left[\left| \frac{I_1}{I_n} \right|^2 \cdot \left| \frac{Z_S}{Z_L} \right| \right] + \left| \frac{R_n}{Z_L} \right| + 1 \right\rangle^{-1}. \quad (2)$$

Combining (1) and (2), it is easily verified that PTE mainly depends on mutual inductances between coupled resonators M_{ij} $(i, j = 1, \dots, n, i \neq j)$, coupled resonator resistances R_i, $i = 1, \dots, n$, source impedance Z_S and load impedance Z_L for a given DCR configuration, namely [62],

$$\text{PTE} = f\left(M_{12}, \dots, M_{(i-1)i}, \dots, M_{(n-1)n}, R_1, \dots, R_i, \dots, R_n, Z_S, Z_L\right), \quad i = 1, \dots, n. \quad (3)$$

It should be noted that the theoretical analysis of DCR configurations has experimentally been verified in [40, 52, 60, 62–64].

3.2. Assumptions Concerning WPT to DCR Configurations. To study the WPT behavior of DCR configurations, three assumptions are needed so that the theoretical analysis of these configurations will be facilitated without, however, losing its theoretical generality, namely the following.

(1) All the used coupled resonators are identical—that is, they consist of the same subcomponents presenting the same geometrical, material, and circuital characteristics—and they are placed in linear path in each DCR configuration—that is, they are coaxial.

(2) All the considered coupled resonators for a given CDR configuration are assumed to have the same self-resonant frequency in order to efficiently exchange power between them while dissipate little power to extraneous objects, namely,

$$2\pi f_R L_i - \frac{1}{2\pi f_R C_i} = 0, \quad i = 1, \dots, n, \quad (4)$$

where f_R is the self-resonance frequency (for more details, see also Section 4.1.5).

(3) Source and load impedances are assumed to be Ohmic resistances.

3.3. Simplified Approximation for Coaxial and Circular DCR Configurations. With reference to Figure 2(a), from (3) and by exploiting well-known Maxwell's equation to calculate mutual inductances of coaxial circular filamentary current loops [57, 62, 104], PTE of a DCR configuration with n

coaxial and circular coupled resonators can be expressed as a function of the distances of every two coupled resonators d_{ik}, $i, k = 1, \ldots, n$, $i < k$, the source impedance and the load impedance, namely [40, 52, 60, 62–64],

$$\mathrm{PTE} = g\left(d_{12}, \ldots, d_{ik}, \ldots, d_{(n-1)n}, Z_S, Z_L\right), \quad i, k = 1, \ldots, n, \; i < k. \tag{5}$$

Actually, as it regards DCR configurations of interest in this paper, equal distance span d_{SPAN} is assumed between adjacent coupled resonators. Thus, PTE of a DCR configuration with n equally spaced coaxial and circular coupled resonators solely depends on the number of coupled resonators, the source impedance and the load impedance for a given transfer distance. Therefore, PTE of these DCR configurations can further be simplified to

$$\mathrm{PTE} = h\left(n, d_{\mathrm{SPAN}}, Z_S, Z_L\right). \tag{6}$$

3.4. Practical DCR Configurations and Approximate CSM Method. Until now, the theoretical analysis presented in Sections 3.1–3.3 can successfully deal with the peculiarities of the behavior of practical DCR configurations. However, its main problem, which is common to all relevant theoretical models of the literature [40, 52, 60, 62–64], is the size of the used mutual inductance matrices that hinders its practical implementation when a great number of coupled resonators needs to be deployed.

So far, several approximate methods with converging results have been used to determine the PTE of WPT systems [14, 21, 24, 26, 73, 105–109]. In this paper, the CSM method is proposed, which constitutes a refined version of the scattering matrix method first presented in [73]. To apply the CSM method, a DCR configuration is separated into segments—DCR modules, each of them comprising the successive coupled resonators encountered across the transfer distance—see Figure 2(b). The model used in the present work is based on the scattering matrix formulation introduced in [73–75] and alternatively used as ABCD matrix formulation [24, 26].

3.4.1. CSM Method and DCR Modules. WPT through the various DCR modules is taken into account through their respective CSMs. As depicted in Figure 2(b), a typical DCR configuration comprises three different types of DCR modules, namely the following.

(1) *Source DCR Module.* It is the first DCR module of a DCR configuration. With reference to Figure 2(a), the CSM of the source DCR module is given by

$$\mathbf{T}_1 = \mathbf{T}_S\left(V_S, Z_S, R_1, L_1, C_1, M_{12}, \ldots, M_{1i}, \ldots, M_{1n}\right) = \begin{bmatrix} T_{S,11} & T_{S,12} \\ T_{S,21} & T_{S,22} \end{bmatrix}, \tag{7}$$

where $T_{S,11}$, $T_{S,12}$, $T_{S,21}$, and $T_{S,22}$ are the elements of the $\mathbf{T}_S$ matrix.

(2) *Load DCR Module.* It is the last DCR module of a DCR configuration. With reference to Figure 2(a), the CSM of the source DCR module is given by

$$\mathbf{T}_n = \mathbf{T}_L\left(Z_L, R_n, L_n, C_n, M_{1n}, \ldots, M_{in}, \ldots, M_{(n-1)n}\right) = \begin{bmatrix} T_{L,11} & T_{L,12} \\ T_{L,21} & T_{L,22} \end{bmatrix}, \tag{8}$$

where $T_{L,11}$, $T_{L,12}$, $T_{L,21}$, and $T_{L,22}$ are the elements of the $\mathbf{T}_L$ matrix.

(3) *Coupled Resonator DCR Modules.* There are $n - 2$ coupled resonator DCR modules, that is, all the other DCR modules exclusive of source and load DCR modules. With reference to Figure 2(a), the CSM of a coupled resonator DCR module i is given by

$$\mathbf{T}_i = \mathbf{T}_i\left(R_i, L_i, C_i, M_{1i}, \ldots, M_{(i-1)i}, \ldots, M_{in}\right), \quad i = 2, \ldots, n-1. \tag{9}$$

Since input and output terminations are assumed matched and having already determined the CSMs of the various DCR modules, the overall CSM of the DCR configuration considered is evaluated in the appropriate cascade rule order from [74, 75, 110, 111]

$$\mathbf{T}_{\mathrm{overall}} = \begin{bmatrix} T_{11} & T_{12} \\ T_{21} & T_{22} \end{bmatrix} = \mathbf{T}_S \cdot \prod_{i=2}^{n-1} \mathbf{T}_i \cdot \mathbf{T}_L = \prod_{i=1}^{n} \mathbf{T}_i, \tag{10}$$

where T_{11}, T_{12}, T_{21}, and T_{22} are the elements of the $\mathbf{T}_{\mathrm{overall}}$ matrix.

The overall end-to-end scattering matrix is obtained from

$$\mathbf{S}_{\mathrm{overall}} = \begin{bmatrix} S_{11} & S_{12} \\ S_{21} & S_{22} \end{bmatrix} = \begin{bmatrix} \dfrac{T_{21}}{T_{11}} & T_{22} - \left(\dfrac{T_{21}T_{12}}{T_{11}}\right) \\ \dfrac{1}{T_{11}} & \dfrac{-T_{12}}{T_{11}} \end{bmatrix}, \tag{11}$$

where S_{11}, S_{12}, S_{21}, and S_{22} are the elements of the $\mathbf{S}_{\mathrm{overall}}$ matrix.

With reference to (11) and Figure 2(a), the PTE of a DCR configuration is given by the S_{21} element of the $\mathbf{S}_{\mathrm{overall}}$ matrix, that is,

$$\mathrm{PTE} = \mathrm{PTE}\left(V_S, Z_S, Z_L, R_1, \ldots, R_n, L_1, \ldots, L_n, C_1, \ldots, C_n, M_{12}, \ldots, M_{(i-1)i}, \ldots, M_{(n-1)n}\right) = \left|\frac{V_{\mathrm{out}}^-}{V_{\mathrm{in}}^+}\right|^2 = \left|S_{21}\right|^2 = \left|T_{11}\right|^{-2}, \tag{12}$$

where V_{out}^- and V_{in}^+ are the reflected load voltage and the incident source voltage wave, respectively,—see Figure 2(a)—when the incident wave a_2 of load module and the reflected wave b_1 of source is equal to zero, respectively, that is, the output and input termination is matched, respectively, which has already been assumed [74, 75, 110, 111].

3.4.2. Approximate CSM Method and Coaxial and Circular DCR Configurations. As it has already been demonstrated in Section 3.4.1, CSM method is a general methodology that can determine the PTE of any DCR configuration in a straightforward way. However, the practicability of (12) remains limited in WPT systems and especially in DCR configurations of this paper due to its complex dependencies on various DCR configuration parameters. To enhance its applicability, approximate CSM method is proposed that comes from CSM method when a number of simplifications are applied, namely the following.

(i) In accordance with assumption 1 of Section 3.2, from (10), the overall CSM is simplified to

$$\mathbf{T}'_{\text{overall}} = \mathbf{T}_S \cdot \mathbf{T}'^{n-2}_i \cdot \mathbf{T}_L. \tag{13}$$

(ii) Due to their structure and with satisfactory accuracy, DCR modules can be approximated as reciprocal (symmetrical), loss-free, and perfectly matched transmission components [24, 26, 73–75, 110, 111]. Thus, the overall CSM of (13) can be further simplified to

$$\begin{aligned}\mathbf{T}'_{\text{overall}} &= \begin{bmatrix} T'_{11} & T'_{12} \\ T'_{21} & T'_{22} \end{bmatrix} = \mathbf{T}_S \cdot \mathbf{T}'^{n-2}_i \cdot \mathbf{T}_L \\ &\cong \begin{bmatrix} T_{S,11} \cdot T'^{n-2}_{i,11} \cdot T_{L,11} & 0 \\ 0 & T_{S,22} \cdot T'^{-(n-2)}_{i,11} \cdot T_{L,22} \end{bmatrix},\end{aligned} \tag{14}$$

where T'_{11}, T'_{12}, T'_{21}, and T'_{22} are the elements of the $\mathbf{T}'_{\text{overall}}$ matrix and $T'_{i,11}$, $T'_{i,12}$, $T'_{i,21}$, and $T'_{i,22}$ are the elements of the $\mathbf{T}'_i$ matrix.

(iii) Based on (12) and the transmission-line-like approximated CSM of (14), the approximated PTE of a DCR configuration with n coupled resonators and distance span d_{SPAN} is determined from

$$\text{PTE}_{\text{approx},n} = \left|T'_{11}\right|^{-2} = \left|T_{S,11} \cdot T'^{n-2}_{i,11} \cdot T_{L,11}\right|^{-2}. \tag{15}$$

(iv) Equation (15) can further be simplified in a flowchart procedure through the following simple cascade rule:

$$\begin{aligned}\text{PTE}_{\text{approx},n} &= \underbrace{\text{PTE}_{\text{approx},2} \cdot \left|T'_{i,11}\right|^{-2\cdot(n-1-2)}}_{\text{PTE}_{\text{approx},n-1}} \cdot \left|T'_{i,11}\right|^{-2}. \\ &= \text{PTE}_{\text{approx},2} \cdot \left|T'_{i,11}\right|^{-2\cdot(n-2)}, \quad n > 2,\end{aligned} \tag{16a}$$

$$\text{PTE}_{\text{approx},2} = \text{PTE}_{\text{approx},1} \cdot \left|T_{L,11}\right|^{-2}, \quad n = 2, \tag{16b}$$

$$\text{PTE}_{\text{approx},1} = \left|T_{S,11}\right|^{-2}, \quad n = 1, \tag{16c}$$

where $|T'_{i,11}|^{-2}$ practically describes the additional PTE that should be taken into consideration when one new DCR module is inserted, thus, expanding the existing DCR configuration.

3.5. DCR Configurations and Other Practical Issues. Although higher voltage levels can be achieved by simply increasing the source excitation voltage of DCR configurations, more effective than usual heat-dissipation methods are needed so as to prevent the source—that is, source transistors—from overheating, package and insulation breakdowns and damage. More specifically, in the case of WPT in HV and in MV applications, the supported power levels require forced air or water cooling on the source coupled resonator so that its temperature remains below 70°C [18, 67]. As it concerns the heat losses of inductances and capacitances, a temperature increase at these components is not critical since they are passive components and are more resilient to the heat. However, the use of insulation oil can mitigate any potential thermal risks.

Except for the aforementioned overheating issues, mid- and long-range WPT technology is not only power lossy but also source and victim of electromagnetic interference (EMI) from and to external resonators, respectively. Since WPT is achieved through DCR configurations, appropriate shielding at the expense of thickness and weight should be adopted. The use of shield permits the system to be more EMI robust in environments where magnetic fields of nearby external resonators may interact with the magnetic field of coupled resonators of DCR configurations. However, shielding is beyond the scope of this paper. Hence, it is assumed that DCR configurations operate in environments free of external resonators that can critically degrade the PTE of DCR configurations [57, 67, 112].

In addition, at high frequencies, the behavior of coupled resonators of DCR configurations becomes different from their ac low-frequency behavior. Actually, skin and proximity effects affect the coupled resonator behavior and are responsible for resonant frequencies. In order to obtain more accurate results applying CSM method at frequencies above several hundred kHz or MHz, the prediction of suitable high-frequency correction terms for the three subcomponents of each DCR module (such as stray capacitance) is crucial for the design of high-frequency DCR configurations [4, 92–103].

Finally, as it has already been mentioned, today's WPT technology mainly utilizes electromagnetic near-field waves while DCR configurations further exploits MRC phenomenon [3, 32, 36, 44–53]. Especially, this phenomenon facilitates the WPT of great quantities of power at mid- and long-range distances. However, this long-range high-power WPT unrests about its harmful results that they could cause to human health [12]. In accordance with [61, 105, 113], WPT should comply with EMI and electromagnetic compatibility (EMC) regulations that impose safety levels with respect to human exposure to the used radio frequency electromagnetic fields.

4. Numerical Results and Discussion

The numerical results of various types of DCR configurations aim at investigating (a) their PTE in relation with specific DCR configuration design parameters; (b) the efficacy and accuracy of describing DCR configurations through

TABLE 1: Nominal parameter values of the practical coupled resonators.

Radius of coupled resonator coils α (mm)	Number of coil turns N_α	Radius of wires R (mm)	Layers of the wire N_d	Load impedance Z_L (Ω)	Self-resonance frequency f_R (kHz)	Copper electrical conductivity σ (MS/m)
155	11	1.2	1	17	530	59.6

TABLE 2: Two DCR configuration scenarios with their nominal parameter values.

	Transfer distance d (m)	Number of coupled resonators n (m)	Distance span d_{SPAN} (m)	Source voltage V_S (V)	Source impedance Z_S (Ω)
Scenario A	0.6	3	0.3	220	3.45
Scenario B	0.9	4	0.3	220	1.42

the approximate CSM method; (c) the potential of power transformer replacement with practical DCR configurations; and (d) minimum and maximum WPT distances that practical DCR configurations can cover for a given TR.

To study the behavior of DCR configurations, practical circular and coaxial coupled resonators have been chosen from the literature. The following theoretical studies are conducted based on the parameters of these practical coupled resonators, which have been experimentally verified in [40, 52, 60, 62–64], that are reported in Table 1.

4.1. Design of High-PTE DCR Configurations. To achieve a high PTE for practical DCR configurations, a wealth of DCR configuration design parameters is required to be appropriately adjusted. Towards that direction, several parameters of them are identified and investigated in terms of their impact on PTE of DCR configurations, namely: the radius of coupled resonator coils, the number of coil turns, the radius of the used wires in coils, the number of coupled resonators for a given transfer distance, the operating frequency, and the source and the load impedances.

As it concerns the simulations of this subsection, except for the nominal parameter values that are detailed in Table 1, two different DCR configuration scenarios—that is, DCR configuration scenario, A and B—are used whose additive nominal parameter values are reported in Table 2.

In accordance with Tables 1 and 2, these two scenarios have been carefully chosen so that the numerical results of the theoretical analysis of Sections 3.1–3.3 will be also experimentally validated by a trial of experimental measurements, which has been presented in [62].

4.1.1. Radius of Coupled Resonator Coils. It is well known that coil dimensions and coil shapes critically affect the magnitude and distribution of their occurred magnetic fields. In fact, these fields are closely related to the PTE of magnetoinductive resonator systems. More specifically, the radius of coupled resonator coils plays significant role towards efficient WPTs in DCR configurations since power is mainly transferred over relatively short distances, which are on the order of this radius [44, 45, 54, 114–117].

In Figure 3, PTE is plotted versus the radius of coupled resonator coils for DCR configuration scenarios A and B. In the same figure, PTE of these two DCR configuration scenarios is also shown with its corresponding experimental measurements when the nominal value of coupled resonator coil radius is considered—that is, $a = 155$ mm.

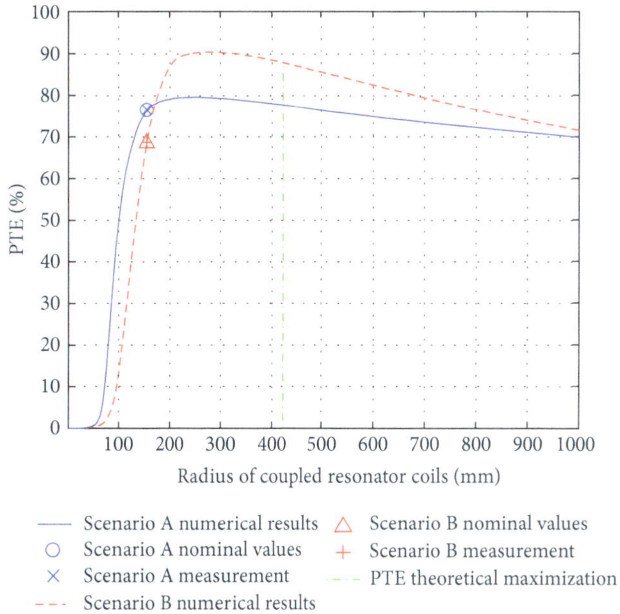

FIGURE 3: PTE versus coupled resonator coil radius for two different DCR configuration scenarios.

It is clearly shown that, when the radius of coupled resonator coils becomes larger, PTE increases till a maximum point and, then, smoothly tends to zero regardless of the adopted DCR configuration. This PTE behavior is explained through the analysis of occurred magnetic fields across DCR configuration transfer path; in a single circular coil with radius α, the magnetic field strength is maximized at distance $\sqrt{2}a/2$ along its coil axis [4, 118]. In the case of DCR configurations, the coil axes of coupled resonators coincide with the DCR configuration transfer path. Similarly to the simple case, PTE maximization occurs when the distance span is approximately equal to this magnetic field maximization distance. Note that this result may imply a reverse design procedure so that the optimum radius of coupled resonator coils can be determined; for example, with reference to Figure 3 and for the nominal distance span ($d_{SPAN} = 0.3$ m), the PTE maximization occurs when the coupled resonator coil radius is approximately equal to $a_{MAX} = 2 \cdot d_{SPAN}/\sqrt{2} \cong 424$ mm, regardless of the adopted DCR configuration scenario.

Furthermore, a physical constraint exists that determines either the upper boundary of coupled resonator coil radius or the elegant PTE transition to zero; when the coupled resonator coil radius increases, its coil circumference tends to become comparable to the wavelength. Hence, above a coil radius threshold, coupled resonator currents present non uniform distributions having negative impact on their PTE performance and explaining the curve form of Figure 3 [57].

4.1.2. Number of Turns of Coupled Resonator Coils. An effective way to increase the useable range of coupled resonator coils and, thus, PTE of a DCR configuration is to add multiple turns to the coupled resonator coils. As shown in [26, 92, 119], increasing the number of coil turns N_{α}, mutual inductance between coupled resonators increases quadratically with N_{α}.

In Figure 4, PTE is plotted versus the number of turns of coupled resonator coils for DCR configuration scenarios A and B. In the same figure, PTE of these two DCR configuration scenarios is also shown with its corresponding measurements when the nominal value of their coupled resonator coil turn number is considered—that is, $N_{\alpha} = 11$.

Observing Figure 4, it is obvious that as the number of turns becomes higher, PTE increases till a maximum point and, then, decreases to zero regardless of the adopted DCR configuration; by only adjusting the number of coil turns in coupled resonators, the DCR configuration PTE can significantly be improved. In practice, this technique can become useful in future since it allows the power utilities to easily design high efficient WPT systems for mid- and long-range applications [120].

Nevertheless, the physical constraint that defines the PTE maximum with respect to the number of coil turns is related with the ohmic losses; as the number of coil turns increases, so do the conductor length and the respective ohmic losses of each coupled resonator coil. This additive loss becomes critical preventing continuous increases in PTE of practical DCR configurations.

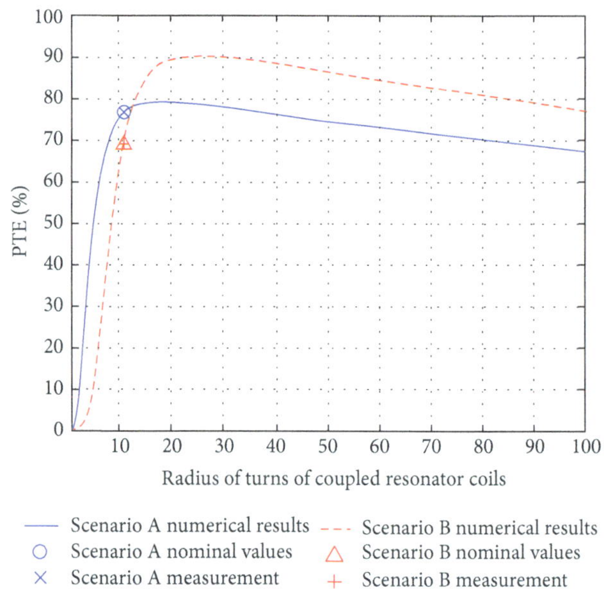

FIGURE 4: PTE versus number of turns of coupled resonator coils for two different DCR configuration scenarios.

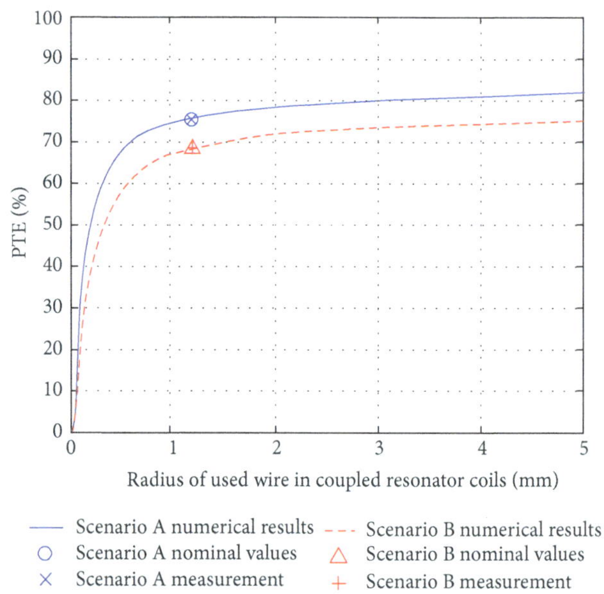

FIGURE 5: PTE versus radius of the used wire in coupled resonator coils for two different DCR configuration scenarios.

4.1.3. Radius of the Used Wires in Coupled Resonator Coils. In accordance with Section 4.1.2, the achievement of admissible PTEs for a given DCR configuration presupposes resistance reduction of used coupled resonator coils. An effective technique towards this reduction of coupled resonator coil resistance is the cross-section extension of their coil wires.

In Figure 5, PTE is plotted versus the radius of the used wires of coupled resonator coils for DCR configuration scenarios A and B. In the same figure, PTE of the two DCR configuration scenarios is also drawn as well as the corresponding measurements when the nominal value of the radius of their coupled resonator coil wires is considered—that is, $R = 1.2$ mm.

From Figure 5, it is evident that the radius increase entails a corresponding PTE increase. However, this PTE increase is marginal after several mm of wire radius due to the skin effect [121]. Actually, except for this PTE marginal increase, the unrestrained increase of wire radius causes enlargement of coupled resonator coil sizes. In many designs of practical DCR configurations, there are certain dimensional limitations that restrict this coupled resonator coil configuration increase. Due to this fact, during the estimation of coupled resonator coil resistance, the coupled resonator coil dimensions must be first specified, and then the resistance of the used coil is derived in relation with (i) radius of coils—see Section 4.1.1; (ii) its number of coil turns—see Section 4.1.2; (iii) coil inductance; and (iv) maximum radius of used coil wires [121].

4.1.4. Number of Coupled Resonators. As it has already been presented in Sections 4.1.1–4.1.3, the WPT characteristics of

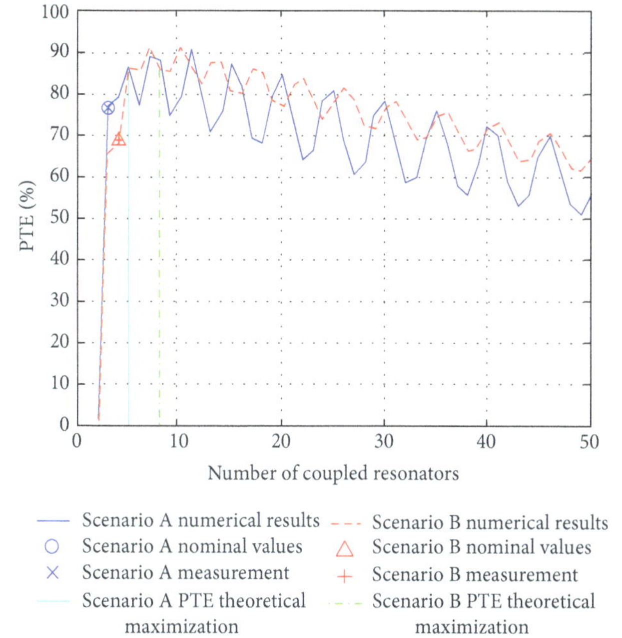

FIGURE 6: PTE versus coupled resonator number for two different DCR configuration scenarios.

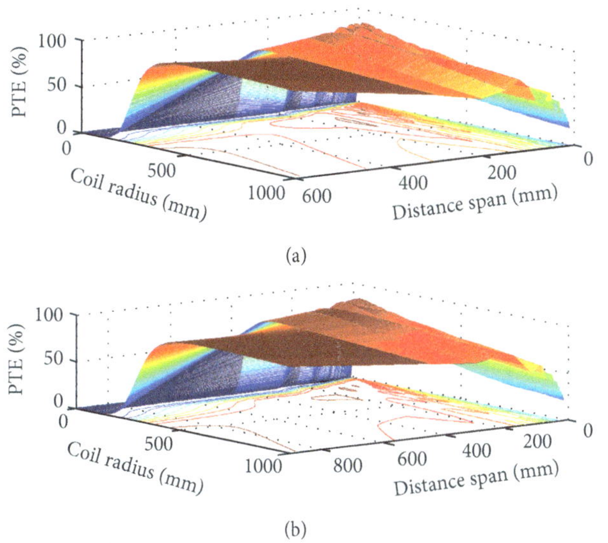

FIGURE 7: PTE versus coupled resonator coil radius and distance span. (a) DCR configuration scenario A. (b) DCR configuration scenario B.

a DCR configuration can be predicted and optimized through the presented theoretical analysis. This is feasible through the appropriate adjustment of the related coupled resonator inherent properties. Nonetheless, the aforementioned PTE findings of Sections 4.1.1–4.1.3 should be taken into consideration during the phase of DCR configuration design. This operation restriction reduces their practicability and applicability in the already installed WPT systems.

In order to overcome the low PTE of mid- and long-range applications without, however, substituting the already installed DCR configurations, a significant number of potential solutions have been proposed [62, 89, 122]. Among these proposals, this that has attracted the recent attention due to its promising PTE results in mid- and long-range applications is the insertion of extra coupled resonators. The additional as well as the existing coupled resonators are rearranged across the existing DCR configuration transfer path so as to achieve higher PTEs.

In Figure 6, PTE is plotted versus the number of coupled resonators for a given transfer distance when the DCR configuration scenarios A and B are applied. In the same figure, PTE of the two DCR configuration scenarios is also plotted with their corresponding measurements when the nominal values of the number of coupled resonators are considered—that is, $n = 3$ and $n = 4$ for DCR configuration scenarios A and B, respectively.

Figure 6 reveals that low PTE can be mitigated by simply adjusting the number of coupled resonators and, thus, the distance span of DCR configurations. Similarly to the PTE theoretical maximization of Figure 3, the PTE maximization occurs when the distance span is equal to the magnetic field maximization distance that is equal to $\sqrt{2}a/2$ or approximately 110 mm, when DCR configuration scenario properties are considered. Since the distance span and the transfer distance are known, this entails that the number of coupled resonators is approximately equal to 5 and 8 for DCR configuration scenarios A and B, respectively. The respective PTE theoretical maximization curves for the two DCR configuration scenarios are also depicted in Figure 6. Note that slight divergences in Figures 3 and 6 between numerical and theoretical results come from the cumulative magnetic field effect of domino arrangements of coupled resonators in contrast with the single resonator coil consideration.

As it is observed from Figure 6, apart from the concativity of PTE curves with respect to the number of coupled resonators, another interesting remark of these curves is the existence of fluctuations on them. In general, an HF-band DCR configuration presents impedance variations because its mutual coupling coefficients significantly change with respect to the distance span; as the distance span decreases, the mutual coupling coefficients averagely increase [123, 124].

Actually, the effect of impedance mismatch averagely increases with respect to the number of coupled resonators. This is due to the fact that the reflected impedance presents important changes depending on the coupling degree that is related to the number of coupled resonators. Therefore, the coupling degree fluctuations cause respective fluctuations on PTE curves. Indeed, based on (6) and assumption 1 of Section 3.2, this PTE behavior is justified because it mainly depends on the number of coupled resonators and on impedance mismatches, which are incorporated into the theoretical model by the presence of source and load impedances.

Comparing Figures 3 and 6, an interesting trade-off between the coil radius of coupled resonators and the number of coupled resonators can be proposed. More specifically, in Figures 7(a) and 7(b), PTE is plotted versus the coil radius of coupled resonators and the distance span of coupled resonators for a given transfer distance when DCR configuration scenarios A and B are examined, respectively. In fact, the mutual coupling decreases when lower coil radii and larger distance spans are applied. This behavior implies a corresponding decrease in PTE performance of DCR configurations. Thus, for given coil radius, PTEs that are greater than 80% can only be achieved if the distance span is slightly larger, equal or slightly smaller than the coil radius ($d_{SPAN}/a \in [0.75\ \ 1.25]$). At very large air gaps ($d_{SPAN}/a \gg 1$), PTE rapidly decreases permitting only a low amount of power to be delivered or a low duty cycle operation to be performed. Nowadays, this low PTE operation, which is also called high-Q coupling, is often adopted at practical today's WPT systems with high number of turns of their coupled resonator coils [2, 3, 52].

In pure electromagnetic terms, PTE behavior can be explained through the analysis of the coupled resonator electric field, namely [54, 125]: (i) *the distance span is smaller than the coil radius*: the electric field follows a Hankel-function form. This function is characterized by its initial exponential-like regime and its long tails compared to the coil radius; (ii) *the distance span is higher than the coil radius*: the function of electric field is characterized by its oscillatory/radiation regime. This entails that power is slowly leaking out of the resonator coil.

Consequently, the distance span can be selected so as to be either shorter or larger than the common coupled resonator coil radius. Nonradiative energy between two coupled resonators can be transferred through the coupling of their resonant-field evanescent tails in both cases [55]. The crucial decision is to select an efficient combination of distance span and wire radius based on practical requirements in order to balance PTE and DCR configuration practicability [58].

4.1.5. Operating Frequency. The nonradiative MRC approach, which is followed in DCR configurations, presents serious potential to deliver power more efficiently in comparison with the traditional lower frequency inductive systems [44, 54, 67, 103, 126–129]. In accordance with a plethora of recent papers and various experimental verifications [3, 36, 54, 68, 126], better PTE performance can be achieved through the use of higher frequencies at kHz or several MHz range.

In Figure 8, PTE is plotted versus the operating frequency for the DCR configuration scenarios A and B. In the same figure, PTE of the two DCR configuration scenarios is also drawn as well as the corresponding extended set of measurements when their self-resonance frequency is considered—that is, $f_R = 530\,\text{kHz}$ [62].

In Figures 9(a) and 9(b), PTE is plotted versus the operating frequency and distance span when DCR configuration scenarios A and B is applied, respectively.

Observing Figures 8, 9(a) and 9(b), several interesting remarks can be pointed out.

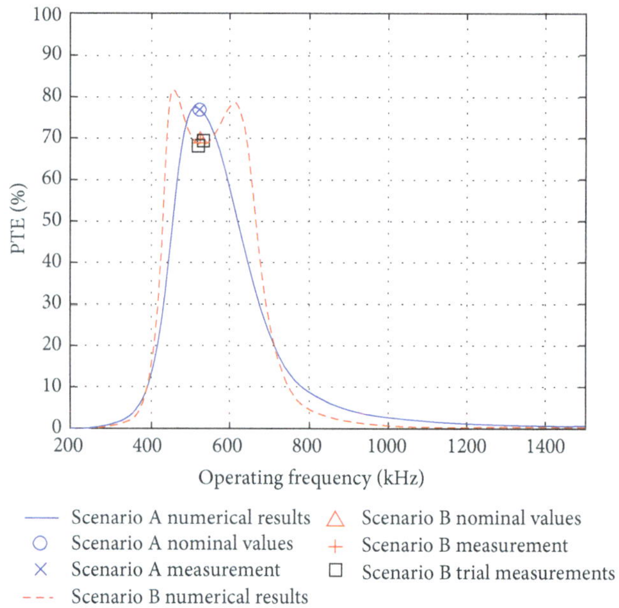

FIGURE 8: PTE versus operating frequency for two different DCR configuration scenarios.

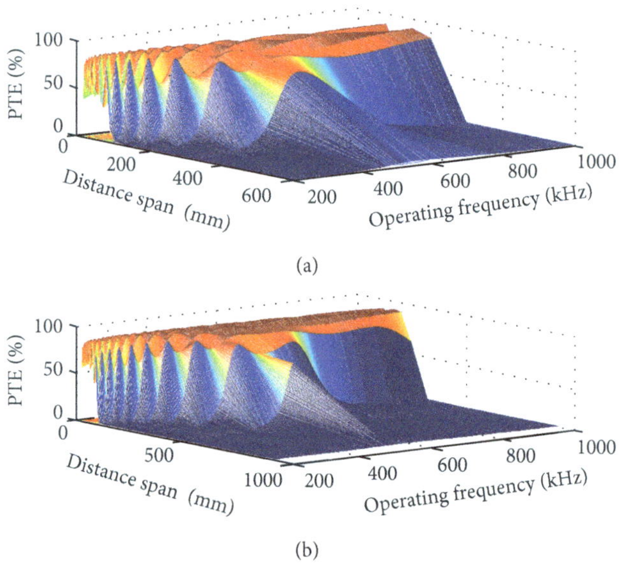

FIGURE 9: PTE versus operating frequency and distance span. (a) DCR configuration scenario A. (b) DCR configuration scenario B.

(i) In the case of traditional WPT systems that use resonant inductive coupling method, operation frequency is fixed with respect to their transfer distance [57]. On the contrary, for the WPT systems that follow MRC method such as DCR configurations studied in this paper, resonant frequency varies with respect to their transfer distance. However, during practical implementations of WPT systems, the available operation frequencies are imposed by the regulation authorities. Hence, in order to maintain resonant frequency stable, winding compensating capacitances should be adaptively tuned.

(ii) Resonant frequency of DCR configurations may differ from the self-resonant frequency of their coupled resonators. In fact, resonant frequency is obtained when the resonant frequency of the odd and the even mode is already known (the two PTE maxima of DCR configuration scenario B of Figure 8) [57].

(iii) As the distance span between coupled resonators increases, the respective coupling decreases. This has as result that the frequency separation that describes the difference between odd and even modes also decreases. This frequency separation reduction continues till the two maxima converge at one (see the sole PTE maximum of DCR configuration scenario A of Figure 8). The operating frequency where the PTE maximization occurs is denoted as the critical coupling point and represents the farthest distance at which maximum PTE still remains achievable—see Figures 9(a) and 9(b) [3].

(iv) When several coupled resonators are deployed and the distance span becomes very small compared to the coupled resonator coil radius, multimode operation is exited in DCR configurations. This operation type is characterized by its great number of PTE maxima at low distance spans in the examined DCR configuration scenarios—see Figures 9(a) and 9(b).

(v) PTE can be improved over a wider range of coupled resonator properties if operating frequency is opted with the other DCR configuration properties. It is noted that the optimal operation frequency can be higher or lower than the resonance frequency depending on DCR configuration properties [63]. Anyway, slight shift from the optimal operation frequency could either increase or decrease PTE critically.

(vi) Combining Figures 7(a) and 7(b) with Figure 8, it can be reported that if the distance span is smaller than the coil radius then DCR configuration operation is considered as overcoupled and it results in maximum PTE. Conversely, when the distance span is higher than the coil radius, DCR configuration operation can be considered as undercoupled having as a result that the amount of power delivered to the load precipitously attenuates with distance (see also Figure 6). With reference to [3], near-constant PTE can be achieved versus distance if the suitable operation frequency is successfully identified for given distance span.

(vii) PTE differences between numerical results and experimental measurements are attributed to the used mathematical formulations and the occurred slight divergences in DCR configuration scenario parameters used in order to determine the subcomponents of coupled resonators.

4.1.6. Source and Load Impedances. Already mentioned in Section 4.1.4, as the number of coupled resonators increases for a given transfer distance, the greater the coupling coefficient and the impedance mismatch become. Conversely, with reference to Figure 8, the profound cause of the existence of more-than-one resonance frequencies, which permits multiple maximum PTEs over very short distance spans, is

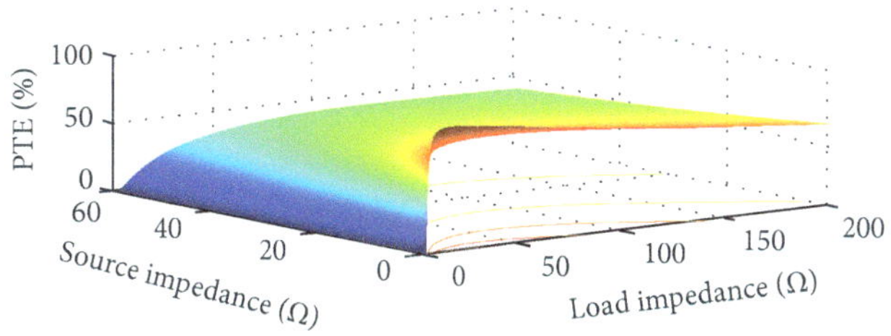

(a)

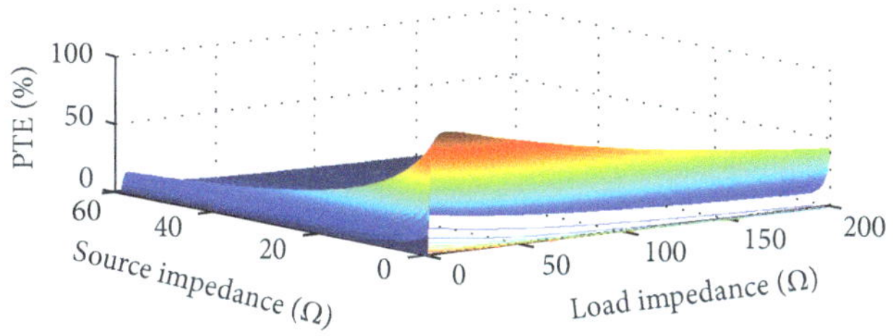

(b)

FIGURE 10: PTE versus source and load impedances. (a) DCR configuration scenario A. (b) DCR configuration scenario B.

the presence of impedance mismatches in DCR configurations [123, 130, 131]. Therefore, PTE of DCR configurations can be optimized by simply regulating impedance mismatches either in source or in load coupled resonators.

In Figures 10(a) and 10(b), PTE is plotted versus the source and load impedances when DCR configuration scenarios A and B are applied, respectively.

From Figures 10(a) and 10(b), certain thoughts can be expressed, namely, as follows.

(i) Similarly to other communications systems, the concept of WPT impedance matching, that is, the adjustment of the ratio between source and load impedance, is conducted using variable capacitances—such as winding compensating capacitances of Figure 2(a)—and inductances in order to maintain high PTE [25].

(ii) In the majority of today's WPT systems, their reported PTE is between 30% and 50% [4, 132]. Recently, interesting feedback approaches that study the effect of loading on maximizing the PTE of WPT systems have been proposed improving PTE performance—that is, PTE can range above 70% [69, 133–135].

(iii) By introducing the concept of impedance matching, new parameters concerning the design of DCR configurations are added. Now, the load problem is beneficially decoupled from the optimization process of the remaining DCR configuration parameters [133, 136]. Actually, there exists an optimum combination of source and load impedances for which PTE is maximized but resorted to numerical methods to find this optimum set. The presented approach which verifies the findings of the theoretical analysis—see (3)—achieves maximum PTEs greater than 80% offering significant performance advantage in practical DCR configuration realizations.

TABLE 3: Design parameters and the two DCR configuration scenarios: PTE and nominal and optimal parameter values.

	Scenario A		Scenario B	
	Nominal values	Optimal values	Nominal values	Optimal values
Radius of coupled resonator coils (Section 4.1.1)/PTE	155 mm/76.38%	230 mm/79.24%	155 mm/68.53%	272/90.47%
Number of turns of coupled resonator coils (Section 4.1.2)/PTE	11/76.38%	18/78.59%	11/68.53%	25/90.27%
Radius of the used wires in coupled resonator coils (Section 4.1.3)/PTE	1.2 mm/76.38%	5 mm/81.45%	1.2 mm/68.53%	5 mm/75.06%
Number of coupled resonator coils (Section 4.1.4)/PTE	3/76.38%	11/90.27%	4/68.53%	7/90.81%
Operating frequency (Section 4.1.5)/PTE	530 kHz/76.38%	521 kHz/76.74%	530 kHz/68.53%	466 kHz/81.34%
Source and load impedance (Section 4.1.6)/PTE	3.45 Ω, 17 Ω/76.38%	0 Ω, 12.6 Ω/92.32%	1.42 Ω, 17 Ω/68.53%	0 Ω, 19.3 Ω/88.95%

(iv) A necessary attribute that characterizes efficient DCR configurations is the ability to operate when a multitude of distance spans and frequencies occur without the need of precise manual tuning in terms of source and load impedances. Similarly to the automatic frequency tuning [3], it is possible to design systems that are automatically adjusted in order to provide maximum PTEs as DCR configurations are expanded within the system's working range.

4.1.7. Discussion on Parameters That Affect PTE of DCR Configurations. The aforementioned analysis has identified the most important design parameters that influence PTE of practical DCR configurations. In Table 3, these parameters as well as their PTE performance are synopsized in the case of considered DCR configuration scenarios A and B when nominal values of Tables 1 and 2 and optimal values of Sections 4.1.1–4.1.6 are adopted.

Examining Table 3, it is clearly shown that DCR configuration parameters of scenario A are very well adjusted whereas scenario B ones are adequately tuned. Anyway, in all the cases examined, their PTEs are relatively close to the optimal values. Therefore, due to these favorable characteristics, these two DCR configurations can be efficiently used in the following theoretical analyses where practical DCR configurations are required to be installed.

Moreover, as it has already been identified and mentioned in [64], the agreement between the theoretical analysis and experimental data is excellent. This result shows that PTE of DCR configurations can be accurately predicted and optimized through the developed models without the need of further experimentation.

4.2. Practical DCR Configurations and Approximate CSM Method. Until now, the theoretical analysis presented in Sections 3.1–3.3 and its respective numerical results demonstrated in Sections 4.1.1–4.1.7 have accurately described PTE behavior of DCR configurations. However, the main problem of the theoretical analysis is the size of the used mutual inductance matrices that hinders its practical implementation when mid- and long-range practical DCR configurations are going to be deployed [40, 52, 60, 62–64]. To save computational resources and time, there is need of a model, such as the approximate CSM method, that gives quick and accurate numerical results for mid- and long-range DCR configurations.

Based on (16a)–(16c), in order to accurately apply the approximate CSM method, the following two-step procedure should be followed.

(1) *Initialization Process.* On the basis of (16b), the initialization process outlines the contribution of source and load modules to PTE. Having applied the theoretical analysis of Section 4.1.4 for $n = 2$, $\mathrm{PTE}_{\mathrm{approx},2}$ is assumed equal to the theoretical PTE.

(2) *Coupled Resonator Contribution.* During this phase, the contribution of coupled resonators to PTE is computed. On the basis of (16a), having applied the theoretical analysis of Section 4.1.4 for n coupled resonators, $\mathrm{PTE}_{\mathrm{approx},n}$ is assumed equal to the theoretical PTE. Since $\mathrm{PTE}_{\mathrm{approx},2}$ and $\mathrm{PTE}_{\mathrm{approx},n}$ are known, the additional PTE degradation due to the insertion of one new DCR module $|T'_{i,11}|^{-2}$ is straightforwardly computed from (16a).

The accuracy of the approximate CSM method is validated through the comparison of its numerical results with the respective ones of the theoretical analysis of Sections 3.1–3.3. Towards that direction, DCR configuration scenario A, whose properties are detailed in Tables 1 and 2, is only considered. Due to configuration and numerical result similarities with DCR configuration scenario A, DCR configuration scenario B is neglected, hereafter.

Across this paper, as it concerns DCR configuration scenario A properties, the number of coupled resonators and the distance span are redefined so that the accuracy of the approximate CSM method will be examined. Note that, in this subsection, the transfer distance of DCR configurations, which are based on DCR configuration scenario A properties, is variable and depends on the number of coupled resonators and the distance span through

$$d = (n-1) \cdot d_{\mathrm{SPAN}}. \tag{17}$$

More specifically, in Figures 11(a) and 11(b), PTE is plotted versus the number of coupled resonators and distance

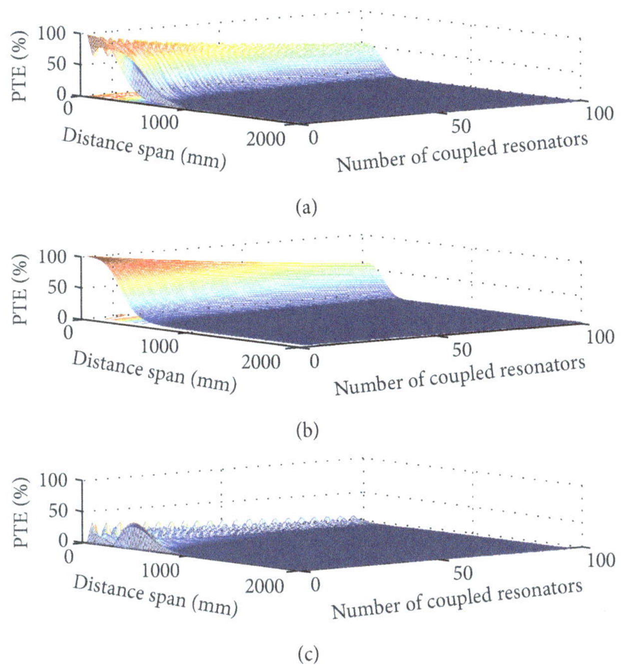

Figure 11: PTE versus the number of coupled resonators and distance span for practical DCR configurations. (a) Theoretical analysis. (b) Approximate CSM method. (c) Absolute difference between theoretical analysis and approximate CSM method.

span when the theoretical analysis of Sections 3.1–3.3 and the approximate CSM method is applied, respectively. In Figure 11(c), the absolute PTE difference between these two methodologies is drawn.

From Figures 11(a)–11(c), it is observed that the theoretical analysis and the approximate CSM method coincide when the number of coupled resonators and/or the distance span are large enough whereas they yield very close results in the case of small DCR configurations (i.e. few coupled resonators, small distance spans). Thus, the approximate CSM method describes PTE accurately either in depth or in extent in the mid- and long-range DCR configurations that are the configurations of interest.

In numerical terms, when practical DCR configurations are deployed, PTE ranges from 4.56×10^{-15}% to 98.33% and from 3.43×10^{-15}% to 97.67% in the case of theoretical analysis and approximate CSM method, respectively. Actually, in the 95% of the practical DCR configurations examined in Figures 11(a)–11(c), the absolute difference between these two methodologies is below 4.92%. Thus, the approximate CSM method offers an accurate representation of the PTE while it is easily adapted to various DCR configurations. The approximate CSM method will be adopted in the rest of the simulations.

Finally, the huge set of available PTE values gives a lot of arrows in the quiver of DCR configuration designers letting a myriad of potential applications to be developed. Some applications among them that are related with power transmission and distribution are going to be highlighted in the following analysis.

4.3. Power Transformers and Practical DCR Configurations. Today's power grids need to become smarter so as to provide an affordable, reliable, and sustainable power supply [80]. Recently, considerable activity has been carried out in order to guide the development of future's smart power grids. However, the majority of these activities emphasizes only the construction, renewable sources and communications side leaving outside the frame state-of-the-art concepts such as WPT technology. This subsection highlights a new vision for the future's smart grids in which the replacement potential of HV/MV, MV/LV and HV/LV transformers (and part of their transmission line branches) is identified. In this vision, each power transformer can be regarded as a practical DCR configuration that functionally consists of a cascade of DCR modules with source voltage equal to power transformer input voltage. The features and functions of DCR configurations, as well as the fertile ground where these features and functions are enabled, are discussed in this subsection.

4.3.1. Overview of HV, MV, and LV Transmission Lines and Their Respective Power Transformers. As it has already been mentioned in Section 2.1, the power grid is mainly divided into three major functions (i.e., power generation, transmission, and distribution) that mainly support three (voltage level) subnetworks, namely [24, 25, 43, 62, 86, 87, 137–139], the following.

(1) *HV Subnetwork.* The ten most typical transmission lines of this subnetwork around the world are those of 1000 kV, 765 kV, 500 kV, 400 kV, 230 kV, 220 kV, 132 kV, 110 kV, 66 kV, and 50 kV.

(2) *MV Subnetwork.* The ten most common types of transmission lines of this subnetwork around the world are those of 35 kV, 33 kV, 22 kV, 15 kV, 11 kV, 10 kV, 6.6 kV, 6 kV, 3.3 kV, and 2.2 kV.

(3) *LV Subnetwork.* The three most typical transmission lines of this subnetwork around the world are those of 400 V, 380 V, and 220 V.

In accordance with Section 2.2 and taking under consideration the voltage levels of the above three subnetworks, three main categories of step-down transformers occur.

(i) *HV/MV and MV/LV Power Transformers.* HV/MV and MV/LV power transformers define the "voltage bridge" between their input—HV and MV, respectively—and their output—MV and LV, respectively—subnetworks. Their TRs depend on these input and output subnetworks.

(ii) *HV/LV power transformers.* Similarly to these two main categories and as it has already been mentioned in Section 2.2, in rare cases of power delivery, HV/LV power transformers can be deployed constituting the third main category of step-down transformers.

Depending on their TRs, $10 \times 10 = 100$ HV/MV, $10 \times 3 = 30$ MV/LV, and $10 \times 3 = 30$ HV/LV power transformers are examined in the rest of this paper. Their input and output voltage levels as well as their respective TRs are reported in Table 4.

Table 4: Operation characteristics of HV/MV, MV/LV, and HV/LV power transformers.

IV	OV	TR	IV	OV	TR	IV	OV	TR	IV	OV	TR	IV	OV	TR
HV/MV power transformers														
1000 kV	35 kV	3.5%	1000 kV	33 kV	3.3%	1000 kV	22 kV	2.2%	1000 kV	15 kV	1.5%	1000 kV	11 kV	1.1%
1000 kV	10 kV	1%	1000 kV	6.6 kV	0.66%	1000 kV	6 kV	0.6%	1000 kV	3.3 kV	0.33%	1000 kV	2.2 kV	0.22%
765 kV	35 kV	4.58%	765 kV	33 kV	4.31%	765 kV	22 kV	2.88%	765 kV	15 kV	1.96%	765 kV	11 kV	1.44%
765 kV	10 kV	1.31%	765 kV	6.6 kV	0.86%	765 kV	6 kV	0.78%	765 kV	3.3 kV	0.43%	765 kV	2.2 kV	0.29%
500 kV	35 kV	7%	500 kV	33 kV	6.6%	500 kV	22 kV	4.4%	500 kV	15 kV	3%	500 kV	11 kV	2.2%
500 kV	10 kV	2%	500 kV	6.6 kV	1.32%	500 kV	6 kV	1.2%	500 kV	3.3 kV	0.66%	500 kV	2.2 kV	0.44%
400 kV	35 kV	8.75%	400 kV	33 kV	8.25%	400 kV	22 kV	5.5%	400 kV	15 kV	3.75%	400 kV	11 kV	2.75%
400 kV	10 kV	2.5%	400 kV	6.6 kV	1.65%	400 kV	6 kV	1.5%	400 kV	3.3 kV	0.83%	400 kV	2.2 kV	0.55%
230 kV	35 kV	15.2%	230 kV	33 kV	14.3%	230 kV	22 kV	9.57%	230 kV	15 kV	6.52%	230 kV	11 kV	4.78%
230 kV	10 kV	4.35%	230 kV	6.6 kV	2.87%	230 kV	6 kV	2.61%	230 kV	3.3 kV	1.43%	230 kV	2.2 kV	0.96%
220 kV	35 kV	15.9%	220 kV	33 kV	15%	220 kV	22 kV	10%	220 kV	15 kV	6.82%	220 kV	11 kV	5%
220 kV	10 kV	4.55%	220 kV	6.6 kV	3%	220 kV	6 kV	2.73%	220 kV	3.3 kV	1.5%	220 kV	2.2 kV	1%
132 kV	35 kV	26.5%	132 kV	33 kV	25%	132 kV	22 kV	16.7%	132 kV	15 kV	11.4%	132 kV	11 kV	8.33%
132 kV	10 kV	7.58%	132 kV	6.6 kV	5%	132 kV	6 kV	4.54%	132 kV	3.3 kV	2.5%	132 kV	2.2 kV	1.67%
110 kV	35 kV	31.8%	110 kV	33 kV	30%	110 kV	22 kV	20%	110 kV	15 kV	13.6%	110 kV	11 kV	10%
110 kV	10 kV	9.09%	110 kV	6.6 kV	6%	110 kV	6 kV	5.45%	110 kV	3.3 kV	3%	110 kV	2.2 kV	2%
66 kV	35 kV	53%	66 kV	33 kV	50%	66 kV	22 kV	33.3%	66 kV	15 kV	22.7%	66 kV	11 kV	16.7%
66 kV	10 kV	15.2%	66 kV	6.6 kV	10%	66 kV	6 kV	9.09%	66 kV	3.3 kV	5%	66 kV	2.2 kV	3.33%
50 kV	35 kV	70%	50 kV	33 kV	66%	50 kV	22 kV	44%	50 kV	15 kV	30%	50 kV	11 kV	22%
50 kV	10 kV	20%	50 kV	6.6 kV	13.2%	50 kV	6 kV	12%	50 kV	3.3 kV	6.6%	50 kV	2.2 kV	4.4%
MV/LV power transformers														
35 kV	400 V	1.14%	35 kV	380 V	1.09%	35 kV	220 V	0.63%	33 kV	400 V	1.21%	33 kV	380 V	1.15%
33 kV	220 V	0.67%	22 kV	400 V	1.82%	22 kV	380 V	1.73%	22 kV	220 V	1%	15 kV	400 V	2.67%
15 kV	380 V	2.53%	15 kV	220 V	1.47%	11 kV	400 V	3.64%	11 kV	380 V	3.45%	11 kV	220 V	2%
10 kV	400 V	4%	10 kV	380 V	3.8%	10 kV	220 V	2.2%	6.6 kV	400 V	6.06%	6.6 kV	380 V	5.76%
6.6 kV	220 V	3.33%	6 kV	400 V	6.67%	6 kV	380 V	6.33%	6 kV	220 V	3.67%	3.3 kV	400 V	12.1%
3.3 kV	380 V	11.5%	3.3 kV	220 V	6.67%	2.2 kV	400 V	18.2%	2.2 kV	380 V	17.3%	2.2 kV	220 V	10%
HV/LV power transformers														
1000 kV	400 V	0.04%	1000 kV	380 V	0.04%	1000 kV	220 V	0.02%	765 kV	400 V	0.05%	765 kV	380 V	0.05%
765 kV	220 V	0.03%	500 kV	400 V	0.08%	500 kV	380 V	0.08%	500 kV	220 V	0.04%	400 kV	400 V	0.1%
400 kV	380 V	0.1%	400 kV	220 V	0.06%	230 kV	400 V	0.17%	230 kV	380 V	0.17%	230 kV	220 V	0.1%
220 kV	400 V	0.18%	220 kV	380 V	0.17%	220 kV	220 V	0.1%	132 kV	400 V	0.30%	132 kV	380 V	0.29%
132 kV	220 V	0.17%	110 kV	400 V	0.36%	110 kV	380 V	0.35%	110 kV	220 V	0.2%	66 kV	400 V	0.61%
66 kV	380 V	0.58%	66 kV	220 V	0.33%	50 kV	400 V	0.8%	50 kV	380 V	0.76%	50 kV	220 V	0.44%

Input voltage level: IV. Output voltage level: OV.

4.3.2. The Potential of Replacing HV/MV, MV/LV, and HV/LV Transformers with Practical DCR Configurations. Already mentioned in the literature [52, 64], a DCR configuration can be interpreted as an expansion of a power transformer, which consists of two coils. Actually, the equations of a power transformer are well-established in textbooks [76–78]. To promote the analysis, from (12), it is reminded that PTE can approximately be assumed equal to the square of the absolute value of its TR for a given power transformer.

Thus, studying Table 4, it is observed that the TR of power transformers ranges from 0.02% to 70%. However, from Section 4.2, it has been validated that practical DCR configuration PTE ranges from 3.43×10^{-15}% to 97.67%. This implies that practical DCR configuration TR approximately ranges from 5.86×10^{-9}% to 98.83%. Therefore, practical DCR configuration TR set defines a superset that includes the respective TR set of power transformers. In order to validate this important finding, in Figure 12(a), practical DCR configuration TR is plotted versus the number of coupled resonators and transfer distance when DCR configuration scenario A properties are adopted. In Figures 12(b) and 12(c), the minimum and maximum power transformer TR contour lines are curved, respectively.

From Figures 12(a)–12(c), it is evident that practical DCR configurations can affordably, reliably, and sustainably replace either HV/MV or MV/LV, or HV/LV power transformers during all their operation cycle since TR of practical DCR configurations overlaps the respective one of power transformers. However, WPT technology defines a new technology solution that demands its time and a lot of research efforts so as to become mature and efficient to substitute already well validated technologies such as power transformer one.

Furthermore, the numerical results of Figures 12(a)–12(c) demonstrate that efficient transfer paths of practical DCR configurations can be long enough reaching distances more

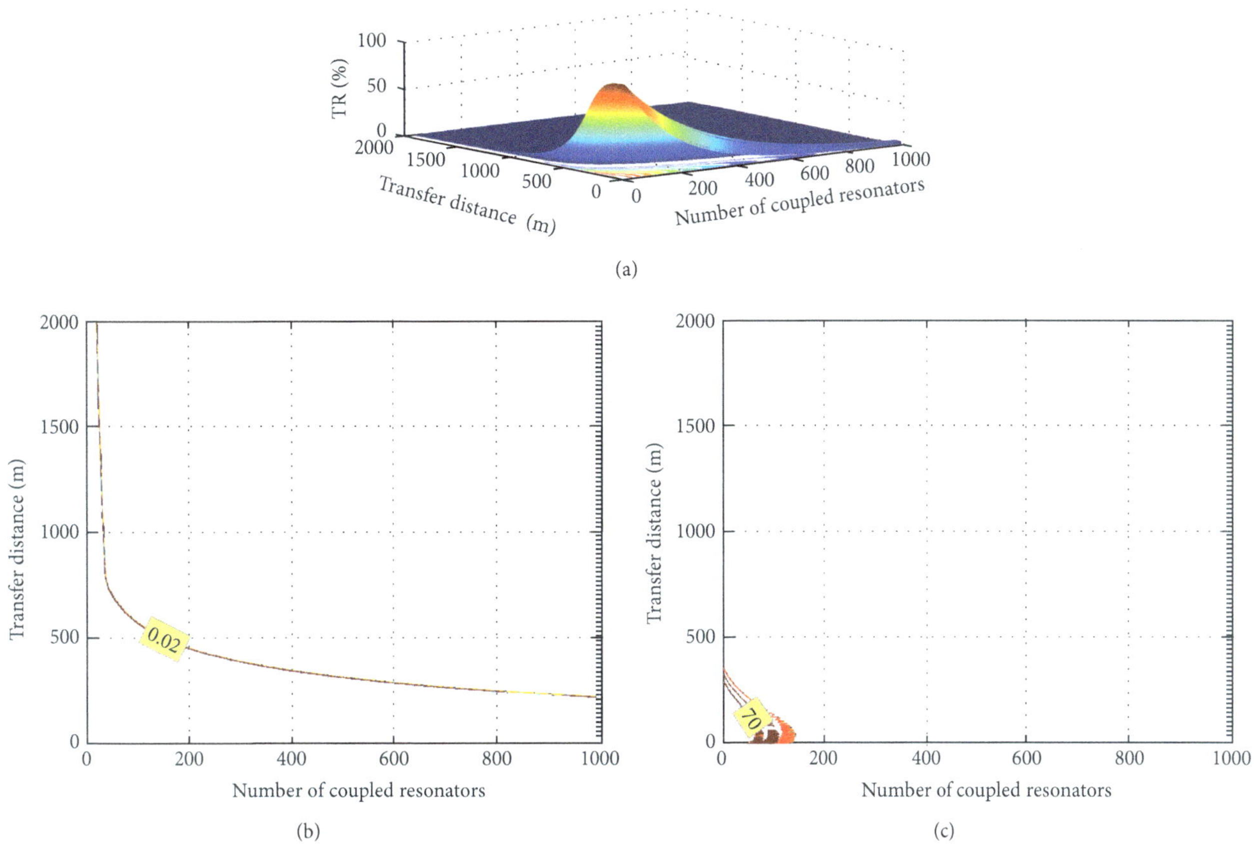

FIGURE 12: TR versus the number of coupled resonators and transfer distance. (a) Practical DCR configurations. (b) Minimum power transformer TR contour lines. (c) Maximum power transformer TR contour lines.

than 1 km by using only few coupled resonators. This order of distances does not only render acceptable the substitution of power transformers but can also permit the replacement of entire transmission line branches. Anyway, plots such as those of Figure 12(a) can operate as rule of thumb in order to design practical DCR configurations given the desired transfer distance and TR.

4.4. Last Cable Problem, Power Transformers, and Practical DCR Configurations. Nowadays, interest in the research and development of WPT technology has skyrocketed. The primary objective of a possible widespread adoption of WPT systems is the solution of last cable problem. To solve this problem, WPT networks can be mixed with the traditional power grid infrastructure. Anyway, last mile and last cable problems as well as their integration open new horizons in the sciences of electrical engineering and sociology.

Prior to investigate the capabilities of practical DCR configurations in terms of minimum and maximum supported WPT distances and emphasize the last cable problem solution, an overview of the parameters that define this problem is given: the lengths of transmission line branches and the dimensions of power transformers.

4.4.1. Exploring the Parameters of Last Cable Problem: Lengths of Transmission Line Branches and Dimensions of Power Transformers. On the basis of the subnetwork taxonomy of Section 4.3.1, the characteristics of the respective transmission lines are the following [24, 25, 43, 62, 86, 87, 137–146].

(1) *HV Subnetwork.* Overhead HV transmission lines generally define the lowest cost delivery method for large bulks of electric power. Utilities mainly employ overhead HV subnetwork for new urban, suburban, and rural installations. Today, thousands of km of overhead HV transmission lines are installed in more than 120 countries. In general, they stretch from approximately 25 km to 190 km from the generation points before reaching critical load destinations. Shorter branches in the range of 10 km to 50 km are used in order to connect overhead HV transmission lines either between them or with HV/MV transformers.

(2) *MV and LV Subnetwork.* Utilities employ either the overhead or the underground distribution power grid for their new urban, suburban, and rural MV and LV installations. Their choice depends on different criteria such as cost requirements, existing grid topology, and plan constraints. Today, in urban areas, about 45%–75% of the underground transmission lines (either MV or LV ones, either main or branch ones) exhibit path lengths shorter than 200 m. Only 1%–9% of them present lengths longer than 500 m. These path

lengths are quite shorter compared to the overhead transmission where path lengths up to 1000 m may be encountered. Actually, in suburban and rural environments, these lengths are significantly longer.

As it regards power transformers, their dimensions are determined by the input voltage, the output voltage, and the supported capacity rating. Among these three factors, the capacity rating often is the key parameter rather than input and output voltages. Based on the aforementioned subnetwork classification, the typical dimensions of the used power transformers are the following [141, 145–147].

(1) *HV/MV and HV/LV Power Transformers.* Due to their high capacity rating and high input voltage, their dimensions are significantly greater than the respective ones of MV/LV transformers. Their typical length, width and height range from 2650 mm to 4400 mm, from 2250 mm to 4250 mm and from 2750 mm to 4900 mm, respectively.

(2) *MV/LV Power Transformers.* Except for the various types of transmission lines, MV/LV power transformers define the most common component of today's power grid. There are different types of MV/LV transformers depending on the use. Their typical length, width and height range from 670 mm to 2400 mm, from 600 mm to 1450 mm and from 885 mm to 2585 mm, respectively.

4.4.2. Approaching the Solution of Last Cable Problem. WPT through DCR configurations can be categorized into two application classes, namely, as follows.

(i) *Short- and Mid-Range Application Class.* To date, this class has high promises as an equipment replacement technology and seems to be particularly suitable for daily and/or urban applications. In fact, the interactions of short- and mid-range DCR configurations with external resonators are reduced due to their simple DCR configuration structure [44, 54, 103, 122, 148]. In order to deliver enhanced power delivery services in short- and mid-range application class, few coupled resonators and/or small distance spans are required.

(ii) *Long-Range Application Class.* This class presents limited implementation potential for consumer applications and urban environments due to the high transferred power and long DCR configurations that are necessary to achieve the levels of power comparable to a power transformer. In order to deliver enhanced power delivery services in long-range applications, several coupled resonators and relatively larger distance spans in comparison with the short- and mid-range application class are required from the installed DCR configurations. Through this class and depending on the subnetwork, DCR configurations can replace both power grid equipment and part of its transmission line branches. In that sense, suburban and rural areas define the most appropriate installation environments, where long-range application class can be applied across.

Table 5: Minimum, median, and maximum TR of HV/MV, MV/LV, and HV/LV power transformers.

Minimum			Median			Maximum		
IV	OV	TR	IV	OV	TR	IV	OV	TR
HV/MV power transformers								
1000 kV	2.2 kV	0.22%	220 kV	10 kV	4.55%	50 kV	35 kV	70%
MV/LV power transformers								
35 kV	220 V	0.63%	6.6 kV	220 V	3.33%	2.2 kV	400 V	18.2%
HV/LV power transformers								
1000 kV	220 V	0.02%	230 kV	400 V	0.17%	50 kV	400 V	0.8%

Input voltage level: IV. Output voltage level: OV.

To investigate the applicability of mid- and long-range application classes and, hence, their contribution to the solution of last cable problem, HV/MV, MV/LV, and HV/LV power transformers are treated separately in the following simulation results. On the basis of Table 4, the minimum, the median and the maximum TR values of HV/MV, MV/LV, and HV/LV power transformers are reported in Table 5. In the same Table, the corresponding input and output voltage levels are also reported.

In Figure 13(a), the transfer distance is plotted versus the number of coupled resonators given the minimum TR of HV/MV power transformers when DCR configuration scenario A properties are adopted. In the same Figure, with respect to Section 4.4.1, the minimum and maximum dimensions of HV/MV transformers are also given. In Figures 13(b) and 13(c), same curves with Figure 13(a) are drawn but for the median and maximum TR, respectively. In Figures 14(a)–14(c) and 15(a)–15(c), same plots with Figures 13(a)–13(c) are curved in the case of MV/LV and HV/LV transformers, respectively.

Observing Figures 13(a)–13(c), 14(a)–14(c), and 15(a)–15(c), several interesting conclusions regarding the interaction between WPT technology and last cable problem can be deduced, more specifically as follows.

(1) Depending on the required TR, there is a plethora of available DCR configurations and corresponding transfer distances. Given the TR and the desired transfer distance, the exact number of coupled resonators can straightforwardly be determined for all the types of today's step-down power transformers.

(2) In general, transfer distance of DCR configurations is a concave function with respect to the number of their coupled resonators for a given TR. Combined with the knowledge of Figure 6, this concave behavior implies that the same transfer distance result can be achieved through two different DCR configurations in the majority of the transfer distance cases. This finding is very useful since the suitable selection among DCR configurations may differentiate depending on frequency and EMI constraints that are imposed by the authorities, the safety operation limitations with

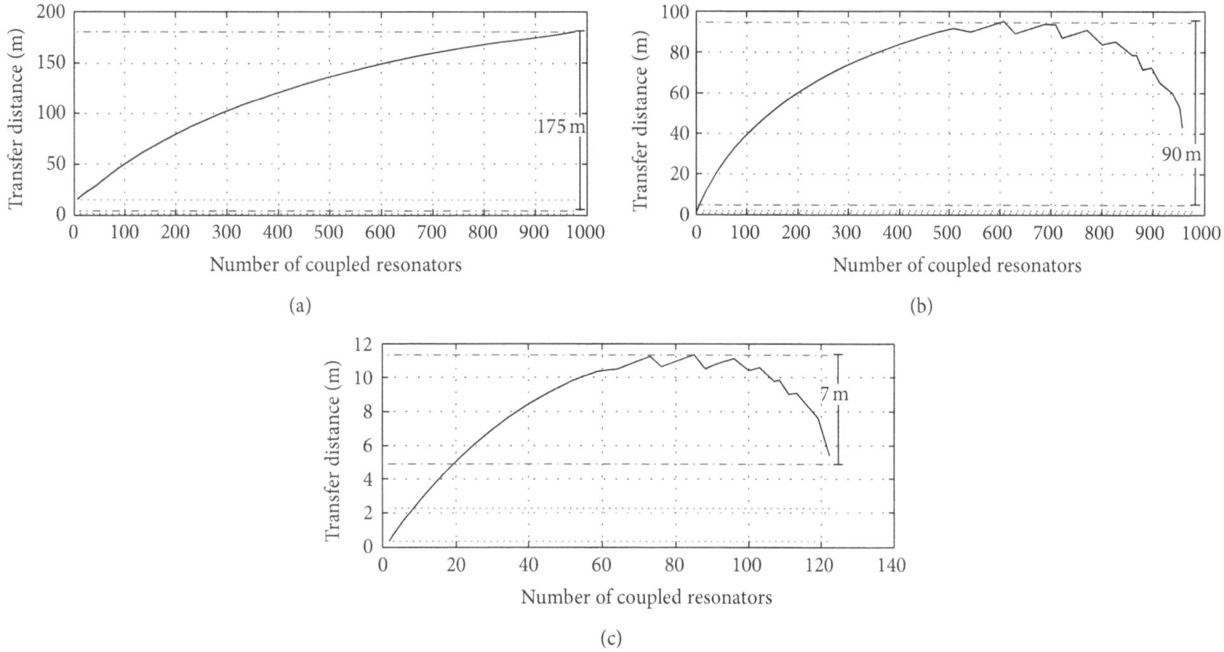

FIGURE 13: Transfer distance versus the number of coupled resonators when DCR configuration scenario A properties (—) are applied for various TRs as well as the minimum (···) and maximum (-·-·) dimensions of HV/MV power transformers. (a) Minimum HV/MV power transformer TR. (b) Median HV/MV power transformer TR. (c) Maximum HV/MV power transformer TR.

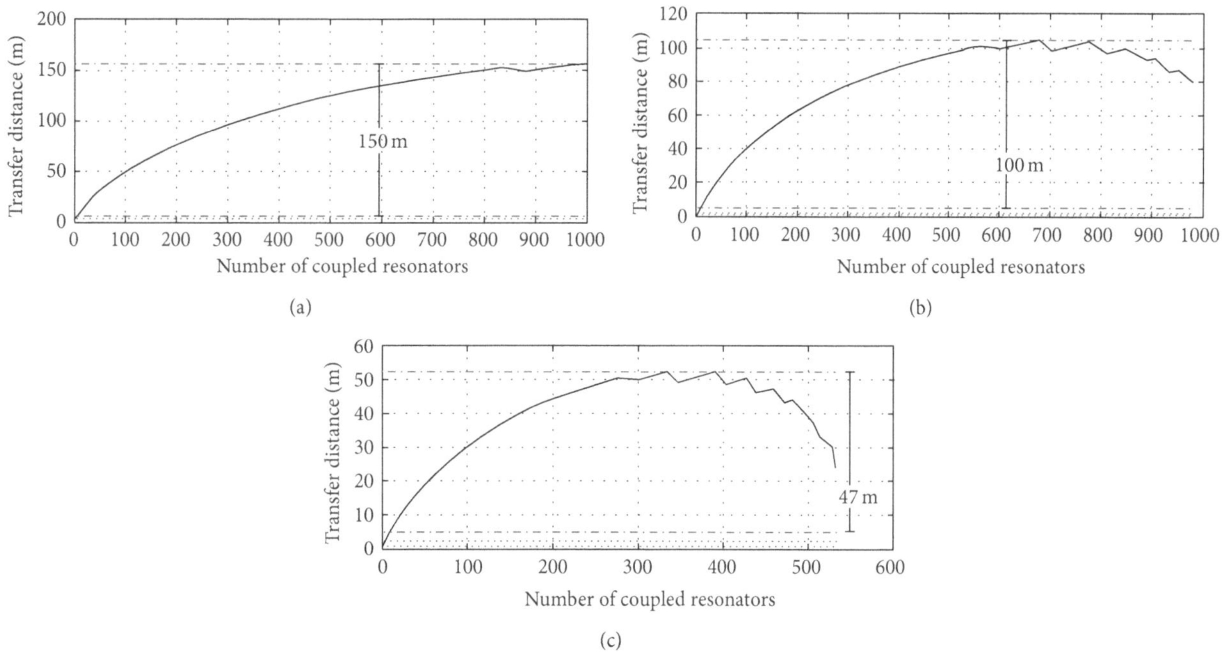

FIGURE 14: Same as Figure 13 but for MV/LV power transformers.

respect to the human presence and health, other application constraints, and so forth.

(3) The transfer distance range primarily depends on the required TR. Lower TRs—for example, HV/MV and HV/LV power transformers—permit longer maximum DCR configuration transfer distances whereas higher TRs—for example, MV/LV power transformers—demand shorter ones. The same results apply in the case of minimum transfer distances.

(4) In all the cases examined, the dimensions of power transformers define small subsets of the corresponding DCR configuration transfer distance sets. Therefore, all the types of today's step-down power

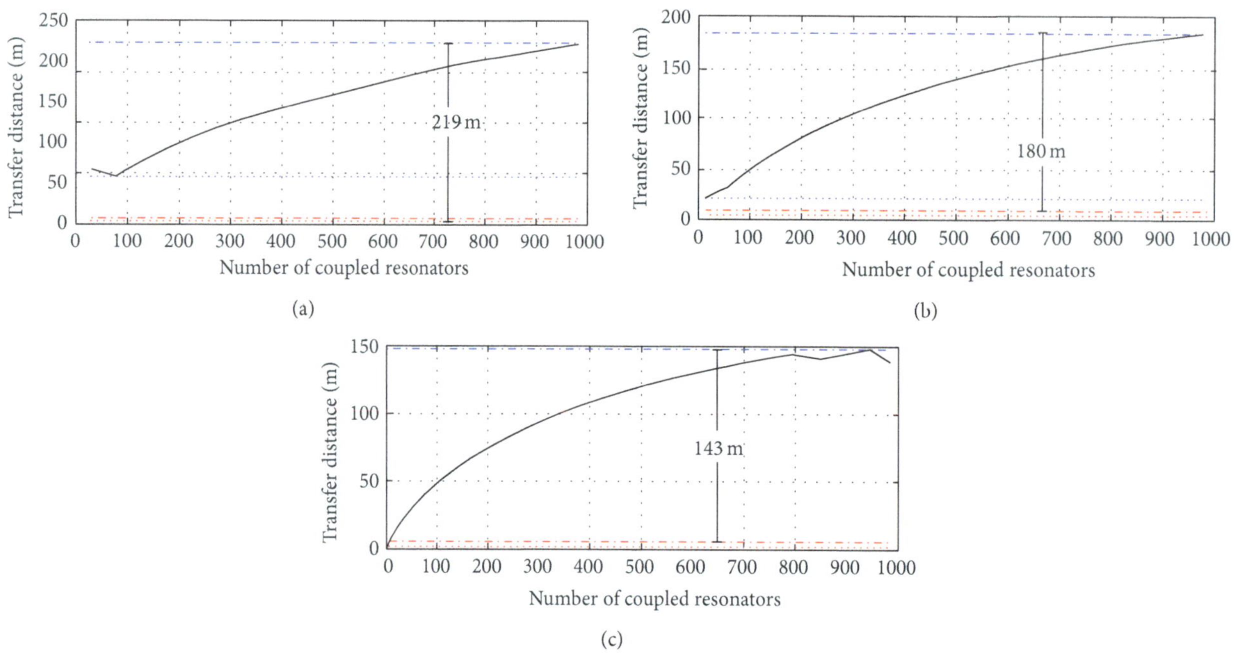

FIGURE 15: Same as Figure 13 but for HV/LV power transformers.

transformers can potentially be replaced by deploying a little number of coupled resonators without, however, exceeding their existing dimensions; say, HV/MV, MV/LV and HV/LV transformers can be replaced with 4–20, 2–11 and 2–6 coupled resonators when DCR configuration scenario A properties are adopted.

(5) Despite the great potential of power transformer replacement, DCR configurations can straightforwardly help towards a partial solution of last cable problem. In fact, in all the cases examined, power transformer maximum dimensions are significantly lower than the maximum transfer distance supported by DCR configurations. On average, this difference between these two maxima is equal to 90 m, 100 m and 180 m in the case of HV/MV, MV/LV, and HV/LV power transformers, respectively. Therefore, except for power transformers, DCR configurations can effectively replace part of their existing branch wiring.

(6) Combining the previous findings with the reports of Section 4.4.1, it is obvious that practical DCR configurations can replace entire transmission line branches in the case of MV and LV subnetworks whereas they can substitute parts of HV subnetwork transmission line branches. In numerical terms and on average, this difference between these two maxima can range from 7 m to 175 m, from 47 m to 150 m and from 143 m to 219 m in the case of HV/MV, MV/LV and HV/LV power transformers, respectively, so that their surrounding transmission line branches can be replaced. Thus, the wiring replacement potential depends on the TR target, the dimensions of the transformer, and the transmission line branch lengths.

(7) According to Section 4.1.7 and Table 3, DCR configuration parameters of scenario A are very well adjusted and, thus, its PTE and TR performance gives a representative image of the capabilities of today's WPT technology in mid- and long-range applications. However, there is enough room for improvements so that DCR configurations will be perfectly tuned and, finally, reveal their optimum capabilities concerning last cable problem solution.

5. Future Work

In reality, there is a long journey ahead for WPT technology and its state-of-the-art products to be affordably, widely, reliably, sustainably, and safely adopted in the society, replacing some of the traditional and well-proven wired systems. However, during these preliminary stages of WPT technology development, theoretical aspects and advancements, visions and proofs of concept are welcome and necessary, as well.

DCR configurations will accelerate practical adaptations of WPT technology in various sectors of daily activity such as power delivery, industry, medicine, transportation and consumer electronics [60, 149]. Especially, in environments where sparking and arching hazards may occur, DCR configurations can be the perfect solution and adopted so as to eliminate any external metallic contacts [67].

Finally, the need for power electronics that have greater compactness, better manufacturability and higher performance triggers the pursuit of dramatic increases in operating frequencies of DCR configurations. Realizing advantages such as miniaturization, more sophisticated power applications, improved PTE and TR performance, and better integration of the passive components of DCR configurations, however, requires devices, passive components and circuit designs that can operate efficiently at the necessary frequencies [150, 151]. Therefore, a series of interconnected and peripheral systems (e.g., rectifiers, cooling systems, source transistors, impedance matching systems, modulators, filters, input and output subsystems, e.t.c.) is required to be redesigned so that the operation optimization of practical DCR configurations will be ensured and, at the same time, the EMI of practical DCR configurations to the other already existing power and telecommunications systems will be reduced [3, 152].

6. Conclusions

In this paper, the replacement potential of step-down power transformers and their transmission line branches with DCR configurations has thoroughly been investigated.

Based on an exhaustive theoretical analysis as well as experimental and numerical assessment procedure, the important replacement potential of step-down power transformers with DCR configurations has been identified. Except for this critical verification, a significant number of secondary findings has been reported: first, the analytical theoretical analysis suitable for the study of DCR configurations has been presented and verified through numerical results and experimental measurements. In addition, it has been demonstrated that practical DCR configurations that have been used during numerical results and for power transformer replacements are well tuned concerning their PTE and TR performance in comparison with maximum achievable PTE and TR, respectively. Second, a great number of parameters that influences PTE and TR performance of DCR configurations has been recognized, namely, the radius of coupled resonator coils, the number of turns of coupled resonator coils, the radius of the used wires in coupled resonator coils, the number of coupled resonator coils, the operating frequency, and the source and load impedances. On the basis of these design parameters and the theoretical analysis outcomes, possible DCR configuration adjustments have been recognized so that the maximization of PTE and TR can occur. Third, the approximate CSM method has been proposed, applied and compared with the respective results of the theoretical analysis giving impressive converging results: in the 95% of the practical DCR configurations examined, the absolute TR difference between these two methodologies is below 4.92%. The main benefit of approximate CSM method is its computational speed in comparison with the respective one of the theoretical analysis when mid- and long-range DCR configurations are deployed. Fourth, in all 160 cases of HV/MV, MV/LV, and HV/LV power transformers used across the world, practical DCR configurations can comfortably replace all of them. Actually, HV/MV, MV/LV, and HV/LV power transformers can be replaced with 4–20, 2–11 and 2–6 coupled resonators when practical DCR configurations are deployed regardless of their power transformer TRs. Fifth, taking into account a thorough report concerning dimensions of transformers and transmission line branches, it has been verified that practical DCR configurations cannot only substitute all step-down power transformers of the today's power grid but can also replace entire transmission line branches too: in the case of HV/MV, MV/LV and HV/LV power transformers, DCR configurations can effectively replace 90 m, 100 m, and 180 m of their existing transmission line wiring, respectively.

Finally, WPT technology and DCR configurations have been proven to be theoretically feasible technological solutions for replacing power transformers and their corresponding wiring. However, in order to become affordable, wide, reliable, sustainable, and safe power grid components, WPT technology and DCR configurations require a lot of academic research, system interaction and business interest in the related research topics.

References

[1] N. Tesla, "Apparatus for transmitting electrical energy," US patent number 1,119,732, December 1914.

[2] D. Kurschner, C. Rathge, and U. Jumar, "Design methodology for high efficient inductive power transfer systems with high coil positioning flexibility," *IEEE Transactions on Industrial Electronics*, vol. 60, no. 1, pp. 372–381, 2013.

[3] A. P. Sample, D. A. Meyer, and J. R. Smith, "Analysis, experimental results, and range adaptation of magnetically coupled resonators for wireless power transfer," *IEEE Transactions on Industrial Electronics*, vol. 58, no. 2, pp. 544–554, 2011.

[4] A. K. RamRakhyani, S. Mirabbasi, and M. Chiao, "Design and optimization of resonance-based efficient wireless power delivery systems for biomedical implants," *IEEE Transactions on Biomedical Circuits and Systems*, vol. 5, no. 1, pp. 48–63, 2011.

[5] G. A. Kendir, W. Liu, G. Wang et al., "An optimal design methodology for inductive power link with Class-E amplifier," *IEEE Transactions on Circuits and Systems I*, vol. 52, no. 5, pp. 857–866, 2005.

[6] S. C. Tang, F. A. Jolesz, and G. T. Clement, "A wireless batteryless deep-seated implantable ultrasonic pulser-receiver powered by magnetic coupling," *IEEE Transactions on Ultrasonics, Ferroelectrics, and Frequency Control*, vol. 58, no. 6, pp. 1211–1221, 2011.

[7] M. P. Theodoridis and S. V. Mollov, "Distant energy transfer for artificial human implants," *IEEE Transactions on Biomedical Engineering*, vol. 52, no. 11, pp. 1931–1938, 2005.

[8] K. Shiba, A. Morimasa, and H. Hirano, "Design and development of low-loss transformer for powering small implantable medical devices," *IEEE Transactions on Biomedical Circuits and Systems*, vol. 4, no. 2, pp. 77–85, 2010.

[9] M. Sawan, S. Hashemi, M. Sehil, F. Awwad, M. Hajj-Hassan, and A. Khouas, "Multicoils-based inductive links dedicated to power up implantable medical devices: modeling, design and experimental results," *Biomedical Microdevices*, vol. 11, no. 5, pp. 1059–1070, 2009.

[10] M. Soma, D. C. Galbraith, and R. L. White, "Radio-frequency coils in implantable devices: misalignment analysis and design procedure," *IEEE Transactions on Biomedical Engineering*, vol. 34, no. 4, pp. 276–282, 1987.

[11] M. Kiani and M. Ghovanloo, "An RFID-based closed-loop wireless power transmission system for biomedical applications," *IEEE Transactions on Circuits and Systems II*, vol. 57, no. 4, pp. 260–264, 2010.

[12] H. Vazquez-Leal, A. Gallardo-Del-Angel, R. Castaneda-Sheissa, and F. J. Gonzalez-Martinez, "The phenomenon of wireless energy transfer: experiments and philosophy," in *Wireless Energy Transfer Based on Electromagnetic Resonance: Principles and Engineering Explorations*, K. Y. Kim, Ed., pp. 978-953, InTech, 2012.

[13] F. Segura-Quijano, J. Garcia-Canton, J. Sacristan, T. Oses, and A. Baldi, "Wireless powering of single-chip systems with integrated coil and external wire-loop resonator," *Applied Physics Letters*, vol. 92, no. 7, Article ID 074102, 3 pages, 2008.

[14] N. A. Keeling, G. A. Covic, and J. T. Boys, "A unity-power-factor IPT pickup for high-power applications," *IEEE Transactions on Industrial Electronics*, vol. 57, no. 2, pp. 744–751, 2010.

[15] M. Dockhorn, D. Kürschner, and R. Mecke, "Contactless power transmission with new secondary converter topology," in *Proceedings of the 13th International Power Electronics and Motion Control Conference (EPE-PEMC '08)*, pp. 1734–1739, September 2008.

[16] M. L. G. Kissin, J. T. Boys, and G. A. Covic, "Interphase mutual inductance in polyphase inductive power transfer systems," *IEEE Transactions on Industrial Electronics*, vol. 56, no. 7, pp. 2393–2400, 2009.

[17] J. De Boeij, E. Lomonova, J. L. Duarte, and A. J. A. Vandenput, "Contactless planar actuator with manipulator," in *Proceedings of the European Conference on Power Electronics and Applications (EPE '07)*, pp. 1–9, Aalborg, Denmark, September 2007.

[18] D. Kürschner and C. Rathge, "Contactless energy transmission systems with improved coil positioning flexibility for high power applications," in *Proceedings of the 39th IEEE Annual Power Electronics Specialists Conference (PESC '08)*, pp. 4326–4332, Rhodes, Greece, June 2008.

[19] U. K. Madawala and D. J. Thrimawithana, "A bidirectional inductive power interface for electric vehicles in V2G systems," *IEEE Transactions on Industrial Electronics*, vol. 58, no. 10, pp. 4789–4796, 2011.

[20] C.-S. Wang, O. H. Stielau, and G. A. Covic, "Design considerations for a contactless electric vehicle battery charger," *IEEE Transactions on Industrial Electronics*, vol. 52, no. 5, pp. 1308–1314, 2005.

[21] H. H. Wu, G. A. Covic, J. T. Boys, and D. J. Robertson, "A series-tuned inductive-power-transfer pickup with a controllable AC-voltage output," *IEEE Transactions on Power Electronics*, vol. 26, no. 1, pp. 98–109, 2011.

[22] Z. Pantic, S. Bhattacharya, and S. Lukic, "Optimal resonant tank design considerations for primary track compensation in inductive power transfer systems," in *Proceedings of the 2nd IEEE Energy Conversion Congress and Exposition (ECCE '10)*, pp. 1602–1609, September 2010.

[23] A. P. Hu, C. Liu, and H. L. Li, "A novel contactless battery charging system for soccer playing robot," in *Proceedings of the 15th International Conference on Mechatronics and Machine Vision in Practice (M2VIP '08)*, pp. 646–650, December 2008.

[24] M. Koizumi, K. Komurasaki, T. Komaru, Y. Mizuno, T. Shibata, and K. Kano, "Mid-range wireless power transmission in space," in *Proceedings of the 8th Annual International Energy Conversion Engineering Conference*, Nashville, Tenn, USA, July 2010.

[25] M. Koizumi, K. Komurasaki, Y. Mizuno, and Y. Arakawa, "Wireless power feeding with strongly coupled magnetic resonance for a flying object," *Wireless Engineering and Technology*, no. 2, pp. 86–89, 2012.

[26] T. Komaru, M. Koizumi, K. Komurasaki, T. Shibata, and K. Kano, "Compact and tunable transmitter and receiver for magnetic resonance power transmission to mobile objects," in *Wireless Power Transfer—Principles and Engineering Explorations*, K. Y. Kim, Ed., InTech, 2012.

[27] R. J. Sedwick, "Long range inductive power transfer with superconducting oscillators," *Annals of Physics*, vol. 325, no. 2, pp. 287–299, 2010.

[28] Y. Jang and M. M. Jovanović, "A contactless electrical energy transmission system for portable-telephone battery chargers," *IEEE Transactions on Industrial Electronics*, vol. 50, no. 3, pp. 520–527, 2003.

[29] C.-G. Kim, D.-H. Seo, J.-S. You, J.-H. Park, and B. H. Cho, "Design of a contactless battery charger for cellular phone," *IEEE Transactions on Industrial Electronics*, vol. 48, no. 6, pp. 1238–1247, 2001.

[30] J. Hirai, T.-W. Kim, and A. Kawamura, "Study on intelligent battery charging using inductive transmission of power and information," *IEEE Transactions on Power Electronics*, vol. 15, no. 2, pp. 335–345, 2000.

[31] D. Ahn and S. Hong, "Effect of coupling between multiple transmitters or multiple receivers on wireless power transfer," *IEEE Transactions on Industrial Electronics*, vol. 60, no. 7, pp. 2602–2613, 2013.

[32] H. Jabbar, Y. S. Song, and T. T. Jeong, "RF energy harvesting system and circuits for charging of mobile devices," *IEEE Transactions on Consumer Electronics*, vol. 56, no. 1, pp. 247–253, 2010.

[33] B. Choi, J. Nho, H. Cha, T. Ahn, and S. Choi, "Design and implementation of low-profile contactless battery charger using planar printed circuit board windings as energy transfer device," *IEEE Transactions on Industrial Electronics*, vol. 51, no. 1, pp. 140–147, 2004.

[34] J. Murakami, F. Sato, T. Watanabe et al., "Consideration on cordless power station—contactless power transmission system," *IEEE Transactions on Magnetics*, vol. 32, no. 5, pp. 5037–5039, 1996.

[35] K. Hatanaka, F. Sato, H. Matsuki et al., "Power transmission of a desk with a cord-free power supply," *IEEE Transactions on Magnetics*, vol. 38, no. 5 I, pp. 3329–3331, 2002.

[36] T. C. Beh, M. Kato, T. Imura, S. Oh, and Y. Hori, "Automated impedance matching system for robust wireless power transfer via magnetic resonance coupling," *IEEE Transactions on Industrial Electronics*, vol. 60, no. 9, pp. 3689–3698, 2013.

[37] T.-W. Yoo and K. Chang, "Theoretical and experimental development of 10 and 35 GHz rectennas," *IEEE Transactions on Microwave Theory and Techniques*, vol. 40, no. 6, pp. 1259–1266, 1992.

[38] J. A. Hagerty, F. B. Helmbrecht, W. H. McCalpin, R. Zane, and Z. B. Popović, "Recycling ambient microwave energy with broadband rectenna arrays," *IEEE Transactions on Microwave Theory and Techniques*, vol. 52, no. 3, pp. 1014–1024, 2004.

[39] E. Waffenschmidt and T. Staring, "Limitation of inductive power transfer for consumer applications," in *Proceedings of the 13th European Conference on Power Electronics and Applications (EPE '09)*, Barcelona, Spain, September 2009.

[40] W. X. Zhong, C. K. Lee, and S. Y. Hui, "Wireless power domino-resonator systems with noncoaxial axes and circular structures," *IEEE Transactions on Power Electronics*, vol. 27, no. 11, pp. 4750–4762, 2012.

[41] M. Dionigi, A. Costanzo, and M. Mongiardo, "Network methods for analysis and design of resonant wireless power transfer systems," in *Wireless Power Transfer—Principles and Engineering Explorations*, K. Y. Kim, Ed., InTech, 2012.

[42] S. Atluri and M. Ghovanloo, "Design of a wideband power-efficient inductive wireless link for implantable biomedical devices using multiple carriers," in *Proceedings of the 2nd International IEEE EMBS Conference on Neural Engineering*, pp. 533–537, March 2005.

[43] R. J. Sedwick, "A fully analytic treatment of resonant inductive coupling in the far field," in *Wireless Power Transfer—Principles and Engineering Explorations*, K. Y. Kim, Ed., InTech, 2012.

[44] A. Kurs, A. Karalis, R. Moffatt, J. D. Joannopoulos, P. Fisher, and M. Soljačić, "Wireless power transfer via strongly coupled magnetic resonances," *Science*, vol. 317, no. 5834, pp. 83–86, 2007.

[45] A. Kurs, R. Moffatt, and M. Soljačić, "Simultaneous mid-range power transfer to multiple devices," *Applied Physics Letters*, vol. 96, no. 4, Article ID 044102, 2010.

[46] J. Kim, H.-C. Son, D.-H. Kim, and Y.-J. Park, "Optimal design of a wireless power transfer system with multiple self-resonators for an LED TV," *IEEE Transactions on Consumer Electronics*, vol. 58, no. 3, pp. 775–780, 2012.

[47] F. Zhang, X. Liu, S. A. Hackworth, R. J. Sclabassi, and M. Sun, "In vitro and in vivo studies on wireless powering of medical sensors and implantable devices," in *Proceedings of IEEE/NIH Life Science Systems and Applications Workshop (LiSSA '09)*, pp. 84–87, April 2009.

[48] A. Kumar, S. Mirabbasi, and M. Chiao, "Resonance-based wireless power delivery for implantable devices," in *Proceedings of IEEE Biomedical Circuits and Systems Conference (BioCAS '09)*, pp. 25–28, November 2009.

[49] T. Imura, T. Uchida, and Y. Hori, "Flexibility of contactless power transfer using magnetic resonance coupling to air gap and misalignment for EV," *World Electric Vehicle Journal*, vol. 3, no. 1, pp. 24–34, 2009.

[50] P. Si, A. P. Hu, S. Malpas, and D. Budgett, "A frequency control method for regulating wireless power to implantable devices," *IEEE Transactions on Biomedical Circuits and Systems*, vol. 2, no. 1, pp. 22–29, 2008.

[51] H. Sugiyama, "Optimal designs for wireless resonant energy link based on nonradiative magnetic field," in *Proceedings of 13th IEEE International Symposium on Consumer Electronics (ISCE '09)*, pp. 428–429, May 2009.

[52] S. Cheon, Y.-H. Kim, S.-Y. Kang, M. L. Lee, J.-M. Lee, and T. Zyung, "Circuit-model-based analysis of a wireless energy-transfer system via coupled magnetic resonances," *IEEE Transactions on Industrial Electronics*, vol. 58, no. 7, pp. 2906–2914, 2011.

[53] C. M. Zierhofer and E. S. Hochmair, "Geometric approach for coupling enhancement of magnetically coupled coils," *IEEE Transactions on Biomedical Engineering*, vol. 43, no. 7, pp. 708–714, 1996.

[54] A. Karalis, J. D. Joannopoulos, and M. Soljačić, "Efficient wireless non-radiative mid-range energy transfer," *Annals of Physics*, vol. 323, no. 1, pp. 34–48, 2008.

[55] S. L. Ho, J. Wang, W. N. Fu, and M. Sun, "A comparative study between novel witricity and traditional inductive magnetic coupling in wireless charging," *IEEE Transactions on Magnetics*, vol. 47, no. 5, pp. 1522–1525, 2011.

[56] S.-H. Lee and R. D. Lorenz, "Development and validation of model for 95%-efficiency 220-W wireless power transfer over a 30-cm air gap," *IEEE Transactions on Industry Applications*, vol. 47, no. 6, pp. 2495–2504, 2011.

[57] H. Hirayama, "Equivalent circuit and calculation of its parameters of magnetic-coupled-resonant wireless power transfer," in *Wireless Power Transfer—Principles and Engineering Explorations*, K. Y. Kim, Ed., InTech, 2012.

[58] J. Wang, S. L. Ho, W. Fu, C. T. Kit, and M. Sun, "Finite-element analysis and corresponding experiments of resonant energy transfer for wireless transmission devices," *IEEE Transactions on Magnetics*, vol. 47, no. 5, pp. 1074–1077, 2011.

[59] R. E. Hamam, A. Karalis, J. D. Joannopoulos, and M. Soljačić, "Efficient weakly-radiative wireless energy transfer: an EIT-like approach," *Annals of Physics*, vol. 324, no. 8, pp. 1783–1795, 2009.

[60] F. Zhang, S. A. Hackworth, W. Fu, C. Li, Z. Mao, and M. Sun, "Relay effect of wireless power transfer using strongly coupled magnetic resonances," *IEEE Transactions on Magnetics*, vol. 47, no. 5, pp. 1478–1481, 2011.

[61] H. Hoang and F. Bien, "Maximizing efficiency of electromagnetic resonance wireless power transmission systems with adaptive circuits," in *Wireless Power Transfer—Principles and Engineering Explorations*, K. Y. Kim, Ed., InTech, 2012.

[62] W. Zhong, C. K. Lee, and S. Y. R. Hui, "General analysis on the use of Tesla's resonators in domino forms for wireless power transfer," *IEEE Transactions on Industrial Electronics*, vol. 60, no. 1, pp. 261–270, 2013.

[63] C. K. Lee, W. X. Zhong, and S. Y. R. Hui, "Effects of magnetic coupling of nonadjacent resonators on wireless power domino-resonator systems," *IEEE Transactions on Power Electronics*, vol. 27, no. 4, pp. 1905–1916, 2012.

[64] S. Cheon, Y. H. Kim, S. Y. Kang, M. L. Lee, and T. Zyung, "Wireless energy transfer system with multiple coils via coupled magnetic resonances," *ETRI Journal*, vol. 34, no. 4, pp. 527–535, 2012.

[65] R. Koma, S. Nakamura, S. Ajisaka, and H. Hashimoto, "Basic analysis of the circuit model using relay antenna in magnetic resonance coupling position sensing system," in *Proceedings of IEEE/ASME International Conference on Advanced Intelligent Mechatronics (AIM '11)*, pp. 25–30, July 2011.

[66] M. K. Watfa, H. AlHassanieh, and S. Selman, "Multi-hop wireless energy transfer in WSNs," *IEEE Communications Letters*, vol. 15, no. 12, pp. 1275–1277, 2011.

[67] Z. N. Low, R. A. Chinga, R. Tseng, and J. Lin, "Design and test of a high-power high-efficiency loosely coupled planar wireless power transfer system," *IEEE Transactions on Industrial Electronics*, vol. 56, no. 5, pp. 1801–1812, 2009.

[68] T. Imura and Y. Hori, "Maximizing air gap and efficiency of magnetic resonant coupling for wireless power transfer using equivalent circuit and Neumann formula," *IEEE Transactions on Industrial Electronics*, vol. 58, no. 10, pp. 4746–4752, 2011.

[69] M. W. Baker and R. Sarpeshkar, "Feedback analysis and design of RF power links for low-power bionic systems," *IEEE Transactions on Biomedical Circuits and Systems*, vol. 1, no. 1, pp. 28–38, 2007.

[70] C. V. Jones, *The Unified Theory of Electrical Machines*, Butterworths, London, UK, 1967.

[71] M. G. Say, *Alternating Current Machines*, Pitman, New York, NY, USA, 4th edition, 1976.

[72] R. R. A. Syms, E. Shamonina, and L. Solymar, "Magneto-inductive waveguide devices," *IEE Proceedings: Microwaves, Antennas and Propagation*, vol. 153, no. 2, pp. 111–121, 2006.

[73] A. Bodrov and S.-K. Sul, "Analysis of wireless power transfer by coupled mode theory (CMT) and practical considerations to increase power transfer efficiency," in *Wireless Power Transfer—Principles and Engineering Explorations*, K. Y. Kim, Ed., InTech, 2012.

[74] G. Gonzalez, *Microwave Transistor Amplifiers: Analysis and Design*, Prentice-Hall, Upper Saddle River, NJ, USA, 2nd edition, 1996.

[75] D. M. Pozar, *Microwave Engineering*, Addison-Wesley, Reading, Mass, USA, 1990.

[76] J. H. Harlow, *Electric Power Transformer Engineering*, CRC Press, 2007.

[77] A. C. Franklin, D. P. Franklin, and S. A. Stigant, *The J & P Transformer Book: A Practical Technology of the Power Transformer*, Butterworths, 1983.

[78] J. Avila-Rosales, *Modeling of the Power Transformer for Electromagnetic Transient Studies in Power Systems*, vol. 1, University of Wisconsin-Madison, 1980.

[79] M. Erol-Kantarci and H. T. Mouftah, "Suresense: sustainable wireless rechargeable sensor networks for the smart grid," *IEEE Wireless Communications*, vol. 19, no. 3, pp. 30–36, 2012.

[80] F. Li, W. Qiao, H. Sun et al., "Smart transmission grid: vision and framework," *IEEE Transactions on Smart Grid*, vol. 1, no. 2, pp. 168–177, 2010.

[81] L. L. Grigsby, *Electric Power Generation, Transmission, and Distribution*, CRC Press, 2012.

[82] M. Chetty and R. Buyya, "Weaving computational grids: how analogous are they with electrical grids?" *Computing in Science and Engineering*, vol. 4, no. 4, p. 61, 2002.

[83] W. Kempton and J. Tomić, "Vehicle-to-grid power implementation: from stabilizing the grid to supporting large-scale renewable energy," *Journal of Power Sources*, vol. 144, no. 1, pp. 280–294, 2005.

[84] T. Flick and J. Morehouse, *Securing the Smart Grid: Next Generation Power Grid Security*, Syngress, 2010.

[85] A. Keyhani, *Design of Smart Power Grid Renewable Energy Systems*, Wiley-IEEE Press, 2011.

[86] U.S. Department of Energy, "Large Power Transformers and the U.S. Electric Grid," Tech. Rep., Infrastructure Security and Energy Restoration, Office of Electricity Delivery and Energy Reliability, June 2012, http://energy.gov/sites/prod/files/Large%20Power%20Transformer%20Study%20-%20June%202012_0.pdf.

[87] L. Xie, Y. Shi, Y. T. Hou, and H. D. Sherali, "Making sensor networks immortal: an energy-renewal approach with wireless power transfer," *IEEE/ACM Transactions on Networking*, vol. 20, no. 6, pp. 1748–1761, 2012.

[88] N. Shinohara, "Power without wires," *IEEE Microwave Magazine*, vol. 12, no. 7, pp. S64–S73, 2011.

[89] F. Zhang, S. A. Hackworth, W. Fu, and M. Sun, "The relay effect on wireless power transfer using witricity," in *Proceedings of the 14th Biennial IEEE Conference on Electromagnetic Field Computation (CEFC '10)*, Chicago, Ill, USA, May 2010.

[90] R. R. A. Syms, E. Shamonina, V. Kalinin, and L. Solymar, "A theory of metamaterials based on periodically loaded transmission lines: interaction between magnetoinductive and electromagnetic waves," *Journal of Applied Physics*, vol. 97, no. 6, Article ID 064909, 2005.

[91] R. R. A. Syms, L. Solymar, I. R. Young, and T. Floume, "Thin-film magneto-inductive cables," *Journal of Physics D*, vol. 43, no. 5, Article ID 055102, 2010.

[92] E. M. Thomas, J. D. Heebl, C. Pfeiffer, and A. Grbic, "A power link study of wireless non-radiative power transfer systems using resonant shielded loops," *IEEE Transactions on Circuits and Systems I*, vol. 59, no. 9, pp. 2125–2136, 2012.

[93] G. Grandi, M. K. Kazimierczuk, A. Massarini, and U. Reggiani, "Stray capacitances of single-layer solenoid air-core inductors," *IEEE Transactions on Industry Applications*, vol. 35, no. 5, pp. 1162–1168, 1999.

[94] A. Massarini, M. K. Kazimierczuk, and G. Grandi, "Lumped parameter models for single- and multiple-layer inductors," in *Proceedings of the 27th Annual IEEE Power Electronics Specialists Conference (PESC '96)*, pp. 295–301, January 1996.

[95] M. Bartoli, N. Noferi, A. Reatti, and M. K. Kazimierczuk, "Modeling Litz-wire winding losses in high-frequency power inductors," in *Proceedings of the 27th Annual IEEE Power Electronics Specialists Conference (PESC '96)*, pp. 1690–1696, January 1996.

[96] Z. Yang, W. Liu, and E. Basham, "Inductor modeling in wireless links for implantable electronics," *IEEE Transactions on Magnetics*, vol. 43, no. 10, pp. 3851–3860, 2007.

[97] New EnglandWire Technologies, "LitzWire—Product N. E. W. Technologies," 2005, http://newenglandwire.com/.

[98] F. Tourkhani and P. Viarouge, "Accurate analytical model of winding losses in round Litz wire windings," *IEEE Transactions on Magnetics*, vol. 37, no. 1, pp. 538–543, 2001.

[99] Q. Yu and T. W. Holmes, "A study on stray capacitance modeling of inductors by using the finite element method," *IEEE Transactions on Electromagnetic Compatibility*, vol. 43, no. 1, pp. 88–93, 2001.

[100] G. S. Dimitrakakis, E. C. Tatakis, and E. J. Rikos, "A new model for the determination of copper losses in transformer windings with arbitrary conductor distribution under high frequency sinusoidal excitation," in *Proceedings of the European Conference on Power Electronics and Applications (EPE '07)*, pp. 1–10, September 2007.

[101] C. D. Sijoy and S. Chaturvedi, "Calculation of accurate resistance and inductance for complex magnetic coils using the finite-difference time-domain technique for electromagnetics," *IEEE Transactions on Plasma Science*, vol. 36, no. 1, pp. 70–79, 2008.

[102] G. Grandi, M. K. Kazimierczuk, A. Massarini, and U. Reggiani, "Stray capacitances of single-layer air-core inductors for high-frequency applications," in *Proceedings of the 31st IAS Annual Meeting on Industry Applications Conference*, pp. 1384–1388, October 1996.

[103] B. L. Cannon, J. F. Hoburg, D. D. Stancil, and S. C. Goldstein, "Magnetic resonant coupling as a potential means for wireless power transfer to multiple small receivers," *IEEE Transactions on Power Electronics*, vol. 24, no. 7, pp. 1819–1825, 2009.

[104] J. C. Maxwell, *A Treatise on Electricity and Magnetism*, Dover, New York, NY, USA, 1954.

[105] IEEE—International Committee on Electromagnetic Safety ICES, "Standard for Safety Levels with Respect to Human Exposure to Radio Frequency Electromagnetic Fields, 3 kHz to 300 GHz," IEEE Std. C95.1-1991, 1992.

[106] "Guidelines for limiting exposure to time-varying electric, magnetic, and electromagnetic fields (up to 300 GHz)," *Health Physics*, vol. 74, no. 4, pp. 494–522, 1998.

[107] F. van der Pijl, P. Bauer, and M. Castilla, "Control method for wireless inductive energy transfer systems with relatively large air gap," *IEEE Transactions on Industrial Electronics*, vol. 60, no. 1, pp. 213–390, 2013.

[108] A. J. Moradewicz and M. P. Kazmierkowski, "Contactless energy transfer system with FPGA-controlled resonant converter," *IEEE Transactions on Industrial Electronics*, vol. 57, no. 9, pp. 3181–3190, 2010.

[109] H. H. Wu, J. T. Boys, and G. A. Covic, "An AC processing pickup for IPT systems," *IEEE Transactions on Power Electronics*, vol. 25, no. 5, pp. 1275–1284, 2010.

[110] J. A. Dobrowolski, *Microwave Network Design Using the Scattering Matrix*, Artech House, Norwood, Mass, USA, 2010.

[111] R. Mavaddat, *Network Scattering Parameters. Advanced Series in Circuits and Systems*, World Scientific, River Edge, NJ, USA, 2nd edition, 1996.

[112] X. Liu and S. Y. R. Hui, "An analysis of a double-layer electromagnetic shield for a universal contactless battery charging platform," in *Proceedings of the 36th IEEE Power Electronics Specialists Conference*, pp. 1767–1772, June 2005.

[113] J. C. Lin, "Safety standards for human exposure to radio frequency radiation and their biological rationale," *IEEE Microwave Magazine*, vol. 4, no. 4, pp. 22–26, 2003.

[114] J. M. Fernandez and J. A. Borras, "Contactless battery charger with wireless control link," US patent number 6,184,651, issued in February 2001.

[115] G. Scheible, B. Smailus, M. Klaus, K. Garrels, and L. Heinemann, "System for wirelessly supplying a large number of actuators of a machine with electrical power," US patent number 6,597,076, July 2003.

[116] L. Ka-Lai, J. W. Hay, and P. G. W. Beart, "Contact-less power transfer," US patent number 7,042,196, May 2006.

[117] J. Wang, J. Li, S. L. Ho et al., "Lateral and angular misalignments analysis of a new PCB circular spiral resonant wireless charger," *IEEE Transactions on Magnetics*, vol. 48, no. 11, pp. 4522–4525, 2012.

[118] U.-M. Jow and M. Ghovanloo, "Design and optimization of printed spiral coils for efficient transcutaneous inductive power transmission," *IEEE Transactions on Biomedical Circuits and Systems*, vol. 1, no. 3, pp. 193–202, 2007.

[119] T. Komaru, M. Koizumi, K. Komurasaki, T. Shibata, and K. Kano, "Parametric evaluation of mid-range wireless power transmission," in *Proceedings of the IEEE-ICIT International Conference on Industrial Technology (ICIT '10)*, pp. 789–792, March 2010.

[120] H. Hoang, S. Lee, Y. Kim, Y. Choi, and F. Bien, "An adaptive technique to improve wireless power transfer for consumer electronics," *IEEE Transactions on Consumer Electronics*, vol. 58, no. 2, pp. 327–332, 2012.

[121] H. Sugiyama, "Performance analysis of magnetic resonant system based on electrical circuit theory," in *Wireless Power Transfer—Principles and Engineering Explorations*, K. Y. Kim, Ed., InTech, 2012.

[122] J. Choi and C. Seo, "High-efficiency wireless energy transmission using magnetic resonance based on metamaterial with relative permeability equal to -1," *Progress in Electromagnetics Research*, vol. 106, pp. 33–47, 2010.

[123] B.-J. Jang, S. Lee, and H. Yoon, "HF-band wireless power transfer system: concept, issues, and design," *Progress in Electromagnetics Research*, vol. 124, pp. 211–231, 2012.

[124] V. Marian, B. Allard, C. Vollaire, and J. Verdier, "Strategy for microwave energy harvesting from ambient field or a feeding source," *IEEE Transactions on Power Electronics*, vol. 27, no. 11, pp. 4481–4491, 2012.

[125] E. Y. Chow, C.-L. Yang, Y. Ouyang, A. L. Chlebowski, P. P. Irazoqui, and W. J. Chappell, "Wireless powering and the study of RF propagation through ocular tissue for development of implantable sensors," *IEEE Transactions on Antennas and Propagation*, vol. 59, no. 6, pp. 2379–2387, 2011.

[126] L. Chen, S. Liu, Y. Zhou, and T. Cui, "An optimizable circuit structure for high-efficiency wireless power transfer," *IEEE Transactions on Industrial Electronics*, vol. 60, no. 1, pp. 339–349, 2013.

[127] C. Zhu, K. Liu, C. Yu, R. Ma, and H. Cheng, "Simulation and experimental analysis on wireless energy transfer based on magnetic resonances," in *Proceedings of IEEE Vehicle Power and Propulsion Conference (VPPC '08)*, pp. 1–4, September 2008.

[128] Z. N. Low, J. J. Casanova, P. H. Maier, J. A. Taylor, R. A. Chinga, and J. Lin, "Method of load/fault detection for loosely coupled planar wireless power transfer system with power delivery tracking," *IEEE Transactions on Industrial Electronics*, vol. 57, no. 4, pp. 1478–1486, 2010.

[129] Y.-H. Kim, S.-Y. Kang, M.-L. Lee, B.-G. Yu, and T. Zyung, "Optimization of wireless power transmission through resonant coupling," in *Proceedings of the 6th International Conference-Workshop on Compatability and Power Electronics (CPE '09)*, pp. 426–431, May 2009.

[130] L. Peng, J. Y. Wang, L. X. Ran, O. Breinbjerg, and N. A. Mortensen, "Performance analysis and experimental verification of mid-range wireless energy transfer through non-resonant magnetic coupling," *Journal of Electromagnetic Waves and Applications*, vol. 25, no. 5-6, pp. 845–855, 2011.

[131] H.-S. Kim, D.-H. Won, and B.-J. Jang, "Simple design method of wireless power transfer system using 13.56MHz loop antennas," in *Proceedings of IEEE International Symposium on Industrial Electronics (ISIE '10)*, pp. 1058–1063, July 2010.

[132] W. H. Ko, S. P. Liang, and C. D. F. Fung, "Design of radio-frequency powered coils for implant instruments," *Medical and Biological Engineering and Computing*, vol. 15, no. 6, pp. 634–640, 1977.

[133] M. Zargham and P. G. Gulak, "Maximum achievable efficiency in near-field coupled power-transfer systems," *IEEE Transactions on Biomedical Circuits and Systems*, vol. 6, no. 3, pp. 228–245, 2012.

[134] R. R. Harrison, "Designing efficient inductive power links for implantable devices," in *Proceedings of IEEE International Symposium on Circuits and Systems (ISCAS '07)*, pp. 2080–2083, May 2007.

[135] K. M. Silay, D. Dondi, L. Larcher et al., "Load optimization of an inductive power link for remote powering of biomedical implants," in *Proceedings of IEEE International Symposium on Circuits and Systems (ISCAS '09)*, pp. 533–536, May 2009.

[136] G. Simard, M. Sawan, and D. Massicotte, "High-speed OQPSK and efficient power transfer through inductive link for biomedical implants," *IEEE Transactions on Biomedical Circuits and Systems*, vol. 4, no. 3, pp. 192–200, 2010.

[137] S. Ray, *Electrical Power Systems: Concepts, Theory and Practice*, Prentice-Hall of India, 2007.

[138] K.-H. Grote and E. K. Antonsson, *Springer Handbook of Mechanical Engineering*, vol. 10, Springer, 2009.

[139] J. Schlabbach and K.-H. Rofalski, *Power System Engineering: Planning, Design, and Operation of Power Systems and Equipment*, Wiley-VCH, 2008.

[140] E. Lakervi and E. J. Holmes, *Electricity Distribution Network Design*, vol. 21, Peter Peregrinus, 1995.

[141] M. V. Bakshi and U. A. Bakshi, *Elements of Power Systems*, Technical Publications, 2009.

[142] A. G. Lazaropoulos, "Towards modal integration of overhead and underground low-voltage and medium-voltage power line communication channels in the smart grid landscape: model

expansion, broadband signal transmission characteristics, and statistical performance metrics (invited paper)," *ISRN Signal Processing*, vol. 2012, Article ID 121628, 17 pages, 2012.

[143] A. G. Lazaropoulos, "Review and progress towards the common broadband management of high-voltage transmission grids: model expansion and comparative modal analysis," *ISRN Electronics*, vol. 2012, Article ID 935286, 18 pages, 2012.

[144] S. Sivanagaraju, *Electric Power Transmission and Distribution*, Pearson Education India, 2008.

[145] U. A. Bakshi and M. V. Bakshi, *Generation, Transmission and Distribution*, Technical Publications Pune, Pune, India, 2001.

[146] J. C. de Sosa, *Analysis and Design of High-Voltage Transmission Lines*, iUniverse Incorporated, Bloomington, Ind, USA, 2010.

[147] MAKSAN, "Product Catalog: Power and Distribution Transformers," Tech. Rep., 2012.

[148] M. C. K. Wiltshire, J. B. Pendry, I. R. Young, D. J. Larkman, D. J. Gilderdale, and J. V. Hajnal, "Microstructured magnetic materials for rf flux guides in magnetic resonance imaging," *Science*, vol. 291, no. 5505, pp. 849–851, 2001.

[149] D. Schneider, "Wireless power at a distance is still far away," *IEEE Spectrum*, vol. 47, no. 5, pp. 34–39, 2010.

[150] D. J. Perreault, J. Hu, J. M. Rivasy et al., "Opportunities and challenges in very high frequency power conversion," in *Proceedings of the 24th Annual IEEE Applied Power Electronics Conference and Exposition (APEC '09)*, pp. 1–14, February 2009.

[151] F. Blaabjerg, Z. Chen, and S. B. Kjaer, "Power electronics as efficient interface in dispersed power generation systems," *IEEE Transactions on Power Electronics*, vol. 19, no. 5, pp. 1184–1194, 2004.

[152] U. Reggiani, A. Massarini, L. Sandrolini et al., "Experimental verification of predicted electromagnetic fields radiated by straight interconnect cables carrying high-frequency current," in *Power Tech Conference Proceedings*, vol. 1, p. 5, Bologna, Italy, June 2003.

22

An Improved Walsh Function Algorithm for Use in Sinusoidal and Nonsinusoidal Power Components Measurement

Saifulnizam Bin Abdul Khalid,[1] Garba Aliyu,[1] Mohd Wazir Mustafa,[1] and Hussein Shareef[2]

[1] *Faculty of Electrical Engineering, Universiti Teknologi Malaysia, 81310 Skudai, Malaysia*
[2] *Faculty of Electrical Engineering and Built Environment, Universiti Kebangsaan Malaysia, 43600 Bangi, Malaysia*

Correspondence should be addressed to Saifulnizam Bin Abdul Khalid; nizam@fke.utm.my

Academic Editor: Antonio Moreno-Munoz

This paper presents an improved Walsh function IWF algorithms as an alternative approach for active and reactive powers measurement in linear and nonlinear, balanced and unbalanced sinusoidal three-phase load system. It takes advantage of Walsh function unified approach, simple algorithm and its intrinsic high level of accuracy as a result of coefficient characteristics and energy behaviour representation. The developed algorithm was modeled on the Matlab Simulink software; different types of load, linear and nonlinear, were also modeled based on practical voltage and current waveforms and tested with the proposed improved Walsh algorithms. The IEEE standard 1459–2000 which is based on fast Fourier transform FFT approach was used as benchmark for the linear load system. The data obtained from laboratory experiment to determine power components in harmonic load systems using Fluke 435 power quality analyzer PQA which complies with IEC/EN61010-1-2001 standard was modeled and used to validate the improved algorithm for nonlinear load measurement. The results showed that the algorithm has the potential to effectively measure three-phase power components under different load conditions.

1. Introduction

The awareness of the problems associated with nonlinear devices and their effects on power components measurement algorithms has considerably increased. The use of these power electronics equipment and appliances with microprocessors has been shown to contribute to the distortion of the voltage and current waveforms of the supply system at the distribution end [1]. This distortion affects the smooth operation of most algorithms use for power components measurement and invariably has negative effect on the quality of the power supply system. Accurate measurement and evaluation of energy consumption are of utmost importance for effective planning, billing, monitoring, maintenance, and further development of the electricity supply system. The deregulation of the power sector has led electricity operators to seek for ways to make the best use of the supply systems. Proper evaluation of the energy consumption is one of the important tasks in electric power industries, especially in electric bill and the electrical energy quality estimation and control [2, 3]. Therefore, several companies are into design and manufacture of energy meters using different algorithms to meet the increasing demand of smart metering [4]. In the pricing of electric bill based on the value of the integral of the load active power measured using kilowatt-hour meter, the electric utility board incurs losses in revenue for energy delivered to current harmonic generating customers and also those that cause current asymmetry [5]. Most times, customers that do not generate harmonics but are supplied with distorted or asymmetrical voltages are made to pay not only for the useful energy but also for the energy which may have caused malfunction of their equipment [6]. There is no universally accepted definition for the measurement of reactive power component in the presence of nonlinear and nonsinusoidal waveform [7, 8]. The IEEE working group on appropriate determination of billing system based on sinusoidal and nonsinusoidal situation defined distortion power in terms of total fundamental and harmonic constituents with less unwieldy theory [8].

The conventional method of determining power consumption is to evaluate using measured values of the voltage (V), current (I), and power factor (pf). The apparent power S, active power P, and reactive power Q are defined as in [9]. This way of power evaluation requires accurate measurement of the root mean square (RMS) values of the voltage and current before performing the multiplication operation so as to determine the apparent power S. Precise determination of RMS values of voltage and current is complex and poses a serious challenge in electrical measurement [10]. For this method of power calculation to be effective in a three phase network the following has to be satisfied.

(i) The RMS voltages and current on all the three phases must be identical.

(ii) Phase angles between voltage and current must be maintained at 120° and are time invariant.

These conditions may not be attainable in the real practical world due to the nature of the loads most of which are nonlinear [11].

This paper proposed an improved algorithm as an alternative for power components measurement/estimation using the Walsh function which has the advantage of better accuracy and reliability in energy measurement as it takes into consideration among others, elimination of the effect of harmonic on reactive power measurement. Walsh function transform algorithm has a unique and essentially high level of accuracy as a result of coefficient characteristics and energy behaviour representation [12]. This proposed algorithm might be integrated into any type of energy meter for measurement.

The remaining of this paper is organized as follows. Section 2 presents Walsh function and its analytical expression, while Section 3 highlights the steps involved in the development of the proposed improved algorithms for measurement using the Walsh function, Section 4 has the modeling and simulation of the improved algorithms, and Section 5 is the simulation with laboratory experiment results. Lastly, Section 6 is the conclusions.

2. Walsh Function (WF) Analytical Expression

The Walsh function method for the evaluation of energy parameters was presented as a mathematical tool to analyze energy meter output behaviour for indepth error detection [9]. The method is faster when compared with other techniques like FFT. It does not require the phase shift of $\pi/2$ between voltage and current and it has less computation requirement. However, when the load becomes nonlinear and nonsinusoidal the voltage and current waveforms become distorted. This leads to harmonic which is a multiple of the fundamental frequencies of the current and/or voltage. The authors exert that harmonic current affects the reactive power measurement so another attempt using modified approach for power components measurement in both sinusoidal and nonsinusoidal condition was made; the approach reduced computational demand, though the influence of harmonics to the measurement results which was not accounted for, was the main setback of the algorithms [13–15].

The attractions to Walsh-function-based technique for use in energy parameter evaluation are the following.

(a) The Walsh transform analyzes signals into rectangular waveform rather than sinusoidal ones and is computed more rapidly when compared with Fourier transform FT.

(b) Walsh-function-based algorithm contains addition and subtractions only and hence results in considerably simplified hardware implementation of power evaluation process.

(c) The IEEE/IEC definition of a phase shift of $\pi/2$ between the voltage and the current signal mainly used for reactive power evaluation is eliminated from signal processing operation when using Walsh function [8].

Generalized Walsh functions and transforms were introduced in 1923 by J. L. Walsh, but their application to engineering and other fields did not happen until recently [16] with some basic and enlightening properties of these functions and transforms considered. This function can be applied among other uses to develop an algorithm that would be applicable to nonlinear loads problem analysis. It is a full orthogonal system with unique properties, which include that it has only two values +1 and −1 over specified normalized period T. This greatly influences the effectiveness of signal processing operation as being related to measurement of power components and characteristics of power distribution system. Analytically the Walsh function is expressed as

$$\mathrm{Wal}\,(n, \beta) = (-1)^{\sum_{k=1}^{m}(n_{m-k+1} \oplus n_{m-k})\beta_k}, \tag{1}$$

where n is the order of the function form $n = 1, 2, 3, \ldots, n_m$ which is the mth coefficients of the n represented in binary code, that is, $n = (n_0, n_1, n_2, n_3 \cdots n_m)_2$, $n_m = 0, 1$. With m being the highest order WF serial number in the system, β is the argument of WF that defines the coefficients of β_k in binary code $\beta = (\beta_k, \beta_k \cdots \beta_k)_2$, $\beta_k = 0, 1$, and $k = 1, 2, 3, \ldots, m$. From (1) the graphical representation of the first sixteenth order Walsh function is generated as shown in Figure 1.

3. Walsh Function Algorithm

The IEEE standard 1459-2000 for the instantaneous voltages (v_a, v_b, v_c) and currents (i_a, i_b, i_c) in a three-phase distribution network with linear balanced or unbalanced load is given as [17]. Instantaneous three-phase sinusoidal voltages are

$$\begin{aligned} v_a &= \sqrt{2}V_a \sin(\omega t), \\ v_b &= \sqrt{2}V_b \sin(\omega t - 120), \\ v_c &= \sqrt{2}V_c \sin(\omega t + 120). \end{aligned} \tag{2}$$

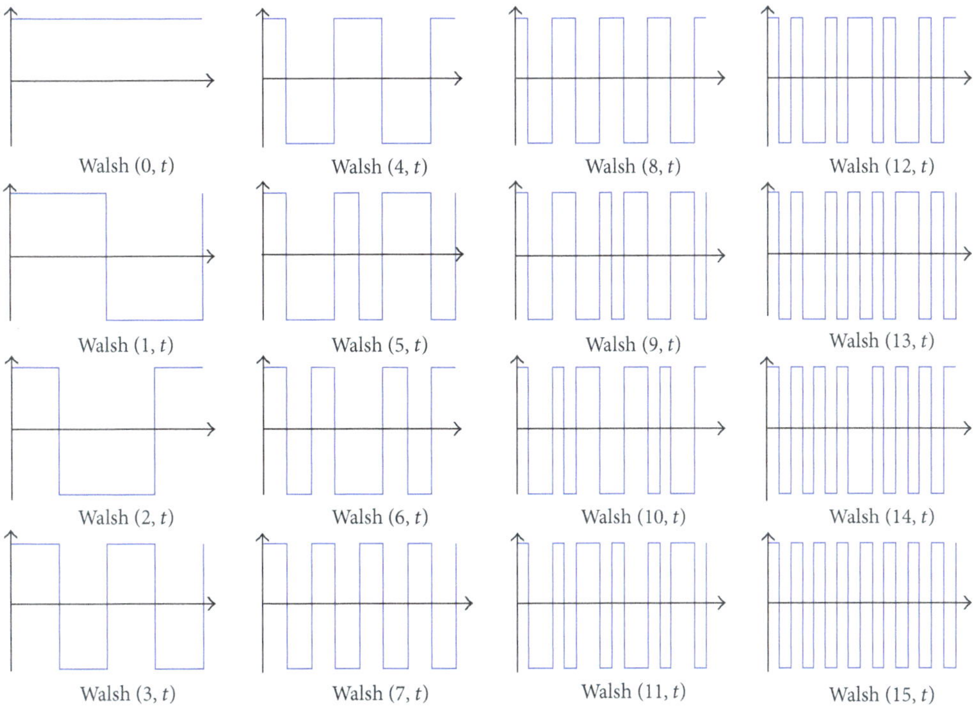

FIGURE 1: The first 16th order of Walsh function.

Instantaneous three-phase currents are

$$\begin{aligned} i_a &= \sqrt{2} I_a \sin(\omega t - \theta_a), \\ i_b &= \sqrt{2} I_b \sin(\omega t - \theta_b - 120), \\ i_c &= \sqrt{2} I_c \sin(\omega t - \theta_c + 120), \end{aligned} \tag{3}$$

where V_a, V_b, V_c and I_a, I_b, I_c are the RMS values of the line to neutral voltages and currents for the phases a, b, and c, respectively. θ_a, θ_b and θ_c are the phase angles between the respective voltage and current. $\omega = 2\pi f$, f = frequency in Hertz, $t = 1/f$.

The instantaneous reactive powers for the three phases represented with say p_a, p_b, and p_c are given by

$$p_a = v_a \times i_a, \qquad p_b = v_b \times i_b, \qquad p_c = v_c \times i_c. \tag{4}$$

Substituting for (v_a, v_b, v_c) and (i_a, i_b, i_c) in the previous expressions and solve trigonometrically the instantaneous powers for phases a, b, and c of the three-phase linear balanced and unbalanced load network are derived as in (5):

$$\begin{aligned} p_a &= P_a - [P_a \cos(2\omega t) - Q_a \sin(2\omega t)], \\ p_b &= P_b + P_b - \sqrt{3} Q_b \cos(2\omega t) + \left(P_b \sqrt{3} + Q_b\right) \sin(2\omega t), \\ p_c &= P_c + \left(P_c + Q_c \sqrt{3}\right) \cos(2\omega t) + \left(Q_c - P_c \sqrt{3}\right) \sin(2\omega t), \end{aligned} \tag{5}$$

where $P_a = V_a I_a \cos\theta_a$, $P_b = V_b I_b \cos\theta_b$, $P_c = V_c I_c \cos\theta_c$, $Q_a = V_a I_a \sin\theta_a$, $Q_b = V_b I_b \sin\theta_b$, and $Q_c = V_c I_c \sin\theta_c$, respectively.

From the instantaneous power of (5) the algorithms for measuring the reactive powers on these three-phase systems say, Q_a, Q_b, and Q_c, using the Walsh function are obtained by multiplying both sides of the equations by the third-order Walsh function, that is, $\mathrm{Wal}(3, t)$ and integrate over the period T of power system frequency. The integral of the third order Walsh function with a constant and also a multiplier of $\cos(2\omega t)$ is equal to zero. Because $\cos(2\omega t)$ is orthogonal (right angled) with the third Walsh function so the result is obtained as follows:

$$\begin{aligned} Q_a &= -\frac{\pi}{2T} \int_0^T p_a \mathrm{Wal}(3, t)\, dt, \\ Q_b &= \frac{\pi}{2T} \int_0^T p_b \mathrm{Wal}(3, t)\, dt - P_b \sqrt{3}, \\ Q_c &= \frac{\pi}{2T} \int_0^T p_c \mathrm{Wal}(3, t)\, dt + P_c \sqrt{3}. \end{aligned} \tag{6}$$

The algorithms for real or active powers in the three phases a, b, and c are determined by multiplying (5) by the zero-order Walsh function, that is, $\mathrm{Wal}(0, t)$ and integrate over the period T. In Walsh algorithm the zero-order function is a constant $\mathrm{Wal}(0, t) = 1$, over the period of T as can be seen from Figure 1, so all the integral terms on the right-hand side of (5) that involve product of $\mathrm{Wal}(0,t)$ with either $\cos(2\omega t)$ or $\sin(2\omega t)$ approach to zero thus giving (7) as

$$\begin{aligned}\frac{1}{T}\int_0^T p_a \mathrm{Wal}(0,t)\,dt &= \frac{1}{T}\int_0^T \mathrm{Wal}(0,t)\,P_a dt,\\ \frac{1}{T}\int_0^T p_b \mathrm{Wal}(0,t)\,dt &= \frac{1}{T}\int_0^T \mathrm{Wal}(0,t)\,P_b dt,\\ \frac{1}{T}\int_0^T p_c \mathrm{Wal}(0,t)\,dt &= \frac{1}{T}\int_0^T \mathrm{Wal}(0,t)\,P_c dt.\end{aligned} \tag{7}$$

Solving the equations further active powers P_a, P_b, and P_c are obtained for three-phase power system as follows:

$$\begin{aligned}P_a &= \frac{1}{T}\int_0^T \mathrm{Wal}(0,t)\,p_a dt,\\ P_b &= \frac{1}{T}\int_0^T \mathrm{Wal}(0,t)\,p_b dt,\\ P_c &= \frac{1}{T}\int_0^T \mathrm{Wal}(0,t)\,p_c dt.\end{aligned} \tag{8}$$

The modeling and simulation of (6) and (8) have been presented earlier in [12] and the authors showed that current harmonic affects the reading of the simulation results. The algorithms have to be further improved as to be able to measure the power components in both linear and nonlinear, sinusoidal nonsinusoidal load conditions. It is worthy to mention, at this juncture, that in AC network source voltages are relatively pure sinusoidal waveforms and that it is mostly at the transmission and distribution end that harmonic and other forms of distortions are introduced. The use of statistical estimators to detect distortions like sags and swells using synthetics analysis that allowed the identification of the variance, skewness, and kurtosis associated with normal sinusoidal waveforms was proposed in [18]. The increase use of nonlinear loads like computers, fluorescent lamps, adjustable speed drive motors, arc furnaces, arc welding machines, electronic control and power converters among others cause harmonics. This lead to overloading of neutral conductors, overheating of transformers, tripping of circuit breakers, power factor correction capacitors over stress and skin effect [19]. Also research has shown that the effect of odd harmonic is more significant in power distribution system while that of even harmonic is negligible [20].

4. Proposed Improved Walsh Function Algorithm for Measurement

Harmonic in the power system does not affect the active power measurement but has a great deal of influence on reactive power measurement which invariably affects the power factor and hence the quality of the supply system. To derive the improved Walsh function IWF algorithm for measurement we assume that the load current is contaminated with third-order current harmonic denoted as i_{a3}, i_{b3}, and i_{c3}, with θ_{a3}, θ_{b3}, and θ_{c3} being the phase angles between the fundamental voltages, and the third-order current harmonic waveforms of the phases, I_{a3}, I_{b3}, and I_{c3} are the RMS values of the third-order current harmonic as given below:

$$\begin{aligned}i_{a3} &= I_{a3}\sin\left(3\omega t - \theta_{a3}\right),\\ i_{b3} &= I_{b3}\sin\left(3\omega t - \theta_{b3} - 120\right),\\ i_{c3} &= I_{c3}\sin\left(3\omega t - \theta_{c3} + 120\right).\end{aligned} \tag{9}$$

The instantaneous powers p_{a3}, p_{b3}, and p_{c3} for the three phases a, b, and c under this condition are derived as follows:

$$\begin{aligned}p_{a3} &= P_a + \left(P_{a3} - P_a\right)\cos(2\omega t) + \left(Q_{a3} - Q_a\right)\sin(2\omega t)\\ &\quad - P_{a3}\cos(4\omega t) - Q_{a3}\sin(4\omega t),\\ p_{b3} &= P_b + \left(P_{b3} + P_b - \sqrt{3}Q_b\right)\cos(2\omega t)\\ &\quad + \left(P_b\sqrt{3} + Q_{b3} - Q_b\right)\sin(2\omega t)\\ &\quad - \left(\sqrt{3}Q_{b3} - P_{b3}\right)\cos(4\omega t) + \left(Q_{b3} + P_b\sqrt{3}\right)\sin(4\omega t),\\ p_{c3} &= P_c + \left(P_{c3} + P_c - \sqrt{3}Q_c\right)\cos(2\omega t)\\ &\quad + \left(P_c\sqrt{3} + Q_{c3} - Q_c\right)\sin(2\omega t)\\ &\quad - \left(\sqrt{3}Q_{c3} - P_{c3}\right)\cos(4\omega t) + \left(Q_{c3} + P_c\sqrt{3}\right)\sin(4\omega t),\end{aligned} \tag{10}$$

where $P_{a3} = V_a I_{a3}\cos\theta_{a3}$, $P_{b3} = V_b I_{b3}\cos\theta_{b3}$, $P_{c3} = V_c I_{c3}\cos\theta_{c3}$, $Q_{a3} = V_a I_{a3}\sin\theta_{a3}$, $Q_{b3} = V_b I_{b3}\sin\theta_{b3}$, and $Q_{c3} = V_c I_{c3}\sin\theta_{c3}$, respectively.

To obtain the improved algorithm for reactive power under this harmonic condition we apply the Walsh function by multiplying (10) with the third-order WF, that is, $\mathrm{Wal}(3,t)$ and integrate over the time T. According to the Walsh functions all the integrals of the right-hand side terms of (10) that involves the multipliers of $\cos(2\omega t)$, $\cos(4\omega t)$, and $\sin(4\omega t)$ and the constant P are all equal to zero [14]. Hence (11) become

$$\begin{aligned}&\frac{1}{T}\int_0^T p_a \mathrm{Wal}(3,t)\,dt\\ &\quad = \frac{1}{T}\int_0^T \left(Q_{a3} - Q_a\right)\sin(2\omega t)\times \mathrm{Wal}(3,t)\,dt,\\ &\frac{1}{T}\int_0^T p_b \mathrm{Wal}(3,t)\,dt\\ &\quad = \frac{1}{T}\int_0^T \left(P_b + \sqrt{3}Q_{b3} - Q_b\right)\sin(2\omega t)\times \mathrm{Wal}(3,t)\,dt,\\ &\frac{1}{T}\int_0^T p_c \mathrm{Wal}(3,t)\,dt\\ &\quad = \frac{1}{T}\int_0^T \left(P_c + \sqrt{3}Q_{c3} - Q_c\right)\sin(2\omega t)\times \mathrm{Wal}(3,t)\,dt.\end{aligned} \tag{11}$$

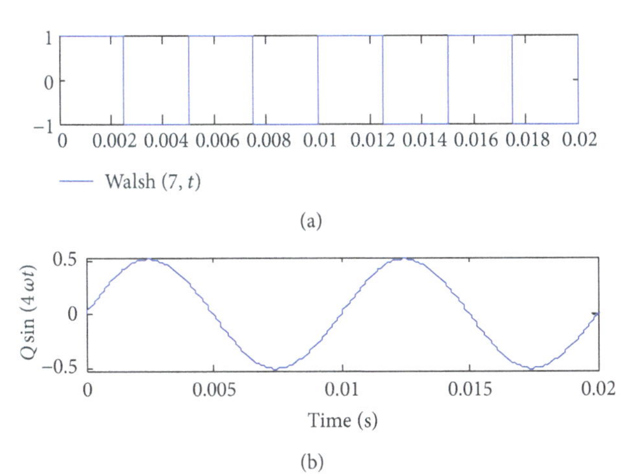

FIGURE 2: (a) 7th order WF, (b) $\sin(4\omega t)$ waveforms.

The product of the 3rd-order WF the expressions, $(Q_{a3} - Q_a)\sin(2\omega t)$, $(P_b\sqrt{3} + Q_{b3} - Q_b)\sin(2\omega t)$, and $(P_c\sqrt{3} + Q_{c3} - Q_c)\sin(2\omega t)$, results in the full-wave rectification of the terms:

$$\frac{1}{T}\int_0^T p_a \mathrm{Wal}(3,t)\,dt = \frac{1}{T}\int_0^T (Q_{a3} - Q_a)\sin(2\omega t)\,dt,$$

$$\frac{1}{T}\int_0^T p_b \mathrm{Wal}(3,t)\,dt = \frac{1}{T}\int_0^T \left(P_b\sqrt{3} + Q_{b3} - Q_b\right)\sin(2\omega t)\,dt,$$

$$\frac{1}{T}\int_0^T p_c \mathrm{Wal}(3,t)\,dt = \frac{1}{T}\int_0^T \left(P_c\sqrt{3} + Q_{c3} - Q_c\right)\sin(2\omega t)\,dt. \quad (12)$$

Solving for Q_a, Q_b, and Q_c

$$Q_a = -\frac{\pi}{2T}\int_0^T p_a \mathrm{Wal}(3,t)\,dt + Q_{a3},$$
$$Q_b = -\frac{\pi}{2T}\int_0^T p_b \mathrm{Wal}(3,t)\,dt + P_b\sqrt{3} + Q_{b3}, \quad (13)$$
$$Q_c = -\frac{\pi}{2T}\int_0^T p_c \mathrm{Wal}(3,t)\,dt + P_c\sqrt{3} + Q_{c3}.$$

Q_{a3}, Q_{b3}, and Q_{c3} in (13) are the reactive components of the distortion power in the phases; these indicate the influence of the third-order current harmonics i_{a3}, i_{b3}, and i_{c3} on the reactive power measurement algorithms. The third-order Walsh function eliminates the effect of the third-order harmonics of the nonlinear load. Harmonics affect the reactive power measurement of a network. The final terms of (10) are the distortion power terms, that is, $Q_{a3}\sin(4\omega t)$, $(Q_{b3} + P_b\sqrt{3})\sin(4\omega t)$, and $(Q_{c3} + P_c\sqrt{3})\sin(4\omega t)$. They are oscillating with the frequency of 4ω which is similar to the oscillating frequency of the 7th-order Walsh function, $\mathrm{Wal}(7,t)$, as can be seen in Figure 2.

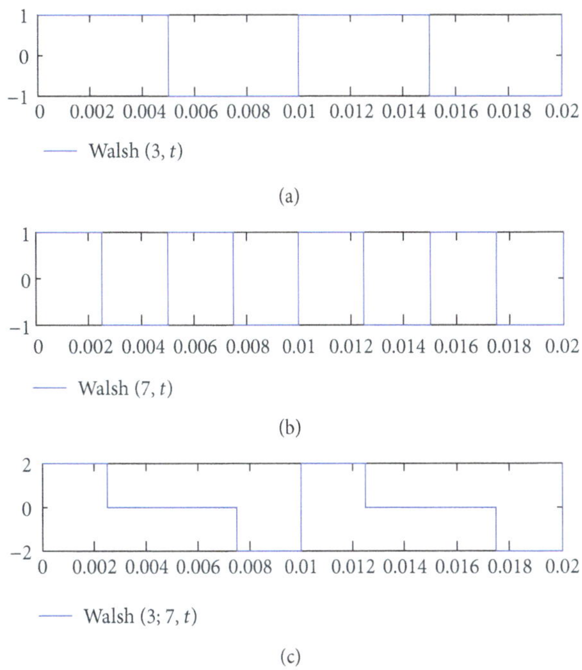

FIGURE 3: (a) $\mathrm{Wal}(3,t)$, (b) $\mathrm{Wal}(7,t)$, and (c) $\mathrm{Wal}(3;7,t)$.

To estimate these distortions power terms both sides of (10) are multiplied by the 7th-order WF and then integrated over the period T and simplified to obtain the following:

$$\frac{1}{T}\int_0^T p_a \mathrm{Wal}(7,t)\,dt = -\frac{1}{T}\int_0^T Q_{a3}\sin(4\omega t)\times \mathrm{Wal}(7,t)\,dt,$$

$$\frac{1}{T}\int_0^T p_b \mathrm{Wal}(7,t)\,dt = \frac{1}{T}\int_0^T \left(Q_{b3} + P_b\sqrt{3}\right)\sin(4\omega t)\times \mathrm{Wal}(7,t)\,dt, \quad (14)$$

$$\frac{1}{T}\int_0^T p_c \mathrm{Wal}(7,t)\,dt = \frac{1}{T}\int_0^T \left(Q_{c3} + P_c\sqrt{3}\right)\sin(4\omega t)\times \mathrm{Wal}(7,t)\,dt.$$

The 7th-order WF is the odd function with the frequency similar to the frequency of the distortion terms. The product of the 7th-order WF with the distortion terms results in their

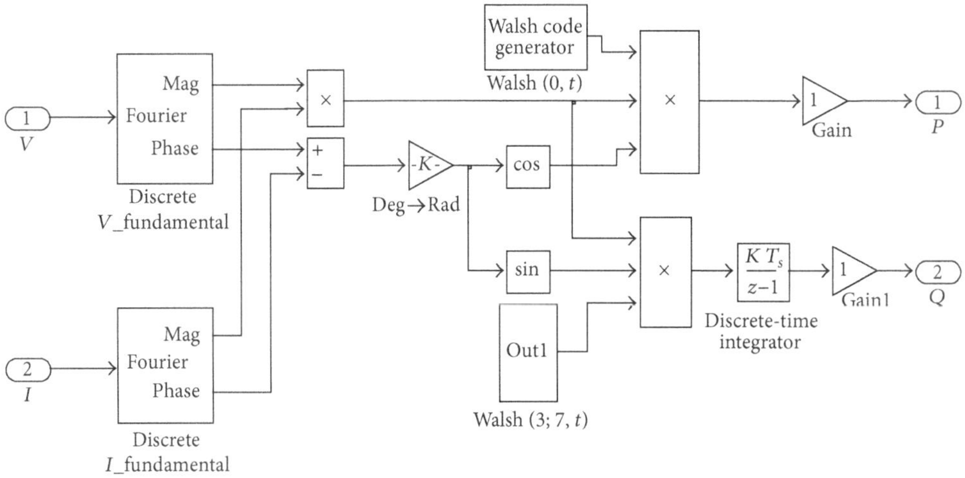

FIGURE 4: The model of the proposed improved Walsh function algorithm.

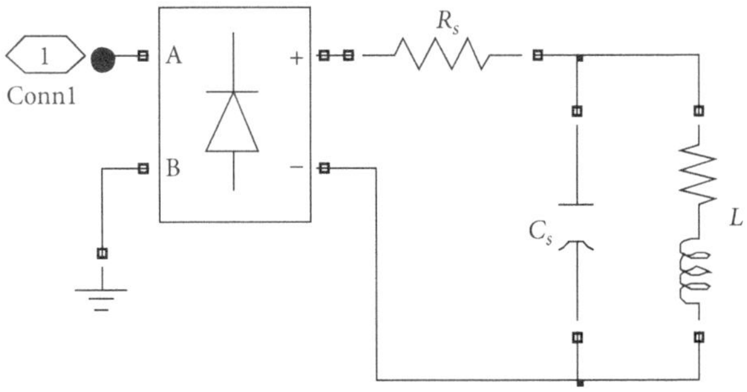

FIGURE 5: Model of nonlinear load.

rectification. So taking cognizance of these rectifying effects, (14) is rewritten as follows:

$$\begin{aligned}
\frac{1}{T}\int_0^T p_a \mathrm{Wal}(7,t)\,dt &= -\frac{1}{T}\int_0^T Q_{a3}\sin(4\omega t)\,dt,\\
\frac{1}{T}\int_0^T p_b \mathrm{Wal}(7,t)\,dt &= \frac{1}{T}\int_0^T \left(Q_{b3}+P_b\sqrt{3}\right)\sin(4\omega t)\,dt,\\
\frac{1}{T}\int_0^T p_c \mathrm{Wal}(7,t)\,dt &= \frac{1}{T}\int_0^T \left(Q_{c3}+P_c\sqrt{3}\right)\sin(4\omega t)\,dt.
\end{aligned} \tag{15}$$

Solving for Q_{a3}, Q_{b3}, and Q_{c3} produces the following:

$$\begin{aligned}
Q_{a3} &= -\frac{\pi}{2T}\int_0^T p_a \mathrm{Wal}(7,t)\,dt,\\
Q_{b3} &= \frac{\pi}{2T}\int_0^T p_b \mathrm{Wal}(7,t)\,dt - P_b\sqrt{3},\\
Q_{c3} &= \frac{\pi}{2T}\int_0^T p_c \mathrm{Wal}(7,t)\,dt - P_c\sqrt{3}.
\end{aligned} \tag{16}$$

Equations (16) are the improved Walsh function algorithms for measuring the distortion power in a three-phase system. Substituting in (13) algorithms for reactive powers are obtained as follows:

$$\begin{aligned}
Q_a &= -\frac{\pi}{2T}\left[\int_0^T p_a \mathrm{Wal}(3,t)\,dt + \int_0^T p_a \mathrm{Wal}(7,t)\,dt\right],\\
Q_b &= \left[\int_0^T p_b \mathrm{Wal}(3,t)\,dt + \int_0^T p_b \mathrm{Wal}(7,t)\,dt\right],\\
Q_c &= \left[\int_0^T p_c \mathrm{Wal}(3,t)\,dt + \int_0^T p_c \mathrm{Wal}(7,t)\,dt\right].
\end{aligned} \tag{17}$$

These algorithms eliminate the effect of the 3rd- to 7th-order harmonics on the reactive power measurement and also essentially reduced the effect of the higher order current harmonics. In other to reduce the computation involved, the 3rd- and 7th-order Walsh are added together which gives the new improved algorithms. The analytical addition of the 3rd and 7th is represented in Figure 3.

From Figure 3 it can be observed that the IWF as a result of the addition of standard 3rd- and the 7th-order WF is defined as follows.

$\mathrm{Wal}(3;7,t)$ is

+1, if t is in the interval $[0;T/8]$, $[T/2;5T/8]$

0, if t is in the interval $[T/8;3T/8]$, $[5T/8;7T/8]$

−1, if t is in the interval $[3T/8;T/2]$, $[7T/8;T]$.

The new improved WF algorithms are derived by multiplying (10), with $\mathrm{Wal}(3;7,t)$ and then integrate over the time T. Now considering the integrals after the equal sign, the p_a, p_b, and p_c are constants so are equal to zero because IWF is a periodic function; 2nd and 4th integrals are also equal to zero as they include cosine functions that are orthogonal with the IWF. The 3rd and 5th integrals comprise the rectification of the sin

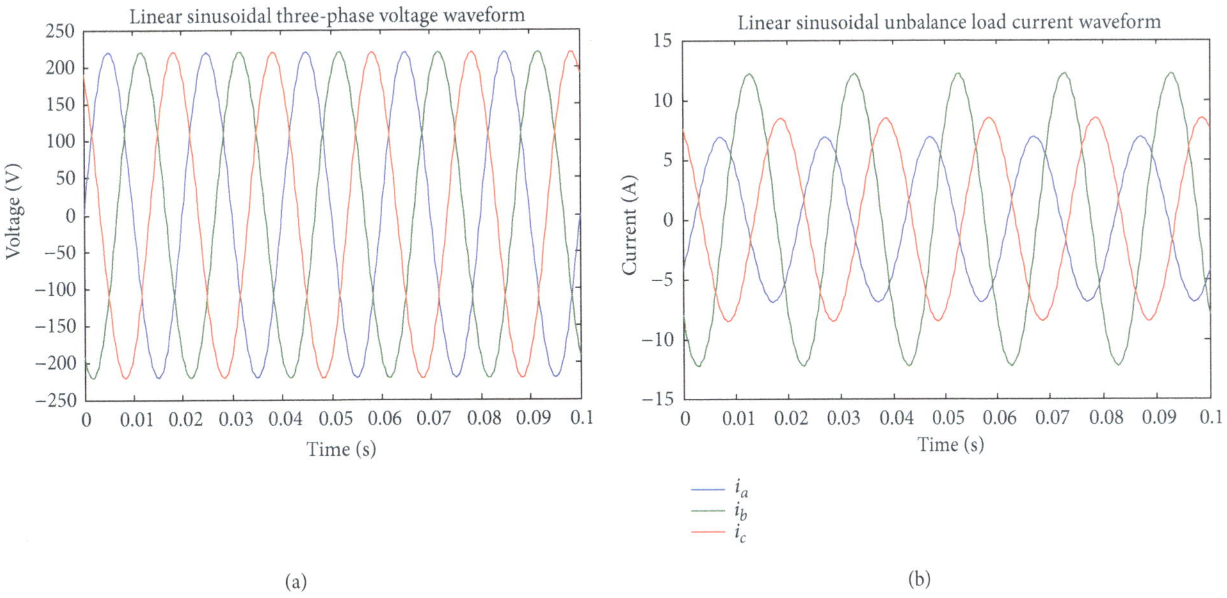

FIGURE 6: (a) 3-phase voltage waveform and (b) 3-phase unbalance current waveform.

functions waveform, so these integrals are not equal to zero. Thus equations are written as follows:

$$\begin{aligned}
&\frac{1}{T}\int_0^T p_a \operatorname{Wal}(3;7,t)\,dt \\
&\quad = \frac{1}{T}\int_0^T \operatorname{Wal}(3;7,t)\left(Q_{a3}-Q_a\right)\sin(2\omega t)\,dt \\
&\quad\quad - \frac{1}{T}\int_0^T \operatorname{Wal}(3;7,t)\,Q_{a3}\sin(2\omega t)\,dt, \\
&\frac{1}{T}\int_0^T p_b \operatorname{Wal}(3;7,t)\,dt \\
&\quad = \frac{1}{T}\int_0^T \operatorname{Wal}(3;7,t)\left(P_b\sqrt{3}+Q_{b3}-Q_b\right)\sin(2\omega t)\,dt \\
&\quad\quad + \frac{1}{T}\int_0^T \operatorname{Wal}(3;7,t)\times\left(Q_{b3}+P_b\sqrt{3}\right)\sin(4\omega t)\,dt, \\
&\frac{1}{T}\int_0^T p_c \operatorname{Wal}(3;7,t)\,dt \\
&\quad = \frac{1}{T}\int_0^T \operatorname{Wal}(3;7,t)\times\left(P_c\sqrt{3}+Q_{c3}-Q_c\right)\sin(2\omega t)\,dt \\
&\quad\quad + \frac{1}{T}\int_0^T \operatorname{Wal}(3;7,t)\times\left(Q_{c3}+P_c\sqrt{3}\right)\sin(4\omega t)\,dt.
\end{aligned} \tag{18}$$

Solving the right-hand side integrals of (18) yields the new improved algorithms in the following:

$$\begin{aligned}
Q_{a,3;7} &= -\frac{\pi}{T}\int_0^T p_a \operatorname{Wal}(3;7,t)\,dt, \\
Q_{b,3;7} &= -\frac{\pi}{T}\int_0^T p_b \operatorname{Wal}(3;7,t)\,dt + \left(P_b\sqrt{3}+Q_{b3}\right), \\
Q_{c,3;7} &= -\frac{\pi}{T}\int_0^T p_c \operatorname{Wal}(3;7,t)\,dt + \left(P_c\sqrt{3}+Q_{c3}\right).
\end{aligned} \tag{19}$$

Equations (7), (16), and (19) are the proposed improved Walsh function IWF algorithms that would be used to measure the active, distortion, and reactive powers, respectively, of a network and also eliminate the effect of higher order harmonic in the three-phase reactive power measurement system. Suffice it to say that in actual cases only lower order harmonics are present in power system signal.

5. Modeling the Proposed IWF Algorithm

Equations (7) and (19) are used to create the model for the active and reactive powers measurement based on the proposed improved algorithms using the Matlab Simulink software tool. Then some of the commonly used nonlinear domestic and industrial loads were modeled along and used in the simulation of the improved Walsh function algorithms for active and reactive powers measurement. Examples are compact fluorescent lamps (CFL) and computers. Loads can be resistive R, inductive L, capacitive C, or a combination, for example, RL, RC, LC, RLC depending on the type of load being modeled. RL was used to model the loads used in this simulation (see Figures 4 and 5).

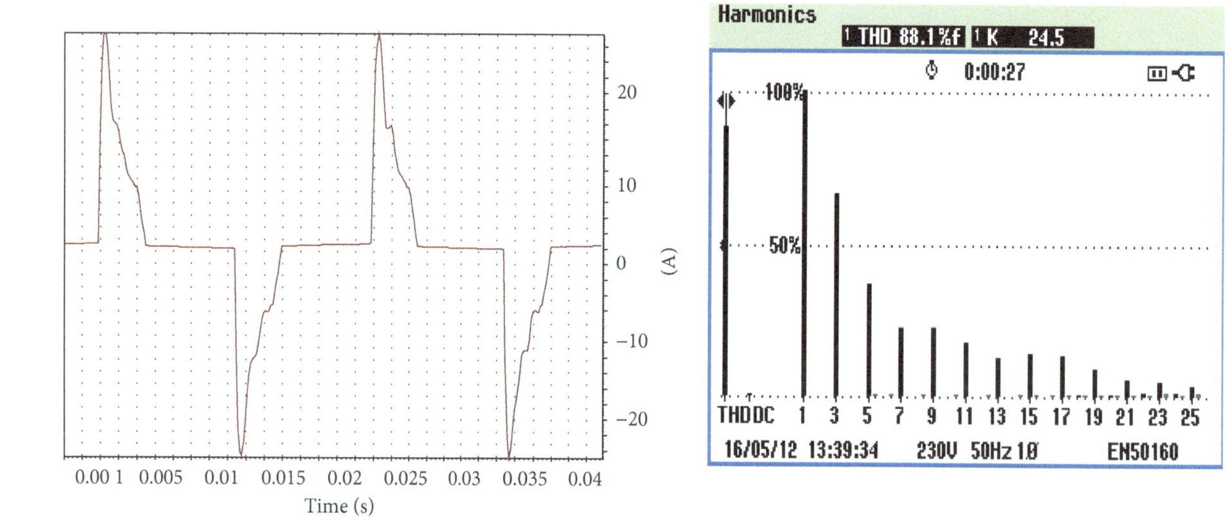

(a) (b)

FIGURE 7: (a) Nonlinear current waveform and (b) harmonic spectrum (experiment).

TABLE 1: The result of case A.

FFT approach	Active power P (watts)	Reactive power Q (var)
Phase a	1181.1	944.8
Phase b	214.1	746.5
Phase c	492.6	1039.1
Total	1887.8	2730.4
Proposed IWF method		
Phase a	1180.7	944.5
Phase b	214.8	746.7
Phase c	493.5	1040.2
Total	1889	2731.4

TABLE 2: The result of case B.

FFT approach	Active power P (watts)	Reactive power Q (var)
Phase a	1027.5	569.5
Phase b	819.3	223.3
Phase c	2098.6	1538.9
Total	3945.4	2331.7
Proposed IWF method		
Phase a	1028	569.6
Phase b	818.6	223.8
Phase c	2099	1539
Total	3945.6	2332.4

The RL nonlinear loads were modeled using universal bridge rectifier; the values for the R_s and C_s were determined using the expressions

$$R_s > \frac{2T_s}{C_s}, \quad C_s < \frac{P_n}{1000\,(2\pi f)\,V_n^2}, \tag{20}$$

where P_n = nominal power of three phase converter VA, V_n = nominal line AC voltage V rms, f = fundamental frequency Hertz, and T_s = sample time.

The R and L of an RL load on dc side of the universal bridge rectifier are calculated using data recorded on the standard 435 fluke power quality analyzer PQA during the laboratory experiment involving measurement of power components of harmonic generating loads.

5.1. Simulation and Discussion of Results. For the simulation of the linear loads a synthetic line to neutral voltages (V phases) is chosen as follows:

$$V_a = 220\angle 0^\circ, \qquad V_b = 220\angle -120^\circ, \qquad V_c = 220\angle 120^\circ. \tag{21}$$

Load impedances for two different cases of unbalanced three-phase systems are chosen as shown in cases A and B below for the simulation to verify the algorithms. Knowing the supply voltage and the impedance of the loads (Z) the resistance (R) and inductive reactance (X_L) of each phase load were calculated and the obtained results for R and X_L were used to configure standard RLC load taken from the Matlab simulink power blocks.

Case A: $Z_a = 36 + j20\Omega$, $Z_b = 55 + j15\Omega$, $Z_c = 15 + j11\Omega$.

Case B: $Z_a = 25 + j20\Omega$, $Z_b = 17 + j60\Omega$, $Z_c = 18 + j38\Omega$.

Figure 6 shows the voltage and current waveform for the three-phase sinusoidal linear unbalanced load system obtained from simulation with the new improved algorithms.

5.2. Linear Load System. For the linear sinusoidal load system, IEEE standard 1459-2000 which is based on FFT approach was used as benchmark for measurement of active and reactive powers of case A for the three phases

TABLE 3: Simulation and experimental results.

S/N	Load type	Fluke 435 PQA meter readings		FFT approach		Proposed IWF algorithm	
		P (watts)	Q (var)	P (watts)	Q (var)	P (watts)	Q (var)
1	CFL lamps (15.8 watts × 45)	711	364.5	710.8	363.4	711	364
2	Desktop computers (60.3 watts × 70)	422.1	20.3	421.9	19.8	422	20
3	Laptop computers (54.6 watts × 8)	436.8	18.5	435.7	18.3	437	18
4	2 ft. fluorescent lamps (27.8 watt × 15)	417	934.5	416.3	933.8	417	934

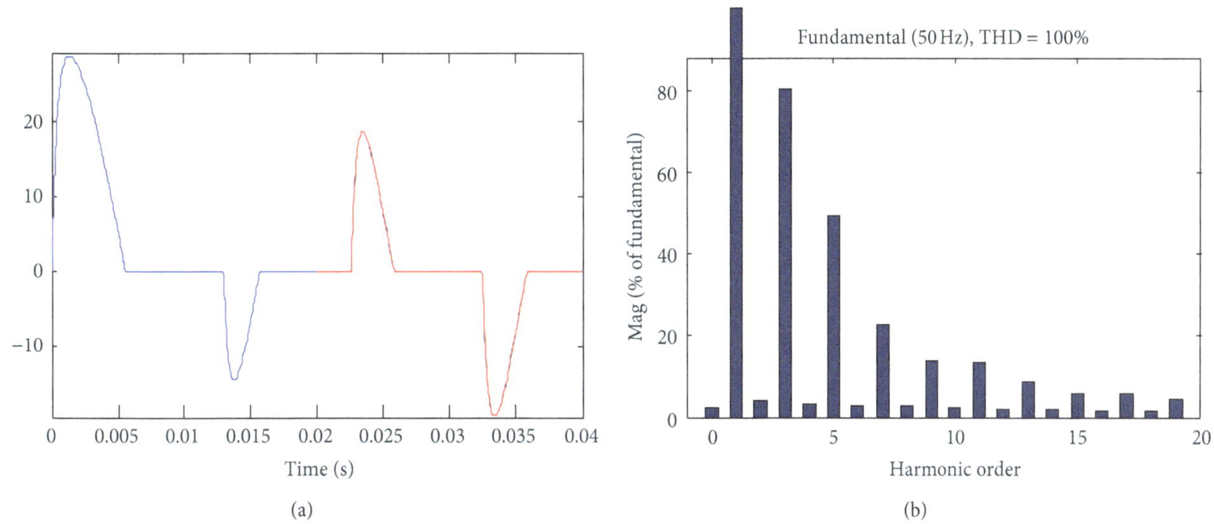

FIGURE 8: (a) Nonlinear current waveform and (b) harmonic spectrum (Simulation).

and compared with the results using the proposed improved algorithms (7) and (19). The results are displayed in Table 1. Similarly for case B results are shown in Table 2. From the results using the IEEE standard 1459-2000 which is based on FFT as reference it can be observed that the proposed Walsh function algorithm has the potential to measure the active and reactive powers with a high degree of accuracy. In terms of computational complexity the proposed Walsh function algorithm has less computational requirement than the FFT.

5.3. Nonlinear Harmonic Load. Experimental results using harmonic load bank and other harmonic generating loads in the laboratory as recorded by fluke PQA meters were used in modeling of the nonlinear loads. The computer and CFL lamps were modeled as RL loads and simulated using both FFT approach and the proposed improved Walsh algorithms for the measurement of active and reactive powers. The laboratory experimental results were used as benchmark. The nonlinear load currents waveform for the experiment using the Fluke PQA meter and that of the model simulation is recorded as shown in Figures 7 and 8. The results of the simulations and experimental measurement of nonlinear loads are presented in Table 3. It can be observed that the standard FFT approach recorded significant error which could be due to the spectral leakage and picket fence effect phenomenon of the FFT approach when used for measurement in nonlinear harmonic load system. On the other hand, the proposed improved Walsh function algorithm for measurement gave a near accurate result as the influence of the harmonics of the nonlinear load has been address in the improved Walsh algorithms. The Fluke power quality analyzer and FFT graphical user interface analyze signals of one phase at a time so the figure shown is for one of the phases and the same is applicable to the other two phases but has been withheld for convenience, clarity, and space. Walsh function transform is a special form of the Fourier transforms [21, 22]. So it is sometimes called Walsh Fourier.

6. Conclusion

A new improved Walsh function algorithm for active and reactive powers measurement was presented. The IEEE standard 1459-2000 fast Fourier transform approach was used as benchmark to validate the algorithm when the load system is linear and sinusoidal. The Fluke 435 power quality analyzer PQA meter was used as reference for nonlinear harmonic load system simulation and measurement. The results show that the proposed improved Walsh function has the potentials of accurately measuring power components under different load conditions. If the algorithm is integrated into measuring instrument it would be able to properly record the active and reactive powers components under sinusoidal and nonsinusoidal load system for proper monitoring, management, and maintenance planning. The approach is simpler with less computation when compared with Fourier transform. The research is continuing in the development of a model for measuring the distortion component of the power system.

Acknowledgments

The authors wish to acknowledge the Ministry of Higher Education, Malaysia (MOHE), for funding of project and Universiti Teknologi Malaysia and Universiti Kebangsaan Malaysia for providing infrastructure and moral support for the research work.

References

[1] E. B. Makram, R. B. Haines, and A. A. Girgis, "Effect of harmonic distortion in reactive power measurement," *IEEE Transactions on Industry Applications*, vol. 28, no. 4, pp. 782–787, 1992.

[2] A. Abiyev, "Measurement of reactive power in unbalanced three-phase power systems by use of walsh functions," *International Journal of Academic Research*, vol. 2, no. 4, pp. 46–56, 2010.

[3] L. S. Czarnecki, "Instantaneous reactive power *p-q* theory and power properties of three-phase systems," *IEEE Transactions on Power Delivery*, vol. 21, no. 1, pp. 362–367, 2006.

[4] S. L. Toral, J. M. Quero, and L. G. Franquelo, "Reactive power and energy measurement in the frequency domain using random pulse arithmetic," *IEE Proceedings*, vol. 148, no. 2, pp. 63–67, 2001.

[5] A. Ozdemir and A. Ferikoglu, "Low cost mixed-signal microcontroller based power measurement technique," *IEE Proceedings*, vol. 151, no. 4, pp. 253–258, 2004.

[6] R. Langella and A. Testa, "A new algorithm for energy measurement at positive sequence of fundamental power frequency, under unbalanced non-sinusoidal conditions," in *Proceedings of the IEEE Conference in Power Technology*, pp. 1558–1563, Lausanne, Switzerland, July 2007.

[7] W. G. Morsi and M. E. El-Hawary, "Defining power components in nonsinusoidal unbalanced polyphase systems: the issues," *IEEE Transactions on Power Delivery*, vol. 22, no. 4, pp. 2428–2438, 2007.

[8] E. Moulin, S. D. Electricite, and S. Dictionary, "Measuring reactive power in energy meters," *Metering International*, vol. 1, no. 1, pp. 52–54, 2002.

[9] T. Lin and A. Domijan Jr., "Recursive algorithm for real-time measurement of electrical variables in power systems," *IEEE Transactions on Power Delivery*, vol. 21, no. 1, pp. 15–22, 2006.

[10] W. M. Lin and T. C. Ou, "Unbalanced distribution network fault analysis with hybrid compensation," *IET Generation, Transmission and Distribution*, vol. 5, no. 1, pp. 92–100, 2011.

[11] W. G. Morsi, C. P. Diduch, L. Chang, and M. E. El-Hawary, "Wavelet-based reactive power and energy measurement in the presence of power quality disturbances," *IEEE Transactions on Power Systems*, vol. 26, no. 3, pp. 1263–1271, 2011.

[12] R. H. Abiyev, A. N. Abiyev, and K. Dimililer, "The Walsh functions based method for reactive power measurement," in *Proceedings of the 33rd Annual Conference of the IEEE Industrial Electronics Society (IECON '07)*, pp. 2337–2342, November 2007.

[13] A. N. Abiyev, "A new electronic reactive power meter based on walsh transform algorithms," in *Proceedngs of the IEEE/PES Conference and Exposition in Transmission and Distribution (PIMS '08)*, April 2008.

[14] A. N. Abiyev and K. Dimililer, "Reactive power measurement in sinusoidal and nonsinusoidal conditions by use of the walsh functions," in *Proceedings of the IEEE International Instrumentation and Measurement Technology Conference (IMTC '08)*, pp. 89–94, May 2008.

[15] J. Vicente, R. Pindado, and I. Martinez, "Design guidelines using selective harmonic elimination advanced method for DC-AC PWM with the Walsh transform," in *Proceedings of the 7th International Conference-Workshop Compatibility and Power Electronics (CPE '11)*, pp. 220–225, June 2011.

[16] M. Maqusi, "On generalized Walsh functions and transform," in *Proceedings of the IEEE Conference on Decision and Control and 11th Symposium on Adaptive Processes*, pp. 499–502, Decembre 1972.

[17] "1459-2000—IEEE Trial Use Standard Definitions for the Measurement of Electric Power Quantities Under Sinusoidal, Non-Sinusoidal, Balanced or Unbalanced Conditions," *IEEE Transaction on Power System*, 2000.

[18] A. Agüera-Pérez, J. C. Palomares-Salas, J. J. G. de la Rosa, J. M. Sierra-Fernández, D. Ayora-Sedeño, and A. Moreno-Muñoz, "Characterization of electrical sags and swells using higher-order statistical estimators," *Measurement*, vol. 44, no. 8, pp. 1453–1460, 2011.

[19] S. Rechka, E. Ngandui, J. Xu, and P. Sicard, "Analysis of harmonic detection algorithms and their application to active power filters for harmonics compensation and resonance damping," *Canadian Journal of Electrical and Computer Engineering*, vol. 28, no. 1, pp. 41–51, 2003.

[20] V. S. Vladimir Terzija, Z. Lazarevic, and M. Popov, "Active and reactive power metering in non-sinusoidal conditions using newton type algorithm," in *Proceedings of the International Conference on Renewable Energies and Power Quality (ICREPQ '97)*, 1997.

[21] T. J. Liang, R. M. O' Connell, and R. G. Hoft, "Inverter harmonic reduction using Walsh function harmonic elimination method," *IEEE Transactions on Power Electronics*, vol. 12, no. 6, pp. 971–982, 1997.

[22] M. D. Vose and A. H. Wright, "The simple genetic algorithm and the Walsh transform part 1, Theory," *Evolutionary Computing*, vol. 6, no. 3, pp. 253–273, 1998.

23

A New Sliding Mode Controller for DC/DC Converters in Photovoltaic Systems

M. Sarvi, I. Soltani, N. NamazyPour, and N. Rabbani

Faculty of Electrical Engineering, Imam Khomeini International University, Qazvin 34149-16818, Iran

Correspondence should be addressed to I. Soltani; i_soltani@ikiu.ac.ir

Academic Editor: Kamaruzzaman Sopian

DC/DC converters are widely used in many industrial and electrical systems. As DC/DC converters are nonlinear and time-variant systems, the application of linear control techniques for the control of these converters is not suitable. In this paper, a new sliding mode controller is proposed as the indirect control method and compared to a simple direct control method in order to control a buck converter in photovoltaic applications. The solar arrays are dependent power sources with nonlinear voltage-current characteristics under different environmental conditions (insolation and temperature). From this point of view, the DC/DC converter is particularly suitable for the application of the sliding mode control in photovoltaic application, because of its controllable states. Simulations are performed in Matlab/Simulink software. The simulation results are presented for a step change in reference voltage and input voltage as well as step load variations. The simulations results of proposed method are compared with the conventional PID controller. The results show the good performance of the proposed sliding mode controller. The proposed method can be used for the other DC/DC converter.

1. Introduction

Modern electronic systems require high-quality, small, lightweight, reliable, and efficient power supplies. So, the DC/DC converters are widely used in many industrial and electrical systems. The most familiar are switching power supplies, DC drives, and photovoltaic systems. The stability is an important aspect in the design of switch mode power supplies; a feedback control is used to achieve the required performance. Ideally the circuit is in steady state, but actually the circuit is affected by line and load variations (disturbances), as well as variation of the circuit component (robustness). These parameters have a severe effect on the behavior of switch mode power supply and may cause instability. Design of controller for these converters is a major concern in power converters design [1–3].

Different control techniques are applied to regulate the DC-DC converters, especially buck converters, in order to obtain a robust output voltage [4, 5].

As DC/DC converters are nonlinear and time-variant systems, the application of linear control techniques for the control of these converters is not suitable. In order to design linear control system using classical linear control techniques, the small signal model is derived by the linearization around a precise operating point from the state space average model [4]. The controllers based on these techniques are simple to implement; however, it is difficult to account the variation of system parameters, because of the dependence of small signal model parameters on the converter operating point [6].

Sliding mode control is a well-known discontinuous feedback control technique which has been exhaustively explored in many books and journal articles. The technique is naturally suited for the regulation of switched controlled systems, such as power electronics devices, in general, and DC/DC power converters, in particular [2]. Many sliding mode controllers have been proposed and used for DC/DC converters [6–9]. These controllers are direct [6] or indirect control method [6, 7]. The direct method is proposed in [6]. In [7], the output capacitor current of DC/DC converter is used to control the output voltage. The differences of the DC/DC output voltage and the reference voltage enter the proportional-Integrator (PI) type controller, and then the output capacitor current of DC/DC converter is decreased from the output of controller [8]. The output voltage and inductor current are used to control of DC/DC converter in [9]. These references [6–9] have not completely investigated the load and line as well as reference regulations.

In this paper, a new sliding mode controller is introduced for DC/DC buck converter as the indirect control method. The proposed controller is compared with a simple direct control method as well as the conventional PID controller. The simulation results are presented for a step change in reference voltage and input voltage as well as for a step load variation. The main contribution of this paper is the presentation of a new indirect sliding mode controller with good accuracy and performance against load and line as well as reference regulations.

2. DC/DC Converters

The DC-DC converters can be divided into two main types: (1) hard-switching pulse width modulated (PWM) converters and (2) resonant and soft-switching converters [1].

Advantages of PWM converters include low component count, high efficiency, constant frequency operation, relatively simple control and commercial availability of integrated circuit controllers, and ability to achieve high conversion ratios for both step-down and step-up applications. The circuit diagram of the DC/DC buck converter is shown in Figure 1. In this figure, the circuit schematic is depicted with the transistor-diode symbols.

By sensing of the DC output and controlling of the switch duty cycle in a negative-feedback loop, the DC output voltage could be regulated against input line and output load changes [3].

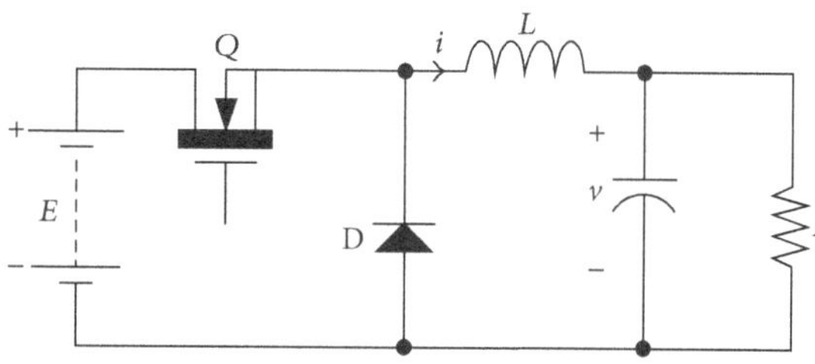

FIGURE 1: Semiconductor realization of the DC/DC buck converter.

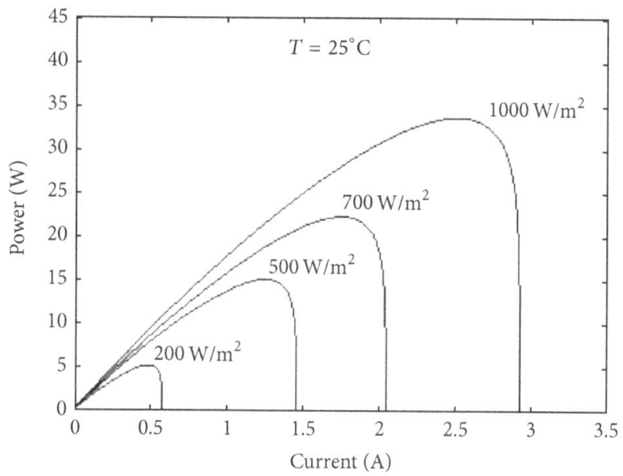

FIGURE 2: *P-I* characteristics of the solar array according to different insolation (for $T = 25^{\circ}$C).

3. Solar Array Characteristic

The solar arrays have nonlinear *V-I* characteristics which depend on the environmental conditions: ambient temperature and insolation.

The nonlinear characteristic of a solar sell is obtained as the following equation [10, 11]:

$$i_{SA} = I_{ph} - I_o \left\{ \exp\left(\left(\frac{q}{AKT} \right) \cdot \left(v_{SA} + R_S i_{SA} \right) \right) - 1 \right\}, \tag{1}$$

where I_{ph} is the generated current under a given insolation condition, I_o is the reverse saturation current, q is the charge of an electron, v_{SA} and i_{SA} are the output voltage and current of the solar cell, respectively, A is the ideality factor for a p-n junction, K is Boltzmann's constant, T is the temperature, and R_S is the series resistance of the solar cell. Figures 2 and 3 indicate the power-current (*P-I*) characteristic of a solar cell for different insolation and temperature, respectively.

4. The State-Space Model of Buck Converter

To obtain the differential equations describing the buck converter, the ideal topology is used as shown in Figure 4. The differential equations describing the DC/DC buck converter dynamics are obtained through the direct application of Kirchoff's current and Kirchoff's voltage laws for each one of the possible circuit topologies arising from the assumed particular switch position function value. Thus, when the switch position function exhibits the value $u = 1$, we obtain the topology corresponding to the nonconducting mode for the diode obtained. Alternatively, when the switch position exhibits the value $u = 0$, the second possible circuit topology corresponding to the conducting mode for the diode is obtained.

The system dynamics is described by the following differential equations.

For $u = 1$,

$$L\frac{di}{dt} = -v + E, \qquad C\frac{dv}{dt} = i - \frac{v}{R}. \tag{2}$$

For $u = 0$,

$$L\frac{di}{dt} = -v, \qquad C\frac{dv}{dt} = i - \frac{v}{R}. \tag{3}$$

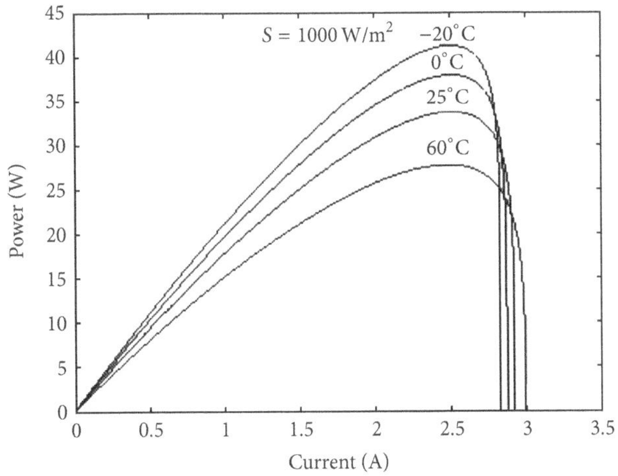

FIGURE 3: *P-I* characteristics of the solar array according to different temperature (for $S = 1000\,\mathrm{W/m^2}$).

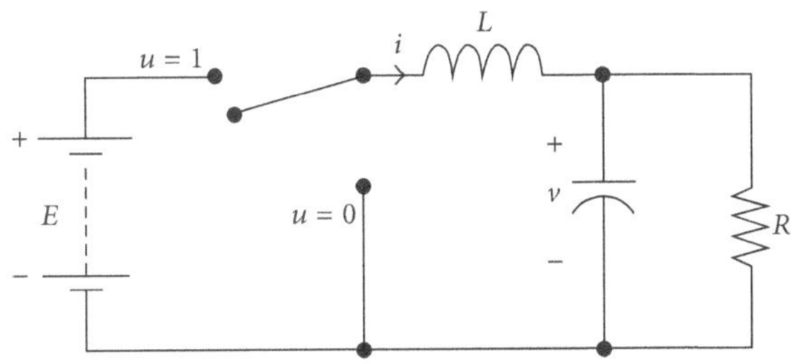

FIGURE 4: Ideal switch representation of the DC/DC buck converter.

By comparing the obtained particular dynamic systems descriptions, the following unified dynamic system model can be obtained:

$$L\frac{di}{dt} = -v + uE,$$
$$C\frac{dv}{dt} = i - \frac{v}{R}. \tag{4}$$

5. Sliding Mode Controller Design

In this section, two methods are presented in order to control the output voltage of a DC/DC buck converter. These methods are direct and indirect sliding mode control.

Direct control is based on the output voltage feedback, where the output voltage (V_C) is directly compared to the reference voltage (V_r).

Sliding surface is as follows:

$$S = V_C - V_r. \tag{5}$$

Indirect control is based on the inductor current control satisfying desired output voltage (V_r). In this method, because of the possible load variation, at first the output impedance (R_o) is determined by dividing the output voltage by the output current, and then the desired load current (I_{ld}) is calculated by dividing the reference voltage by the output impedance (R_o). By adding a coefficient of capacitor current (k) to I_{ld}, the desired inductor current (I_{lr}) is determined. So,

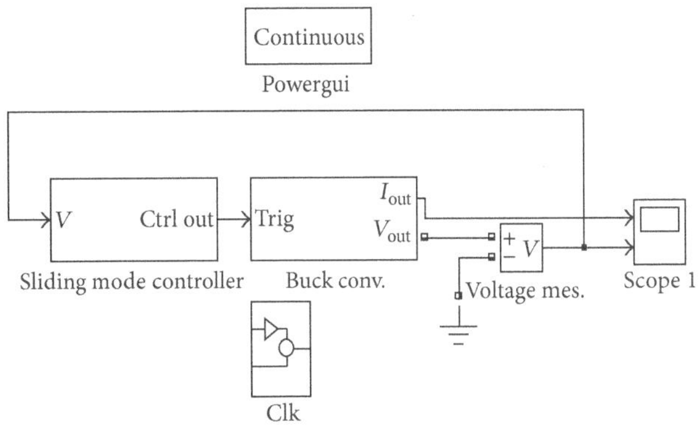

FIGURE 5: Simulink block diagram of direct sliding mode control.

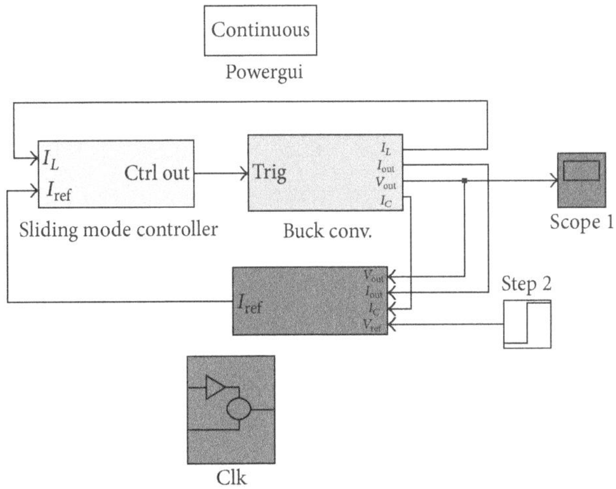

FIGURE 6: Simulink block diagram of indirect sliding mode control.

the sliding surface is defined by comparison between actual indirect current (I_l) and I_{lr} as the following.

Sliding surface can be computed as:

$$S = I_l - I_{lr}. \tag{6}$$

6. Simulation of DC/DC Converter with Sliding Mode Control

In order to investigate the proposed controller performance and accuracy, Matlab/Simulink and its facilities are used.

Simulations are performed on a typical buck converter with the following parameter values: $E = 10\,\mathrm{V}$, $L = 1\,\mathrm{mH}$, $C = 2200\,\mu\mathrm{F}$, $R = 25\,\mathrm{W}$.

Figures 5 and 6 show the Simulink block diagram of direct and indirect sliding mode control of the proposed buck converter, respectively.

7. Results and Discussions

For the evaluation and validation of the proposed controller accuracy and performance, simulation results are presented in three subsections:

(a) analysis of the proposed sliding mode controller and determining the best sliding mode controller response;

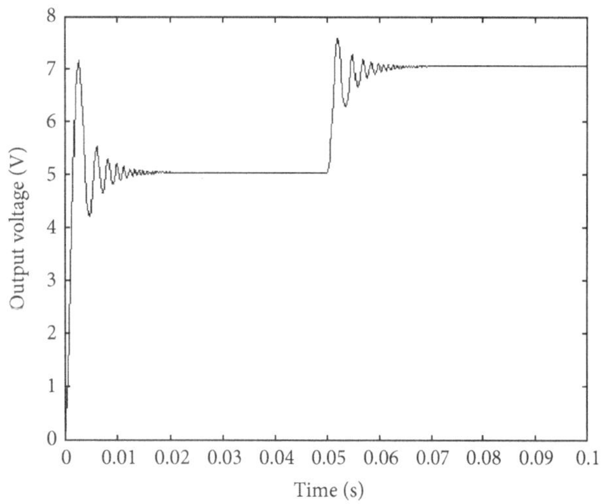

FIGURE 7: The DC/DC output voltage for a step change in reference voltage in direct control method.

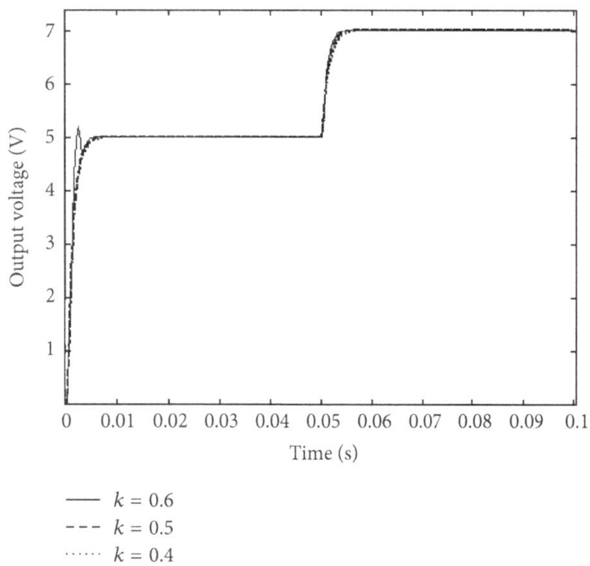

FIGURE 8: The DC/DC output voltage for a step change in reference voltage with respect to different values of k in indirect control method.

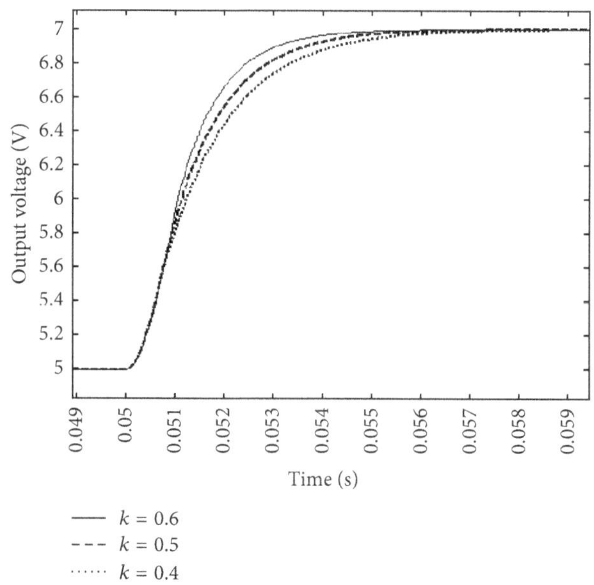

FIGURE 9: The DC/DC output voltage variation about reference variation time.

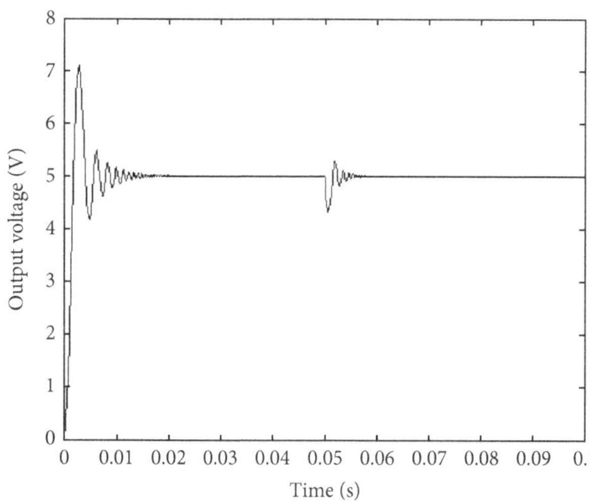

FIGURE 10: The DC/DC output voltage for step load variation in direct control method.

(b) comparison of the proposed sliding mode controller with the conventional PID controller;

(c) investigation of the proposed siding mode controller under environmental (insolation and temperature) conditions.

7.1. Simulation Results of the Proposed Sliding Mode Controller. Figures 7–9 show the output voltage during a change in reference voltage from $V = 5$ to $V = 7$ at $t = 0.05$ sec in direct and indirect control methods, respectively. Figure 7 shows the DC/DC output voltage for a step change in reference voltage (from $V = 5$ to $V = 7$ at $t = 0.05$ sec) in direct method. Figure 8 is plotted with respect to different values of k, where k is capacitor current coefficient.

Figure 9 indicates a zoom window about the output voltage variation.

In order to evaluate the robustness of sliding mode control, the load is changed from $1\,\Omega$ to $0.5\,\Omega$ at $t = 0.05$ sec. The results are indicated in Figures 10–12 for direct and indirect control methods, respectively. Figure 10 shows the DC/DC output voltage for a step change in load (from $1\,\Omega$ to $0.5\,\Omega$ at $t = 0.05$ sec) in direct method. Figure 11 is plotted with respect to different values of k. Figure 12 indicates a zoom window about the output voltage variation.

Figures 13–15 show the input voltage regulation, where the line voltage changes from $E = 10$ to $E = 20$ in $t = 0.05$ sec. Figure 13 shows the DC/DC output voltage for a step change in load (from $1\,\Omega$ to $0.5\,\Omega$ at $t = 0.05$ sec) in direct method. Figure 14 is plotted with respect to different values

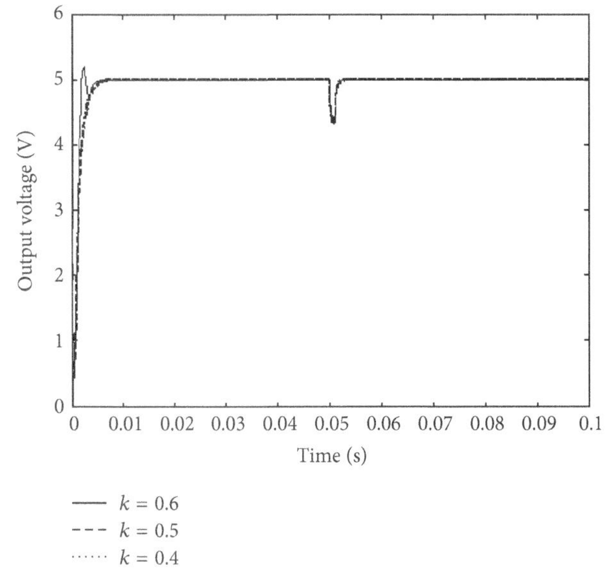

FIGURE 11: The DC/DC output voltage for step load variation with respect to different values of k in indirect control method.

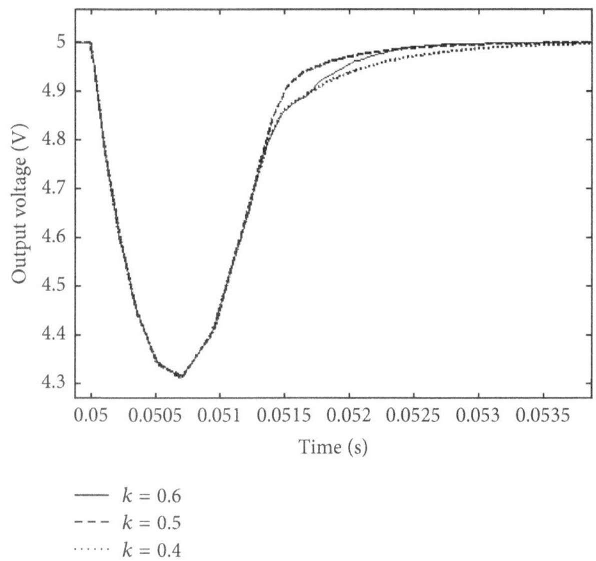

FIGURE 12: The DC/DC output voltage variation about step load variation time.

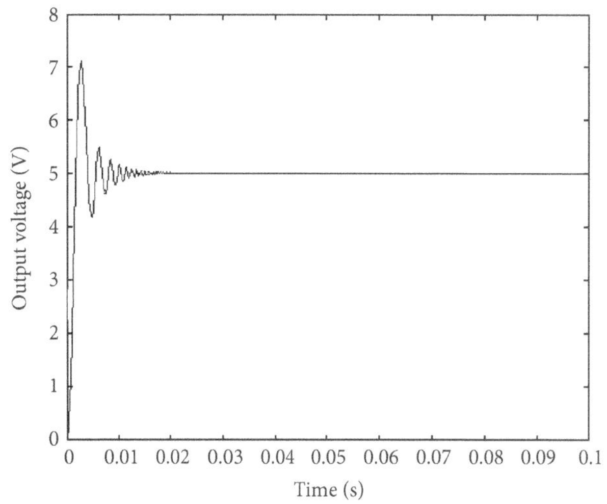

FIGURE 13: The DC/DC output voltage variation for a step change in line voltage in direct control method.

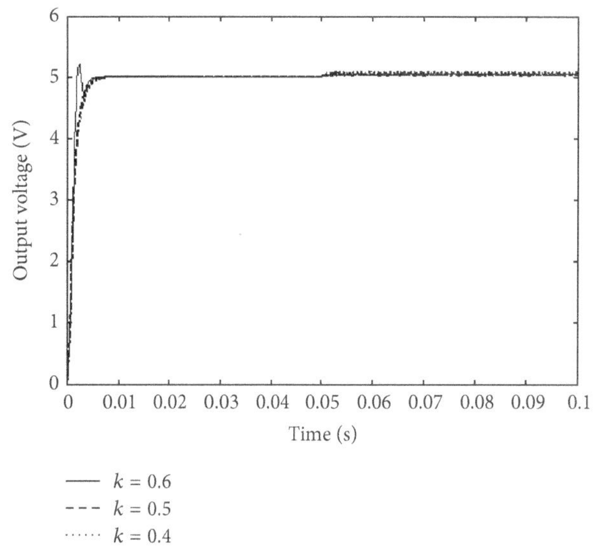

FIGURE 14: The DC/DC output voltage variation for a step change in Line voltage with respect to different values of k in indirect control method.

of k. Figure 15 indicates a zoom window about the output voltage variation.

These results (Figures 7–15) show that the direct control method does not have a good performance. On the other hand, indirect control performance is good but it is related to the k value. By the selection of suitable value of k, the indirect method performance is very good. The results show that the best performance is obtained for $k = 0.5$.

7.2. Comparison of Proposed Sliding Mode and PID Controller. For the validation of the proposed sliding mode controller, the best simulated result compares with conventional PID controller.

The best result is the one with the wave forms of indirect sliding mode control with capacitor current coefficient (k) equal to 0.5.

The noninteracting structure of PID controller is used. This is described by the following equation:

$$G_{\mathrm{PID}} = k_P\left(1 + \frac{1}{sT_i} + sT_d\right), \qquad (7)$$

where k_P is proportional gain and T_i, T_d are integrator and derivative time constants, respectively.

The simulation results of comparison between the proposed sliding mode controller and PID controller are indicated in Figures 16–19.

Figures 16 and 17 show the output voltage of buck converter with sliding mode and PID controller with a step

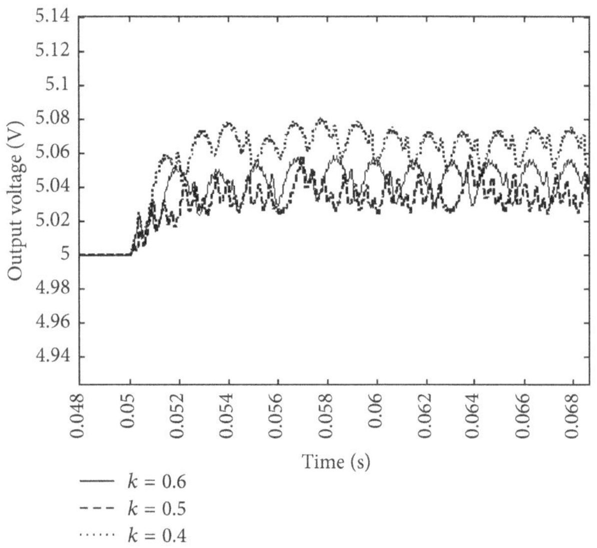

FIGURE 15: The DC/DC output voltage variation about line variation time.

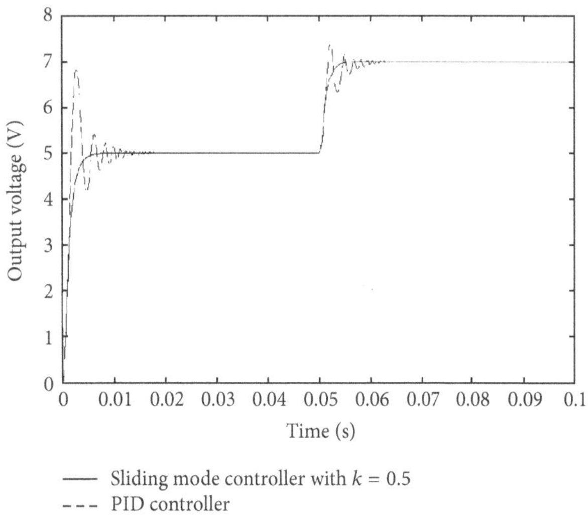

FIGURE 16: The DC/DC output voltage variation with a step change in reference voltage.

change in reference voltage from 5 V to 7 V at $t = 0.05$ sec. Figure 15 indicates a zoom window about the output voltage variation.

Finally, the load resistance is changed from $1\,\Omega$ to $0.5\,\Omega$ at $t = 0.05$ sec. The output voltage variation is indicated in Figure 18. Figure 19 shows a zoom window about the output voltage variation.

7.3. Investigation of the Proposed Sliding Mode Controller under Environmental (Insolation and Temperature) Conditions. In order to investigate of the accuracy of the proposed controller, a system includes a solar array, a DC/DC buck converter, and sliding mode controller. For testing the output voltage and current of photovoltaic array and then the output voltage of buck converter versus changing the environmental parameters instabilities, system was examined by changing

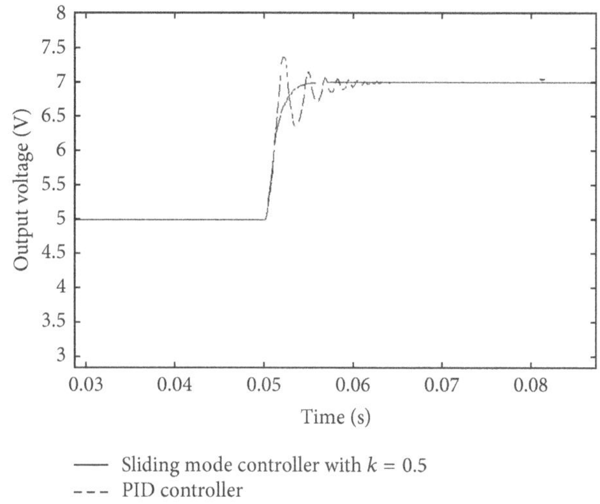

FIGURE 17: The DC/DC output voltage variation about reference voltage variation time.

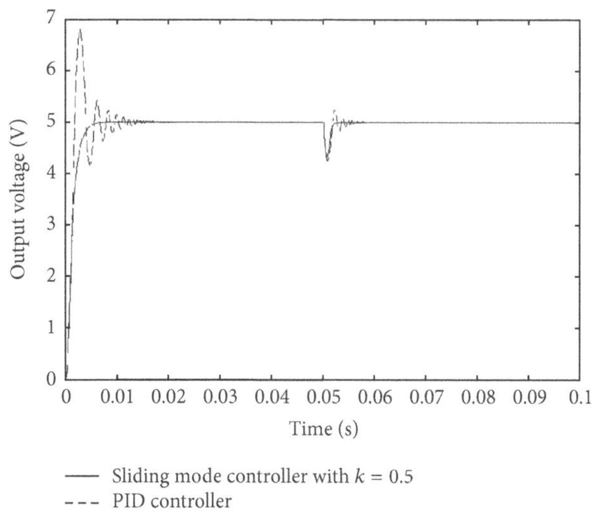

FIGURE 18: The DC/DC output voltage variation with a step change in load.

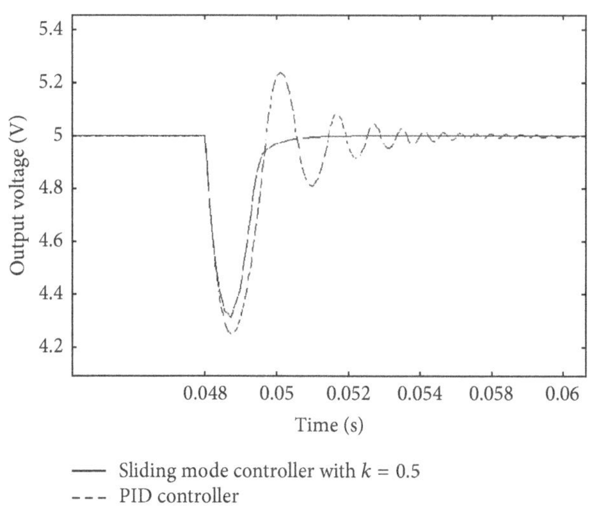

FIGURE 19: The DC/DC output voltage variation about load variation time.

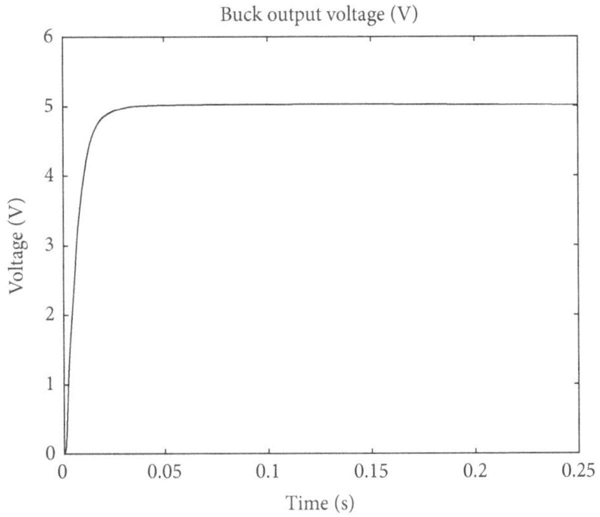

FIGURE 20: Output voltage of DC/DC converter for a step change in insolation from 900 (W/m^2) to 600 (W/m^2) in 30°C.

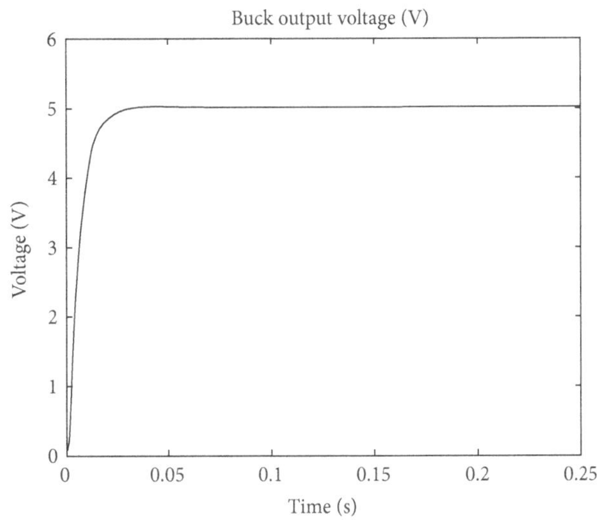

FIGURE 21: Output voltage of DC/DC converter for a step change in ambient temperature from 30°C to 20°C in constant insolation (900 (W/m^2)).

the insolation and ambient temperature of array. The results are illustrated in following figures.

In Figure 20, the insolation is changed from 900 (W/m^2) to 300 (W/m^2) in 30°C ambient temperature at $t = 0.025$ sec. As indicated in these figures, the voltage variation in the output voltage of buck converter is not visible.

In Figure 21, the results of a step change in ambient temperature from 30°C to 20°C in constant insolation (900 (W/m^2)) at $t = 0.025$ sec are illustrated.

8. Conclusion

In this paper, a new indirect sliding mode controller for DC/DC converters is presented. The results of analysis and simulation of proposed method are compared by direct sliding mode as well as conventional PID controller. The main conclusions are the following.

(i) The proposed indirect sliding mode controller has more good performance as compared with direct sliding mode as well as PID control.

(ii) The performance of the proposed sliding mode controller is related to suitable selection of its parameter (k).

(iii) The proposed controller is robust to load and line as well as reference variations.

(iv) The proposed controller can be used to all types of DC/DC converters.

References

[1] M. H. Rashid, *Power Electronics Handbook*, Academic Press, New York, NY, USA, 2001.

[2] J. Clerck Maxwell, *Control Design Techniques in Power Electronics Devices*, Springer, New York, NY, USA, 2006.

[3] A. I. Pressman, *Switching Power Supply Design*, McGrew Hill, New York, NY, USA, 2nd edition, 1998.

[4] J. H. Su, J. J. Chen, and D. S. Wu, "Learning feedback controller design of switching converters via MATLAB/SIMULINK," *IEEE Transactions on Education*, vol. 45, no. 4, pp. 307–315, 2002.

[5] J. A. A. Qahouq, O. Abdel-Rahman, L. Huang, and I. Batarseh, "On load adaptive control of voltage regulators for power managed loads: Control schemes to improve converter efficiency and performance," *IEEE Transactions on Power Electronics*, vol. 22, no. 5, pp. 1806–1819, 2007.

[6] P. Mattavelli, L. Rossetto, G. Spiazzi, and P. Tenti, "General-purpose sliding-mode controller for DC/DC converter applications," in *Proceedings of the IEEE 24th Annual Power Electronics Specialist Conference (PESC '93)*, pp. 609–615, Seattle, Wash, USA, June 1993.

[7] H. Guldemir, "Sliding mode control of DC-DC converters," *Journal of Applied Sciences*, vol. 5, no. 3, pp. 558–592, 2005.

[8] S. C. Tan, Y. M. Lai, and C. K. Tse, "Indirect sliding mode control of power converters via double integral sliding surface," *IEEE Transactions on Power Electronics*, vol. 23, no. 2, pp. 600–611, 2008.

[9] M. Ahmed, M. Kuisma, and P. Silventoinen, "Implementing simple procedure for controlling switch mode power supply using sliding mode control technique," in *Proceedings of the 13th International Symposium on Electrical Apparatus And Technologies*, pp. 9–14, Plovdive, Bulgaria, May 2003.

[10] Z. M. Salameh, B. S. Borowy, and A. R. A. Amin, "Photovoltaic module-site matching based on the capacity factors," *IEEE Transactions on Energy Conversion*, vol. 10, no. 2, pp. 326–331, 1995.

[11] M. A. S. Masoum and M. Sarvi, "Voltage and current based MPPT of solar arrays under variable insolation and temperature conditions," in *Proceedings of the 43rd International Universities Power Engineering Conference (UPEC '08)*, pp. 1–5, Padua, Italy, September 2008.

Permissions

The contributors of this book come from diverse backgrounds, making this book a truly international effort. This book will bring forth new frontiers with its revolutionizing research information and detailed analysis of the nascent developments around the world.

We would like to thank all the contributing authors for lending their expertise to make the book truly unique. They have played a crucial role in the development of this book. Without their invaluable contributions this book wouldn't have been possible. They have made vital efforts to compile up to date information on the varied aspects of this subject to make this book a valuable addition to the collection of many professionals and students.

This book was conceptualized with the vision of imparting up-to-date information and advanced data in this field. To ensure the same, a matchless editorial board was set up. Every individual on the board went through rigorous rounds of assessment to prove their worth. After which they invested a large part of their time researching and compiling the most relevant data for our readers. Conferences and sessions were held from time to time between the editorial board and the contributing authors to present the data in the most comprehensible form. The editorial team has worked tirelessly to provide valuable and valid information to help people across the globe.

Every chapter published in this book has been scrutinized by our experts. Their significance has been extensively debated. The topics covered herein carry significant findings which will fuel the growth of the discipline. They may even be implemented as practical applications or may be referred to as a beginning point for another development. Chapters in this book were first published by Hindawi Publishing Corporation; hereby published with permission under the Creative Commons Attribution License or equivalent.

The editorial board has been involved in producing this book since its inception. They have spent rigorous hours researching and exploring the diverse topics which have resulted in the successful publishing of this book. They have passed on their knowledge of decades through this book. To expedite this challenging task, the publisher supported the team at every step. A small team of assistant editors was also appointed to further simplify the editing procedure and attain best results for the readers.

Our editorial team has been hand-picked from every corner of the world. Their multi-ethnicity adds dynamic inputs to the discussions which result in innovative outcomes. These outcomes are then further discussed with the researchers and contributors who give their valuable feedback and opinion regarding the same. The feedback is then collaborated with the researches and they are edited in a comprehensive manner to aid the understanding of the subject.

Apart from the editorial board, the designing team has also invested a significant amount of their time in understanding the subject and creating the most relevant covers. They scrutinized every image to scout for the most suitable representation of the subject and create an appropriate cover for the book.

The publishing team has been involved in this book since its early stages. They were actively engaged in every process, be it collecting the data, connecting with the contributors or procuring relevant information. The team has been an ardent support to the editorial, designing and production team. Their endless efforts to recruit the best for this project, has resulted in the accomplishment of this book. They are a veteran in the field of academics and their pool of knowledge is as vast as their experience in printing. Their expertise and guidance has proved useful at every step. Their uncompromising quality standards have made this book an exceptional effort. Their encouragement from time to time has been an inspiration for everyone.

The publisher and the editorial board hope that this book will prove to be a valuable piece of knowledge for researchers, students, practitioners and scholars across the globe.

List of Contributors

Piotr Bujlo
Electrotechnical Institute Wroclaw Division of Technology and Materials Science, M. Sklodowskiej-Curie 55/61, 50-369Wroclaw, Poland
Hydrogen South Africa (HySA) Systems and Validation Centre, University of the Western Cape, Robert Sobukwe Road, Bellville, Cape Town 7535, South Africa

Grzegorz Pasciak, Jacek Chmielowiec and Andrzej Sikora
Electrotechnical Institute Wroclaw Division of Technology and Materials Science, M. Sklodowskiej-Curie 55/61, 50-369Wroclaw, Poland

Pinar Tokat, Torbjörn Thiringer and Peiyuan Chen
Division of Electric Power Engineering, Department of Energy and Environment, Chalmers University of Technology, 412 58 Gothenburg, Sweden

Xunmin Ou and Xiliang Zhang
Institute of Energy, Environment and Economy (3E), Tsinghua University, Beijing 100084, China
China Automotive Energy Research Center, Tsinghua University, Beijing 100084, China

Djéssica Tatiana Raspe and Camila da Silva
Department of Technology, Maring´a State University (UEM), 87506-370 Umuarama, PR, Brazil

C. K. Pandey and A. K. Katiyar
Department of Applied Sciences, Institute of Engineering & Technology, Gautam Buddh Technical University, Lucknow 226021, India

Sakaros Bogning Dongue, Donatien Njomo and Lessly Ebengai
Environmental Energy Technologies Laboratory (EETL), Faculty of Sciences,University of Yaound´e 1, P.O. Box812, Yaound´e, Cameroon

Neda Janzadeh and Mojtaba Aghajani Delavar
Islamic Azad University Science and Research Ayatollah Amoli Branch, Amol, Iran

Vidhya Prabhudessai, Anasuya Ganguly and Srikanth Mutnuri
Applied and Environmental Biotechnology Laboratory, Department of Biological Sciences, Birla Institute of Technology and Science, Pilani, K. K. Birla Goa Campus, Goa 403726, India

Delele Worku Ayele
Department of Chemical Engineering, National Taiwan University of Science and Technology, Taipei 106, Taiwan

Wein-Nien Su
Graduate Institute of Engineering, National Taiwan University of Science and Technology, Taipei 106, Taiwan

Bing-Joe Hwang
Department of Chemical Engineering, National Taiwan University of Science and Technology, Taipei 106, Taiwan
National Synchrotron Radiation Research Center, Hsinchu 300, Taiwan

Huaxiao Yan Hui Zhao Yan Zhang and Yuanyu Tian
College of Chemical and Environmental Engineering, Shandong University of Science and Technology, Qingdao 266590, China
Key Laboratory of Low-carbon Energy Chemical Engineering in Universities of Shandong, Shandong University of Science and Technology, Qingdao 266590, China

Kechang Xie
Key Laboratory of Low-carbon Energy Chemical Engineering in Universities of Shandong, Shandong University of Science and Technology, Qingdao 266590, China

A. O. Adelaja and B. I. Babatope
The Department of Mechanical Engineering, University of Lagos, Akoka, Yaba, Lagos 101017, Nigeria

Prakash C. Ghosh
Department Energy Science and Engineering, Indian Institute of Technology Bombay, Mumbai 400076, India

Seyed Mahmood Kazemi and Masoud Rabbani
Department of Industrial Engineering, College of Engineering, University of Tehran, P. O. Box 11155-4563, Tehran, Iran

Christine Cleetus
Department of Mechanical Engineering, NIT Calicut, Kerala 673601, India

Shijo Thomas and Soney Varghese
School of Nano Science and Technology, NIT Calicut, Kerala 673601, India

Daniel F. Moreira
CEHIDRO, Instituto Superior T´ecnico, 1049-001 Lisbon, Portugal

Helena M. Ramos
CEHIDRO, Instituto Superior T´ecnico, 1049-001 Lisbon, Portugal
Instituto Superior T´ecnico, 1049-001 Lisbon, Portugal

S. T. Wara and S. E. Abe
Department of Electrical and Electronics Engineering, College of Engineering, Federal University of Agriculture Abeokuta, Ogun State, Nigeria

Ram Kumar Agrawal
Department of Mechanical Engineering, Kautilya Institute of Technology & Engineering, Jaipur Rajasthan 302022, India

Kamal Kishore Khatri
Department of Mechanical Engineering, Mody Institute of Technology & Science, Lakshmangarh, Sikar Rajasthan 332311, India

Raad Z. Homod
Department of Mechanical Engineering, University of Basrah, Qarmat Ali Campus, Basrah 61004, Iraq

M. A. A. Mohammed, A. Salmiaton, W. A. K. G. Wan Azlina and M. S. Mohamad Amran
Department of Chemical & Environmental Engineering, Faculty of Engineering, Universiti Putra Malaysia, 43400 UPM Serdang, Selangor, Malaysia

Y. H. Taufiq-Yap
Department of Chemistry, Centre of Excellence for Catalysis Science and Technology, Faculty of Science, Universiti Putra Malaysia, 43400 Serdang, Selangor, Malaysia

Ra Git Ata
Department of Electrical & Electronic Engineering, Celal Bayar University, 45140 Manisa, Turkey

Athanasios G. Lazaropoulos
School of Electrical and Computer Engineering, National Technical University of Athens, 9 Iroon Polytechniou Street, Zografou, 15780 Athens, Greece

Saifulnizam Bin Abdul Khalid, Garba Aliyu, Mohd. Wazir Mustafa
Faculty of Electrical Engineering, Universiti Teknologi Malaysia, 81310 Skudai, Malaysia

Hussein Shareef
Faculty of Electrical Engineering and Built Environment, Universiti Kebangsaan Malaysia, 43600 Bangi, Malaysia

M. Sarvi, I. Soltani, N. Namazy Pour and N. Rabbani
Faculty of Electrical Engineering, Imam Khomeini International University, Qazvin 34149-16818, Iran